Dan Romik
Topics in Complex Analysis

Also of Interest

Topics in Complex Analysis
Joel L. Schiff, 2022
ISBN 978-3-11-075769-9, e-ISBN (PDF) 978-3-11-075782-8
in: De Gruyter Studies in Mathematics
ISSN 0179-0986

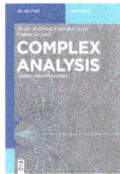

Complex Analysis
Theory and Applications
Teodor Bulboacă, Santosh B. Joshi, Pranay Goswami, 2019
ISBN 978-3-11-065782-1, e-ISBN (PDF) 978-3-11-065786-9

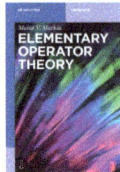

Elementary Operator Theory
Marat V. Markin, 2020
ISBN 978-3-11-060096-4, e-ISBN (PDF) 978-3-11-060098-8

Real Analysis
Measure and Integration
Marat V. Markin, 2019
ISBN 978-3-11-060097-1, e-ISBN (PDF) 978-3-11-060099-5

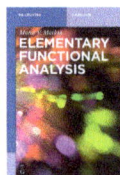

Elementary Functional Analysis
Marat V. Markin, 2018
ISBN 978-3-11-061391-9, e-ISBN (PDF) 978-3-11-061403-9

Dan Romik

Topics in Complex Analysis

—

DE GRUYTER

Mathematics Subject Classification 2020
Primary: 30-01, 11-01; Secondary: 52C07, 52C17

Author
Prof. Dan Romik
Department of Mathematics
University of California
One Shields Ave
Davis CA 95616
USA
romik@math.ucdavis.edu

ISBN 978-3-11-079678-0
e-ISBN (PDF) 978-3-11-079681-0
e-ISBN (EPUB) 978-3-11-079688-9
DOI https://doi.org/10.1515/9783110796810

Library of Congress Control Number: 2023935854

Bibliographic information published by the Deutsche Nationalbibliothek
The Deutsche Nationalbibliothek lists this publication in the Deutsche Nationalbibliografie;
detailed bibliographic data are available on the Internet at http://dnb.dnb.de.

© 2023 the author(s), published by Walter de Gruyter GmbH, Berlin/Boston. The book is published open
access at www.degruyter.com.
Cover image: Guy Kindler and Dan Romik
Typesetting: VTeX UAB, Lithuania
Printing and binding: CPI books GmbH, Leck

www.degruyter.com

To Abigail

Contents

Preface —— XI

0   **Prerequisites and notation** —— **1**
0.1   Prerequisites —— **1**
0.2   Notation —— **1**
        Exercises for Chapter 0 —— **3**

1   **Basic theory** —— **4**
1.1   Motivation: why study complex analysis? —— **4**
1.2   The fundamental theorem of algebra —— **9**
1.3   Holomorphicity, conformality, and the Cauchy–Riemann equations —— **12**
1.4   Additional consequences of the Cauchy–Riemann equations —— **18**
1.5   Power series —— **20**
1.6   Contour integrals —— **22**
1.7   The Cauchy, Goursat, and Morera theorems —— **28**
1.8   Simply connected regions and the general version of Cauchy's theorem —— **32**
1.9   Consequences of Cauchy's theorem —— **36**
1.10   Zeros, poles, and the residue theorem —— **46**
1.11   Meromorphic functions, holomorphicity at ∞, and the Riemann sphere —— **50**
1.12   Classification of singularities and the Casorati–Weierstrass theorem —— **52**
1.13   The argument principle and Rouché's theorem —— **53**
1.14   The open mapping theorem and maximum modulus principle —— **57**
1.15   The logarithm function —— **58**
1.16   The local behavior of holomorphic functions —— **60**
1.17   Infinite products and the product representation of the sine function —— **63**
1.18   Laurent series —— **68**
        Exercises for Chapter 1 —— **71**

2   **The prime number theorem** —— **82**
2.1   Motivation: analytic number theory and the distribution of prime numbers —— **82**
2.2   The Euler gamma function —— **83**
2.3   The Riemann zeta function: definition and basic properties —— **89**
2.4   A theorem on the zeros of the Riemann zeta function —— **97**
2.5   Proof of the prime number theorem —— **99**
        Exercises for Chapter 2 —— **110**

3   **Conformal mapping —— 118**
3.1   Motivation: classifying complex regions up to conformal equivalence —— **118**
3.2   First singleton conformal equivalence class: the complex plane —— **121**
3.3   Second singleton conformal equivalence class: the Riemann sphere —— **123**
3.4   The Riemann mapping theorem —— **124**
3.5   The unit disc and its automorphisms —— **126**
3.6   The upper half-plane and its automorphisms —— **129**
3.7   The Riemann mapping theorem: a more precise formulation —— **131**
3.8   Proof of the Riemann mapping theorem, part I: technical background —— **132**
3.9   Proof of the Riemann mapping theorem, part II: the main construction —— **137**
3.10   Annuli and doubly connected regions —— **140**
    Exercises for Chapter 3 —— **145**

4   **Elliptic functions —— 146**
4.1   Motivation: elliptic curves —— **146**
4.2   Doubly periodic functions —— **149**
4.3   Poles and zeros; the order of a doubly periodic function —— **151**
4.4   Construction of the Weierstrass $\wp$-function —— **154**
4.5   Eisenstein series and the Laurent expansion of $\wp(z)$ —— **158**
4.6   The differential equation satisfied by $\wp(z)$ —— **159**
4.7   A recurrence relation for the Eisenstein series —— **160**
4.8   Half-periods; factorization of the associated cubic —— **161**
4.9 $\wp(z)$ and $\wp'(z)$ generate all doubly periodic functions —— **163**
4.10 $\wp(z)$ as a conformal map for rectangles —— **165**
4.11   The discriminant of a cubic polynomial —— **168**
4.12   The discriminant of a lattice —— **170**
4.13   The J-invariant of a lattice —— **170**
4.14   The modular variable τ: from elliptic functions to elliptic modular functions —— **171**
4.15   The classification problem for complex tori —— **172**
4.16   Equivalence between complex tori and elliptic curves —— **177**
    Exercises for Chapter 4 —— **179**

5   **Modular forms —— 182**
5.1   Motivation: functions of lattices —— **182**
5.2   The modular group $\Gamma = \mathrm{PSL}(2, \mathbb{Z})$ —— **184**
5.3   The modular group as a group of Möbius transformations —— **185**
5.4   The fundamental domain and the modular surface $\mathbb{H}/\Gamma$ —— **186**
5.5   The classification problem for complex tori, part II —— **190**
5.6   The point at $i\infty$, premodular forms, and their Fourier expansions —— **191**
5.7   Fourier expansions and number-theoretic identities —— **194**

5.8   Modular functions — **199**
5.9   Klein's *J*-invariant — **205**
5.10   The *J*-invariant as a conformal map — **208**
5.11   The classification problem for complex tori, part III — **209**
5.12   Modular forms — **209**
5.13   Examples of modular forms — **214**
5.14   Infinite products for modular forms — **218**
         Exercises for Chapter 5 — **228**

6   **Sphere packing in 8 dimensions — 233**
6.1   Motivation: the sphere packing problem in d dimensions — **233**
6.2   A high-level overview of the proof — **236**
6.3   Preparation: some remarks on Fourier eigenfunctions — **237**
6.4   The $(+1)$-Fourier eigenfunction — **239**
6.5   The (-1)-Fourier eigenfunction — **250**
6.6   A modular form inequality — **256**
6.7   Proof of Theorem 6.1 — **263**
         Exercises for Chapter 6 — **265**

A   **Appendix: Background on sphere packings — 267**
A.1   Sphere packings and their densities — **267**
A.2   Lattices and lattice packings — **268**
A.3   Periodic sphere packings — **268**
A.4   Lattice covolume — **269**
A.5   Dual lattices — **269**
A.6   The Poisson summation formula for lattices — **270**
A.7   Construction of the lattice E_8 — **271**
A.8   The Cohn–Elkies sphere packing bounds — **276**
A.9   Magic functions — **278**
A.10   Radial functions and their Fourier transforms — **279**
A.11   Structural properties of E_8 magic functions — **281**
A.12   Summary — **284**
         Exercises for Appendix A — **286**

Bibliography — 289

Web Bibliography — 291

Index — 293

Preface

This book covers the basic theory of complex analysis and a selection of advanced topics. It evolved out of lecture notes from two quarter-long graduate classes that I taught several times at the University of California, Davis in 2016–2021. The book is primarily aimed at graduate students, advanced undergraduate students, and postgraduate mathematics researchers. It is suited for self-study or as a primary reference material for approximately two semester-long graduate-level university courses.

The advanced topics covered in Chapters 2–5 are classical and are discussed in many other places. It is my hope that my own exposition advances the pedagogy of the subject, if only ever so slightly, by simplifying the explanations, logical arguments, notation, etc, as much as it has been within my power to do.

The last chapter, Chapter 6, is more modern in content and covers Maryna Viazovska's spectacular application of modular forms to the solution of the sphere packing problem in dimension 8. Published in 2016, this work was until now only accessible to learn about from the primary literature [71] and from a few expository papers [12, 13, 20, 52]. The detailed exposition of Viazovska's work in Chapter 6, and the accompanying Appendix A covering the relevant background material on sphere packing, should be useful to students and researchers wishing to get up to speed about these beautiful recent developments, which are at the forefront of much ongoing research.

The choice of topics you will find in this work is idiosyncratic and reflects my own mathematical taste, interests, and biases. I make no claim that they are the most important parts of the vast theory that is complex analysis; only that they are beautiful, that they relate to many topics and theories that are of interest to a broad section of pure mathematicians, and that they are, broadly speaking, a fine set of mathematical ideas, one could devote one's time to studying and thinking about. I hope some readers will agree.

I am grateful to Guy Kindler for help with the book cover design and to Christopher Alexander, Jennifer Brown, Brynn Caddel, Keith Conrad, Bo Long, Anthony Nguyen, Jianping Pan, and Brad Velasquez for helpful comments on versions of the lecture notes the book evolved from.

Davis   Dan Romik
March 2023

0 Prerequisites and notation

0.1 Prerequisites

This book assumes knowledge of the following subjects, roughly at the level covered by advanced undergraduate courses in the United States:
- Real analysis and multivariable calculus
- Topology of $\mathbb{R}^n$ (mostly for $n = 2$)
- Complex numbers and their basic properties
- The transcendental functions e^z, $\sin z$, $\cos z$ of a complex variable

In a few places, some familiarity with Fourier analysis is needed to fully understand the material. Specifically, in Chapter 2 the Poisson summation formula is derived from basic properties of Fourier series, and this is used to prove some of the fundamental properties of the Riemann zeta function. Chapter 6 and Appendix A assume knowledge of the Fourier transform in $\mathbb{R}^n$ and its basic properties.

Starting in Chapter 3, and increasingly in Chapter 5, knowledge of the basic language of group theory may be needed to fully understand some of the topics being discussed. No results from group theory are used beyond the definition of a quotient group.

0.2 Notation

The following notation is used throughout the book.
- $\mathbb{R}$ — the real numbers
- $\mathbb{C}$ — the complex numbers
- $\mathbb{Z}$ — the integers
- i — the imaginary unit
- $\text{Re}(z)$ — the real part of a complex number z
- $\text{Im}(z)$ — the imaginary part of a complex number z
- $\bar{z}$ — the complex conjugate of a complex number z
- $|z|$ — the modulus of a complex number z
- $\arg z$ — the argument of a complex number z
- $D_R(z)$ — the open disc of radius R centered at z
- $D_{\leq R}(z)$ — the closed disc of radius R centered at z
- $C_R(z)$ — the circle of radius R centered at z
- $\text{cl}(E)$ — the topological closure of a set $E \subset \mathbb{C}$
- $\mathbb{D}$ — the open unit disc $D_1(0)$
- $\mathbb{H}$ — the upper half-plane: $\{z \in \mathbb{C} : \text{Im}(z) > 0\}$
- Ω — a complex region (open and connected subset of $\mathbb{C}$)

Big-O notation and asymptotic equality. In a few places, the standard **big-O notation** is used. Formally, the statement "$F = O(G)$," where F, G are complex-valued quantities that depend on one or more variables, means that $|F| \leq C|G|$ when the variable or variables in question range over some specified set of values (usually a neighborhood of some limiting point). Big-O expressions can also be combined in various ways in formulas, e. g., "$f(t) = O(e^{-t}) + O(t^2)$ as $t \to \infty$" means that $f(t)$ can be expressed as a sum of two quantities of the forms $O(e^{-t})$ and $O(t^2)$, respectively, as $t \to \infty$.

The statement $F \sim G$ (read as "F is asymptotically equal to G") means that F/G converges to 1 in some limiting sense, which is either specified explicitly or inferred from the context. For example,

$$\sin(x) \sim x \quad \text{as } x \to 0$$

states an asymptotic equality, as does

$$\frac{(2n)!}{(n!)^2} \sim \frac{4^n}{\sqrt{\pi n}} \quad \text{as } n \to \infty.$$

Exercises for Chapter 0

0.1 **Important formulas.** Below there is a list of basic formulas in complex analysis. Review each of them, making sure that you understand what it says and why it is true; that is, if it is a theorem, then prove it, or if it is a definition, then make sure you understand it.

In the formulas below, a, b, c, d, t, x, y denote arbitrary real numbers, and w, z denote arbitrary complex numbers.

a. $i^2 = -1$

b. $(a + bi)(c + di)$
   $= (ac - bd) + (ad + bc)i$

c. $\frac{1}{i} = -i$

d. $z = \text{Re}(z) + i\,\text{Im}(z)$

e. $\bar{z} = \text{Re}(z) - i\,\text{Im}(z)$

f. $\text{Re}(z) = \frac{z+\bar{z}}{2}$

g. $\text{Im}(z) = \frac{z-\bar{z}}{2i}$

h. $|z|^2 = z\bar{z}$

i. $\frac{1}{z} = \frac{\bar{z}}{|z|^2}$

j. $\frac{1}{x+iy} = \frac{x-iy}{x^2+y^2}$

k. $\overline{w \cdot z} = \overline{w} \cdot \overline{z}$

l. $|wz| = |w| \cdot |z|$

m. $||w| - |z|| \le |w + z| \le |w| + |z|$

n. $e^{x+iy} = e^x(\cos(y) + i\sin(y))$

o. $|e^z| = e^{\text{Re}(z)}$

p. $|e^z| \le e^{|z|}$

q. $e^{it} = \cos(t) + i\sin(t)$

r. $|e^{it}| = 1$

s. $\cos(t) = \frac{e^{it}+e^{-it}}{2}$

t. $\sin(t) = \frac{e^{it}-e^{-it}}{2i}$

u. $e^{\pi i} = -1$

v. $e^{\pm\pi i/2} = \pm i$

w. $e^{2\pi i} = 1$

0.2 **Reminder of basic analysis concepts.** Remind yourself of the definitions of the following terms in real and complex analysis and the topology of $\mathbb{C}$, referring to other textbooks or online sources if necessary.

a. real part

b. imaginary part

c. complex conjugate

d. modulus

e. argument

f. open set (in $\mathbb{C}$)

g. closed set

h. closure

i. connected set

j. bounded set

k. compact set

l. region

m. convergent sequence

n. Cauchy sequence

o. limit point

p. accumulation point

q. continuous function

1 Basic theory

What is unpleasant here, and indeed directly to be objected to, is the use of complex numbers. ψ is surely fundamentally a real function.

Erwin Schrödinger, June 6, 1926 letter to Hendrik Lorentz

1.1 Motivation: why study complex analysis?

This book is about **complex analysis**, the area of mathematics that studies holomorphic functions of a complex variable and their properties. Although this may sound a bit specialized, there are (at least) two excellent reasons why all mathematicians should learn about complex analysis. First, it is, in my humble opinion, one of the most beautiful areas of mathematics. One way of putting it is that complex analysis seems to have a very high ratio of theorems to definitions (i. e., a very low "entropy"): you get a lot more as "output" than you put in as "input."

The second reason is that complex analysis and, more generally, complex numbers, have a large number of applications in both the pure mathematics and applied mathematics senses of the word. Moreover, many of these applications are to problems that a priori look like they ought to have little to do with complex numbers. Here are a few examples, including some that will be discussed later in the book:

- **Solving polynomial equations.** In 1545, the Italian thinker Gerolamo Cardano published the famous formula for solving cubic equations, after learning of the solution found earlier by Scipione del Ferro. Historically, this appears to have been the first problem in mathematics to be solved using complex numbers. One surprising aspect of Cardano's formula is that it sometimes requires taking operations in the complex plane as an intermediate step to get to the final answer, even when the cubic equation being solved has only real roots.
- **Proving asymptotic formulas.** A well-known approximation to the factorial function $n!$ is given by **Stirling's formula**, which states that the behavior of the factorial function for large values of n is given by

$$n! \sim \sqrt{2\pi n}\left(\frac{n}{e}\right)^n \tag{1.1}$$

(using the notation of Section 0.2). Another famous asymptotic formula is the Hardy–Ramanujan formula, which states that the number $p(n)$ of integer partitions of n behaves for large n like

$$p(n) \sim \frac{1}{4\sqrt{3}n}e^{\pi\sqrt{2n/3}}. \tag{1.2}$$

A standard approach to proving these types of results uses complex analysis, as discussed, for example, in [28].

- **Counting prime numbers.** Let $\pi(n)$ denote the number of primes less than or equal to n. This function is known as the **the prime-counting function**. The **prime number theorem** states that

$$\pi(n) \sim \frac{n}{\log n} \quad \text{as } n \to \infty.$$

This is one of the most celebrated asymptotic formulas (and, indeed, one of the most famous theorems) in mathematics. Because it deals with prime numbers, it stands apart from the more general class of asymptotic formulas, such as (1.1)–(1.2) mentioned above, and its proof requires more specialized techniques. A standard path to a proof of the prime number theorem goes through complex analysis, and this is the subject of Chapter 2.
- **Evaluation of complicated definite integrals.** Complex analysis offers a set of techniques for evaluating definite integrals that are difficult or impossible to derive using standard calculus methods. An example is the integral

$$\int_0^\infty \sin(t^2)\, dt = \frac{\sqrt{\pi}}{2\sqrt{2}}$$

(known as one of the **Fresnel integrals**). See Exercise 1.47 at the end of this chapter for additional examples.
- **Solving partial differential equations.** Complex-analytic techniques are very useful for solving several kinds of partial differential equation, particularly those arising in various applied physics problems in hydrodynamics, heat conduction, electrostatics, and more.
- **Analyzing alternating current electrical networks.** Electrical engineers learn that the usefulness of Ohm's law can be greatly extended by generalizing the notion of electrical resistance to that of **electrical impedance**, a complex-valued quantity. Complex analysis also has many other important applications in electrical engineering, signal processing, and control theory.
- **Solution of the sphere packing problem in 8 and 24 dimensions.** It was proved in 2016 that the optimal densities for packing unit spheres in 8 and 24 dimensions are $\frac{\pi^4}{384}$ and $\frac{\pi^{12}}{12!}$, respectively. The proofs make use of complex analysis in a fundamental way. The proof for the case of 8 dimensions is presented in Chapter 6.
- **Applications in probability and combinatorics.** Over the last few decades, complex analysis has been applied in spectacular ways to prove asymptotic results in probability and combinatorics. One such application is a proof of the **Cardy–Smirnov formula** in percolation theory, which answers the following question: consider a parallelogram-shaped section of cells in the honeycomb lattice with m rows of cells, each containing n cells. Each cell is colored either black or white according to the outcome of a fair coin toss, independently of all other cells (Fig. 1.1(a)).

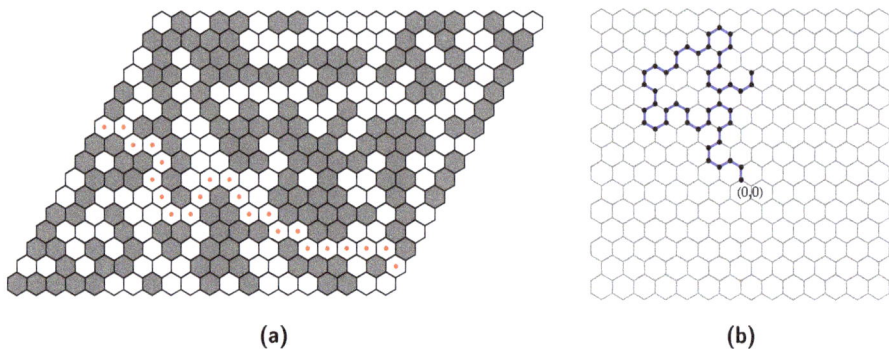

(a)   (b)

Figure 1.1: (a) Percolation on a honeycomb: the Cardy–Smirnov formula gives the asymptotic probability of a left-to-right crossing event. In this sample configuration, a left-to-right crossing has occurred, as illustrated by the trail of red dots representing one possible crossing path. (b) A self-avoiding walk of length 45 on the hexagonal lattice.

A **left-to-right crossing event** is the event that we can find a contiguous path of white-colored cells connecting the left edge of the parallelogram to the right edge. What is the asymptotic probability of this event in the limit as the side lengths of the parallelogram grow to infinity but its shape tends toward a parallelogram with a fixed aspect ratio?

Specifically, let $P(m, n)$ denote the probability of a left-to-right crossing event. Cardy conjectured [10] and Smirnov proved [64] the following result.

Theorem 1.1 (Cardy–Smirnov formula). *As $m, n \to \infty$ with the aspect ratio m/n converging to a fixed value $\lambda \in (0, \infty)$, the probabilities $P(m, n)$ have the limiting behavior*

$$P(m, n) \xrightarrow[\substack{m,n \to \infty \\ m/n \to \lambda}]{} \Phi(\lambda)$$

for an explicit function $\Phi(\lambda)$.

A detailed account of Smirnov's proof can be found in [34, 73]. The function $\Phi(\lambda)$ is most naturally defined as a certain geometric invariant associated with the parallelogram with corners $0, 1, (\frac{1+\sqrt{3}i}{2})\lambda$, and $(\frac{1+\sqrt{3}i}{2})\lambda + 1$ and can be written down explicitly in terms of modular forms [43] and other special functions from complex analysis.

A second example of a recent application of complex analysis to probability and combinatorics is the evaluation of the **connective constant of the hexagonal lattice**. Let c_n denote the number of self-avoiding walks of length n in the hexagonal lattice that start at the origin; that is, hexagonal lattice paths that do not intersect themselves; see Fig. 1.1(b). Without the condition of the path being self-avoiding, the number of such paths would be exactly equal to 3^n. The sequence $(c_n)_{n=1}^\infty$, with initial values $1, 3, 6, 12, 24, 48, 90, 174, 336, \dots$ [W1], is much more mysterious, and its rate of growth (as well as the rates of growth of similar sequences associated with the square lattice and other natural lattices) have been the subject of much study. From general considerations it is fairly easy to see that the sequence grows roughly exponentially, that is, there exists a constant $\mu > 0$ such that $c_n^{1/n} \to \mu$ as $n \to \infty$. The constant μ is known as the connective constant of the hexagonal lattice. Nienhuis [51] conjectured in 1982 and Duminil-Copin and Smirnov [65] proved in 2010 the following remarkable result concerning the value of μ.

Theorem 1.2 (Duminil-Copin–Smirnov theorem). *The connective constant of self-avoiding walks in the hexagonal lattice is equal to* $\sqrt{2 + \sqrt{2}} \approx 1.84776$, *that is, the numbers* c_n *satisfy*

$$\lim_{n\to\infty} c_n^{1/n} = \sqrt{2 + \sqrt{2}}.$$

- **Running the universe.** Nature uses complex numbers in the fundamental laws of physics, Schrödinger's equation and quantum field theory. This is not a mere mathematical convenience or sleight-of-hand, but appears to be a built-in feature of the very equations describing our physical universe. Why? No one knows.[1] (But it is a fun topic for debate; see, e. g., [42], [W2], [W3].)
- **Conformal maps.** A conformal map is a mapping from one planar region to another that preserves angles. This notion, which comes up in purely geometric applications where the algebraic or analytic structure of complex numbers seems irrelevant, are in fact deeply tied to complex analysis. Conformal maps were used by the Dutch artist M. C. Escher (though he had no formal mathematical training) to create amazing art and used by others to better understand, and even to improve on, Escher's work. See Fig. 1.2 and [21, 59] for more on the connection of Escher's work to mathematics. We discuss conformal maps in detail in Chapter 3.
- **Proving number-theoretic identities.** Lagrange proved in 1770 a classic result in number theory, which states that every positive integer can be represented as a sum of four squares of integers. Jacobi later proved a more precise fact: if we denote by $r_4(n)$ the *number* of distinct ways in which a positive integer n can be represented as a sum of four squares (with different orderings counting as distinct), then we have the remarkable identity

$$r_4(n) = 8 \sum_{d\,|\,n,\ 4\nmid d} d. \tag{1.3}$$

(In words: eight times the sum of divisors of n that are not divisible by 4.) This beautiful identity and many others like it with a number-theoretic flavor can be proved

---

1 Schrödinger himself appeared dissatisfied with the idea that his equation uses complex numbers to describe physical reality. See the epigraph at the beginning of this chapter and [42] for further discussion.

Figure 1.2: *Print Gallery*, a lithograph by M. C. Escher from 1956, which was discovered in the early 2000s to be based on a mathematical structure related to a certain complex-analytic map, although it was constructed by Escher purely using geometric intuition. See the paper [21] and the websites [W4], [W5], which have animated versions of Escher's lithograph brought to life using the mathematics of complex analysis. M. C. Escher's "Print Gallery" © 2023 The M.C. Escher Company-The Netherlands. All rights reserved. www.mcescher.com

using complex analysis; see Chapter 5 (and Exercise 5.21 at the end of that chapter for the particular application to proving (1.3)).

– **Complex dynamics.** Iteration of complex-analytic maps can be used to generate beautiful fractals with remarkable properties. A famous example is the iconic **Mandelbrot set** (Fig. 1.3) defined as the set of complex numbers $c \in \mathbb{C}$ for which the sequence of functional iterates $f_c^{(n)}(0)$ of the map $f_c(z) = z^2 + c$ starting from the point $z = 0$ remains bounded.

This has been just a short and necessarily very incomplete survey on the importance of complex analysis. There are many other intriguing applications and connections of complex analysis to other areas of mathematics.

In the next section, I will begin our journey into the subject by proving a famous theorem about polynomials over the complex numbers.

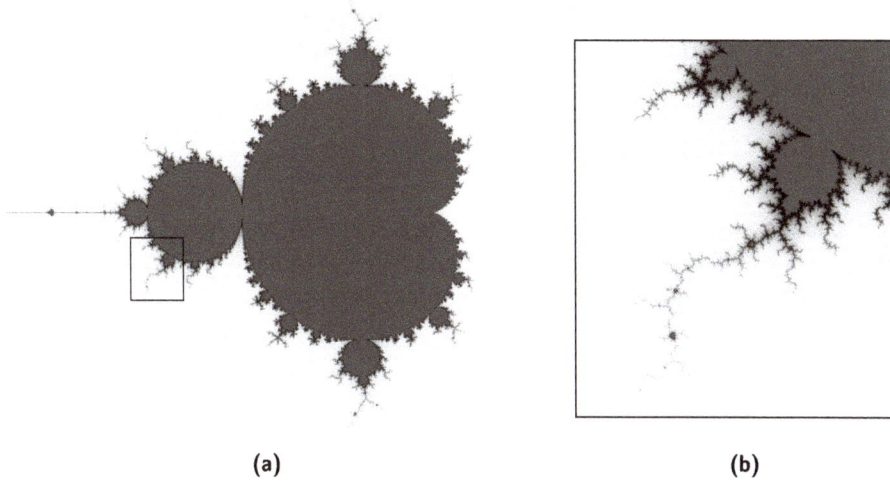

(a)   **(b)**

Figure 1.3: (a) The Mandelbrot set. (b) Magnified details of a small region.

1.2 The fundamental theorem of algebra

One of the most famous results about complex numbers is the **fundamental theorem of algebra**. Although the statement of the theorem is indeed very fundamental to algebra, most of its known *proofs* rely on complex analysis in an essential way. Looking at a few of these proofs seems like a fitting place to start our journey into the theory.

Theorem 1.3 (Fundamental theorem of algebra). *Every nonconstant polynomial*

$$p(z) = a_n z^n + a_{n-1} z^{n-1} + \cdots + a_0 \quad (n \geq 1) \tag{1.4}$$

with complex coefficients has a complex root.

The fundamental theorem of algebra is a striking and subtle result and has many beautiful proofs. I will show you three of them.

First proof: analytic proof. Let $p(z)$ be as in (1.4), and consider where $|p(z)|$ attains its infimum.

First, note that the infimum cannot be attained as $|z| \to \infty$, since

$$|p(z)| = |z|^n \cdot \left(|a_n + a_{n-1} z^{-1} + a_{n-2} z^{-2} + \cdots + a_0 z^{-n}| \right)$$

and, in particular,

$$\lim_{|z| \to \infty} \frac{|p(z)|}{|z|^n} = |a_n|, \tag{1.5}$$

so for large $|z|$, it is guaranteed that $|p(z)| \geq |p(0)| = |a_0|$. Now fix some radius $R > 0$ for which $|z| > R$ implies $|p(z)| \geq |a_0|$, and choose a complex number z_0 in the disc $D_R(0)$ for which $|p(z_0)| = \min_{|z| \leq R} |p(z)|$. (The minimum exists because $p(z)$ is a continuous function on the disc.) We then have that

$$m_0 := \inf_{z \in \mathbb{C}} |p(z)| = \inf_{|z| \leq R} |p(z)| = \min_{|z| \leq R} |p(z)| = |p(z_0)|.$$

Denote $w_0 = p(z_0)$, so that $m_0 = |w_0|$. We now claim that $m_0 = 0$. Indeed, assume by contradiction that this is not the case. The idea is now to examine the local behavior of $p(z)$ around z_0. Expanding $p(z)$ in powers of $z - z_0$, we can write

$$p(z) = w_0 + \sum_{j=1}^{n} c_j (z - z_0)^j$$

for some complex coefficients $c_1, \ldots, c_n$. This can also be written as

$$p(z) = w_0 + c_k (z - z_0)^k + \cdots + c_n (z - z_0)^n, \tag{1.6}$$

where we denote by k the minimal positive index for which $c_j \neq 0$. Now imagine starting at the initial point $z = z_0$ and then making a small perturbation away from z_0 in the direction of some unit vector $e^{i\theta}$. We estimate the way that such a perturbation affects the value $p(z)$. Expansion (1.6) gives

$$p(z_0 + re^{i\theta}) = w_0 + c_k r^k e^{ik\theta} + c_{k+1} r^{k+1} e^{i(k+1)\theta} + \cdots + c_n r^n e^{in\theta}. \tag{1.7}$$

When r (the magnitude of the perturbation) is very small, the power r^k dominates the other terms r^j with $k < j \leq n$; that is, (1.7) can be rewritten as

$$p(z_0 + re^{i\theta}) = w_0 + r^k \left(c_k e^{ik\theta} + c_{k+1} re^{i(k+1)\theta} + \cdots + c_n r^{n-k} e^{in\theta} \right)$$
$$= w_0 + c_k r^k e^{ik\theta} (1 + g(r, \theta)), \tag{1.8}$$

where we denote

$$g(r, \theta) = \sum_{j=k+1}^{n} \frac{c_j}{c_k} r^{j-k} e^{i(j-k)\theta}.$$

Note that $g(r, \theta)$ satisfies a bound of the form

$$|g(r, \theta)| < Ar \tag{1.9}$$

for all $r \in [0, 1]$ and some constant $A > 0$.

To reach a contradiction, we now choose θ, the angle of the perturbation, to be such that the vector $c_k r^k e^{ik\theta}$ "points in the opposite direction" from w_0, that is, such that

$$\frac{c_k r^k e^{ik\theta}}{w_0} \in (-\infty, 0).$$

This is clearly possible: take $\theta = \frac{1}{k}(\arg w_0 - \arg(c_k) + \pi)$. The idea in doing this is that for this choice of θ, the expression $w_0 + c_k r^k e^{ik\theta}$ that forms the dominant term in (1.8) will have a smaller magnitude than w_0 if r is chosen small enough.

To make this precise, choose a number $r \in [0, 1]$ smaller than the minimum of the two numbers $1/(2A)$ (where A is the constant in (1.9)) and $(|w_0|/|c_k|)^{1/k}$. This choice ensures the two inequalities

$$|c_k r^k e^{ik\theta}| < |w_0| \quad \text{and} \quad |g(r, \theta)| < \frac{1}{2}.$$

With those choices for θ and r, we have that

$$|p(z_0 + re^{i\theta})| = |w_0 + c_k r^k e^{ik\theta}(1 + g(r, \theta))| \le |w_0 + c_k r^k e^{ik\theta}| + |c_k r^k g(r, \theta)|$$

$$= |w_0| - |c_k|r^k + |c_k|r^k|g(r, \theta)| < |w_0| - \frac{1}{2}|c_k|r^k < |w_0| = |p(z_0)|.$$

This is in contradiction to the defining property of z_0 and completes the proof. □

Second proof: topological proof. If the constant coefficient $a_0 = p(0)$ of $p(z)$ is equal to 0, then we are done, since 0 is a complex root of $p(z)$. Otherwise, consider the image under p of the circle $|z| = r$. Note that, on the one hand, for sufficiently small values of r, the image is contained in a neighborhood of w_0, so it cannot "go around" the origin.

On the other hand, for r very large, we have

$$p(re^{i\theta}) = a_n r^n e^{in\theta}\left(1 + \frac{a_{n-1}}{a_n}r^{-1}e^{-i\theta} + \cdots + \frac{a_0}{a_n}r^{-n}e^{-in\theta}\right)$$

$$= a_n r^n e^{in\theta}(1 + h(r, \theta)),$$

where $h(r, \theta)$ is a function that satisfies $\lim_{r \to \infty} h(r, \theta) = 0$ (uniformly in θ). As θ goes from 0 to 2π, this is a closed curve that goes around the origin n times (in an approximately circular path, which becomes closer and closer to a circle as $r \to \infty$).

As we gradually increase r from 0 to a very large number, to transition from a curve that does not go around the origin to a curve that goes around the origin n times, there has to be a value of r for which the curve crosses 0. This means that the circle $|z| = r$ contains a point z such that $p(z) = 0$, which was the claim. □

The argument presented in the topological proof is imprecise. It can be made rigorous in a couple of ways—one way we will see a bit later is using Rouché's theorem (see Section 1.13 and Exercise 1.30 at the end of the chapter). The difficulty of making these sorts of arguments precise, in spite of their appealing intuitive nature, gives a hint as to the importance of subtle topological arguments in complex analysis.

As another remark, the topological proof should be compared to the standard calculus proof that any odd-degree polynomial over the reals has a real root. That argument is also "topological", although much more elementary.

Third proof: typical textbook proof (or: "hocus-pocus" proof). This is a one-liner of a proof that assumes some complex analysis knowledge. Recall that an **entire function** is a function $f : \mathbb{C} \to \mathbb{C}$ that is everywhere holomorphic. Recall the well-known **Liouville's theorem**, which states that any bounded entire function is constant.

Assuming this result (which we will prove in Section 1.9), if $p(z)$ is a polynomial with no root, then $1/p(z)$ is an entire function. Moreover, it is bounded, since our earlier observation (1.5) implies that $\lim_{|z| \to \infty} 1/p(z) = 0$. By Liouville's theorem it follows that $1/p(z)$ is a constant, which then has to be 0, leading to a contradiction. $\square$

To summarize this section, we saw three proofs of the fundamental theorem of algebra. They are all beautiful—the "hocus-pocus" proof certainly packs a punch, which is why it is a favorite of complex analysis textbooks—but personally I like the first one best since it is fully rigorous while being completely elementary and not requiring the use of either Cauchy's theorem or any of its consequences, or of subtle topological concepts. Moreover, it employs a "local" argument based on understanding how a polynomial behaves locally, where by contrast the other two proofs can be characterized as "global." It is a general principle in mathematical analysis (that has analogies in other areas of mathematics, such as number theory and graph theory) that local arguments are conceptually easier than global ones.

Suggested exercises for Section 1.2. 1.1, 1.2.

1.3 Holomorphicity, conformality, and the Cauchy–Riemann equations

In this section, we begin to build the theory in a systematic way by laying its most basic cornerstone, the definition of holomorphicity, along with some of the useful ways to think about this fundamental concept.

1.3.1 Definition of holomorphicity

A function $f(z)$ of a complex variable is called **holomorphic** at z_0 if the limit

$$\lim_{h \to 0} \frac{f(z_0 + h) - f(z_0)}{h} \tag{1.10}$$

exists. In this case, we denote this limit by $f'(z_0)$ and call it the **derivative of** f **at** z_0. A function of a complex variable defined on all of the complex plane that is everywhere holomorphic is called an **entire function**.

The terms **analytic**, **differentiable**, and **complex-differentiable** are synonyms for "holomorphic." Some books will make a somewhat pedantic distinction between "analytic" and "holomorphic" as two distinct concepts that are defined in a priori different ways but are then shown to be equivalent soon afterward, at which point the distinction ceases to have any real importance. In this book, we do not follow that approach.

The following are basic properties of complex derivatives.

Lemma 1.4. *Under appropriate assumptions (see Exercise* 1.4*), we have the relations*

$$(f + g)'(z) = f'(z) + g'(z), \tag{1.11}$$

$$(fg)'(z) = f'(z)g(z) + f(z)g'(z), \tag{1.12}$$

$$\left(\frac{1}{f}\right)' = -\frac{f'(z)}{f(z)^2}, \tag{1.13}$$

$$\left(\frac{f}{g}\right)' = \frac{f'(z)g(z) - f(z)g'(z)}{g(z)^2}, \tag{1.14}$$

$$(f \circ g)'(z) = f'(g(z))g'(z). \tag{1.15}$$

Proof. Exercise 1.4. $\square$

The concept of the derivative in complex analysis is clearly at the heart of the subject, and there are several helpful ways to think about its meaning. Assume that $f(z)$ is holomorphic at z_0. In the discussion below, we make the further assumption that $f'(z_0) \neq 0$.

1.3.2 First interpretation of holomorphicity: local geometric behavior

If we write the polar decomposition $f'(z_0) = re^{i\theta}$ of the derivative, then for points z that are close to z_0, we will have the approximate equality

$$\frac{f(z) - f(z_0)}{z - z_0} \approx f'(z_0) = re^{i\theta}$$

or, equivalently,

$$f(z) \approx f(z_0) + re^{i\theta}(z - z_0) + \text{[lower-order terms]},$$

where "lower-order terms" refers to a quantity that is much smaller in magnitude that $|z - z_0|$ when z is close to z_0. Geometrically, this means that to compute $f(z)$, we start from $f(z_0)$ and move by a vector that results by taking the displacement vector $z - z_0$, rotating it by an angle of θ, and then scaling it by a factor of r (which corresponds to a magnification if $r > 1$, a shrinking if $0 < r < 1$, or no scaling if $r = 1$). This idea can be summarized by the slogan:

> Holomorphic functions behave locally as a rotation composed with a scaling.

The local behavior of analytic functions in the case $f'(z) = 0$ is more subtle; see Section 1.16.

1.3.3 Second interpretation of holomorphicity: the Cauchy–Riemann equations

Next, we interpret holomorphicity from the point of view of real analysis. Remembering that complex numbers are vectors that have real and imaginary components, we can denote $z = x + iy$, where x and y are the real and imaginary parts of the complex number z, and $f = u + iv$, where u and v are real-valued functions of z (or, equivalently, of x and y) that return the real and imaginary parts, respectively, of f. Now if f is holomorphic at z, then the limit (1.10) exists as a complex limit, that is, independently of the way h approaches 0 as a complex number. In particular, we can evaluate the limit in two ways by considering two specific ways of letting h approach 0, as a pure real number or as a pure imaginary number. For the first of those possibilities, we have

$$
\begin{aligned}
f'(z) &= \lim_{h \to 0} \frac{f(z+h) - f(z)}{h} \\
&= \lim_{h \to 0,\, h \in \mathbb{R}} \frac{u(x+h+iy) - u(x+iy)}{h} + i \frac{v(x+h+iy) - v(x+iy)}{h} \\
&= \frac{\partial u}{\partial x} + i \frac{\partial v}{\partial x}.
\end{aligned}
$$

Similarly, for the second method of approaching 0, we get that

$$
\begin{aligned}
f'(z) &= \lim_{h \to 0} \frac{f(z+h) - f(z)}{h} \\
&= \lim_{h \to 0,\, h \in i\mathbb{R}} \frac{u(x+h+iy) - u(x+iy)}{h} + i \frac{v(x+h+iy) - v(x+iy)}{h} \\
&= \lim_{h \to 0,\, h \in \mathbb{R}} \frac{u(x+iy+ih) - u(x+iy)}{ih} + i \frac{v(x+iy+ih) - v(x+iy)}{ih} \\
&= -i \frac{\partial u}{\partial y} - i \cdot i \frac{\partial v}{\partial y} = \frac{\partial v}{\partial y} - i \frac{\partial u}{\partial y}.
\end{aligned}
$$

Since these limits are equal, by equating their real and imaginary parts we get a celebrated system of partial differential equations, the **Cauchy–Riemann equations**:

$$
\frac{\partial u}{\partial x} = \frac{\partial v}{\partial y}, \quad \frac{\partial v}{\partial x} = -\frac{\partial u}{\partial y}. \tag{1.16}
$$

We have proved that if f is holomorphic at $z = x + iy$, then the components u and v of f satisfy the Cauchy–Riemann equations (1.16). A kind of converse to this is also true but requires additional assumptions. Assume that $f = u + iv$ is continuously differentiable at $z = x + iy$ (in the sense that each of u and v is a continuously differentiable function of x, y as defined in ordinary real analysis) and satisfies the Cauchy–Riemann equations there. This implies that f has a differential at z; that is, in the notation of vector calculus, if we denote f, z, and Δz as the column vectors

$$f = \begin{pmatrix} u \\ v \end{pmatrix}, \quad z = \begin{pmatrix} x \\ y \end{pmatrix}, \quad \Delta z = \begin{pmatrix} h_1 \\ h_2 \end{pmatrix},$$

then we have

$$f(z + \Delta z) = \begin{pmatrix} u(z) \\ v(z) \end{pmatrix} + \begin{pmatrix} \frac{\partial u}{\partial x} & \frac{\partial u}{\partial y} \\ \frac{\partial v}{\partial x} & \frac{\partial v}{\partial y} \end{pmatrix} \begin{pmatrix} h_1 \\ h_2 \end{pmatrix} + E(h_1, h_2),$$

where $E(\Delta z)$ is a function of Δz that satisfies

$$\lim_{\Delta z \to 0} \frac{|E(\Delta z)|}{|\Delta z|} = 0.$$

Now by the assumption that the Cauchy–Riemann equations hold, we also have

$$\begin{pmatrix} \frac{\partial u}{\partial x} & \frac{\partial u}{\partial y} \\ \frac{\partial v}{\partial x} & \frac{\partial v}{\partial y} \end{pmatrix} \begin{pmatrix} h_1 \\ h_2 \end{pmatrix} = \begin{pmatrix} \frac{\partial u}{\partial x} h_1 + \frac{\partial u}{\partial y} h_2 \\ -\frac{\partial u}{\partial y} h_1 + \frac{\partial u}{\partial x} h_2 \end{pmatrix},$$

which is the vector calculus notation for the complex number

$$\left(\frac{\partial u}{\partial x} - i \frac{\partial u}{\partial y} \right)(h_1 + i h_2) = \left(\frac{\partial u}{\partial x} - i \frac{\partial u}{\partial y} \right) \Delta z.$$

So we have shown that (again, in complex analysis notation)

$$\lim_{\Delta z \to 0} \frac{f(z + \Delta z) - f(z)}{\Delta z} = \lim_{\Delta z \to 0} \left(\frac{\partial u}{\partial x} - i \frac{\partial u}{\partial y} + \frac{E(\Delta z)}{\Delta z} \right) = \frac{\partial u}{\partial x} - i \frac{\partial u}{\partial y}.$$

This proves that f is holomorphic at z with derivative given by $f'(z) = \frac{\partial u}{\partial x} - i \frac{\partial u}{\partial y}$. We summarize the above discussion with the following proposition.

Proposition 1.5 (Cauchy–Riemann equations). *Let $f = u + iv$ be a function of a complex variable z with real and imaginary parts u and v, respectively. If f is holomorphic at z, then the Cauchy–Riemann equations (1.16) are satisfied at z. Conversely, if equations (1.16) are satisfied at z and if u and v are continuously differentiable functions at z, then f is holomorphic at z.*

1.3.4 Third interpretation of holomorphicity: conformal maps

Going back to a more geometric way of thinking about holomorphicity, a further interpretation of the meaning of this property is that holomorphic functions are **conformal mappings** where their derivatives do not vanish. More precisely, assume as before that $f(z)$ is holomorphic at z_0 and $f'(z_0) \neq 0$. Let $\gamma_1, \gamma_2 : (a, b) \to \mathbb{C}$ be two differentiable parameterized planar curves defined on some interval (a, b) containing 0, such that $\gamma_1(0) = \gamma_2(0) = z_0$. The tangent vectors to the curves γ_1 and γ_2 at z_0 are the complex numbers v_1 and v_2 defined by

$$v_1 = \gamma_1'(0), \quad v_2 = \gamma_2'(0). \tag{1.17}$$

Similarly, the tangent vectors to the curves $f \circ \gamma_1$ and $f \circ \gamma_2$ at $f(z_0)$ are

$$w_1 = (f \circ \gamma_1)'(0) \quad w_2 = (f \circ \gamma_2)'(0),$$

which, by a version of the chain rule from vector calculus adapted to complex-analytic notation (Exercise 1.6), can be rewritten as

$$w_1 = f'(\gamma_1(0))\gamma_1'(0) = f'(z_0)\gamma_1'(0), \tag{1.18}$$
$$w_2 = f'(\gamma_2(0))\gamma_2'(0) = f'(z_0)\gamma_2'(0). \tag{1.19}$$

It follows that we can write the inner products (in the ordinary sense of planar vector geometry) between the complex number pairs v_1, v_2 and w_1, w_2 as

$$\langle v_1, v_2 \rangle = \mathrm{Re}(v_1 \overline{v_2}),$$
$$\langle w_1, w_2 \rangle = \mathrm{Re}(w_1 \overline{w_2}) = \mathrm{Re}(\overline{(f'(z_0)\gamma_1'(0))(f'(z_0)\gamma_2'(0))})$$
$$= f'(z_0)\overline{f'(z_0)}\, \mathrm{Re}(v_1 \overline{v_2}) = |f'(z)|^2 \langle v_1, v_2 \rangle. \tag{1.20}$$

If we denote by θ and φ the angle between v_1, v_2 and the angle between w_1, w_2, respectively, we then get using (1.17)–(1.20) that

$$\cos \varphi = \frac{\langle w_1, w_2 \rangle}{|w_1|\,|w_2|} = \frac{|f'(z_0)|^2 \langle v_1, v_2 \rangle}{|f'(z_0)v_1|\,|f'(z_0)v_2|} = \frac{\langle v_1, v_2 \rangle}{|v_1|\,|v_2|} = \cos \theta.$$

So we have shown that under the assumption that $f'(z_0) \neq 0$, the function $f(z)$ maps two curves meeting at an angle θ at z_0 to two curves that meet at the same angle at $f(z_0)$. A function with this property is said to be **conformal** at z_0; see Fig. 1.4.

We can also prove that, under additional assumptions, the converse to the fact that holomorphicity with a nonvanishing derivative implies conformality also holds, making holomorphicity and conformality into nearly equivalent concepts. An important additional condition is that the conformal map needs to be **orientation-preserving**; this

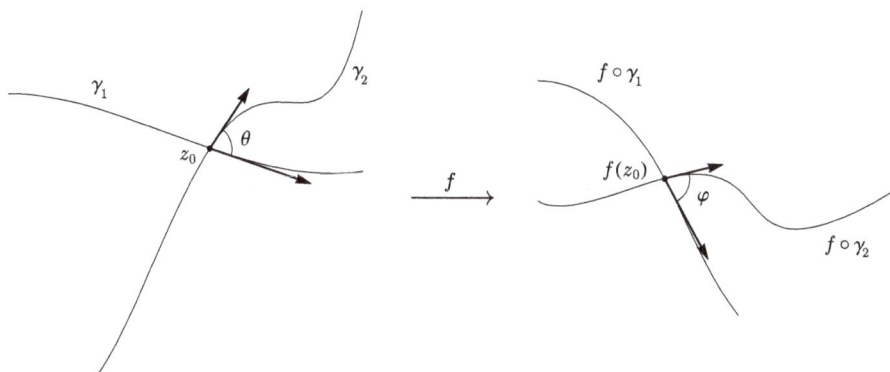

Figure 1.4: A conformal map f preserves the angle between curves crossing at a point: $\theta = \varphi$.

condition can be seen to be necessary by considering the map $f(z) = \bar{z}$, which is conformal but not holomorphic. Recall from vector calculus that for a differentiable vector planar map $f : U \to \mathbb{R}^2$ (where U is some open set in $\mathbb{R}^2$), the **Jacobian matrix of** f is the matrix of partial derivatives,

$$J_f = \begin{pmatrix} \frac{\partial u}{\partial x} & \frac{\partial u}{\partial y} \\ \frac{\partial v}{\partial x} & \frac{\partial v}{\partial y} \end{pmatrix}. \tag{1.21}$$

If $\det J_f > 0$, then we say that f **preserves orientation**.

Theorem 1.6. *If* $f = u + iv$ *is holomorphic at* z_0 *and* $f'(z_0) \neq 0$, *then* f *is conformal at* z_0. *Conversely, if* f *is conformal at* z_0, *continuously differentiable at* z_0 *in the real analysis sense, and preserves orientation at* z_0, *then* f *is holomorphic at* z_0.

The first claim of the theorem was already proved above. The converse direction is proved with the help of the Cauchy–Riemann equations. First, we will need the following simple lemma about linear transformations in the plane.

Lemma 1.7 (Linear conformal maps). *Assume that* $A = \begin{pmatrix} a & b \\ c & d \end{pmatrix}$ *is a* 2×2 *real matrix. The following are equivalent:*
(a) *A preserves orientation (that is,* $\det A > 0$*) and is a* **linear conformal map***, that is, satisfies*

$$\frac{\langle Aw_1, Aw_2 \rangle}{|Aw_1| \, |Aw_2|} = \frac{\langle w_1, w_2 \rangle}{|w_1| \, |w_2|} \quad (w_1, w_2 \in \mathbb{R}^2 \setminus \{(0,0)\}). \tag{1.22}$$

(b) *A takes the form*

$$A = \begin{pmatrix} a & b \\ -b & a \end{pmatrix} \quad \text{for some } a, b \in \mathbb{R} \text{ with } a^2 + b^2 > 0. \tag{1.23}$$

(c) *A takes the form*

$$A = r \begin{pmatrix} \cos\theta & -\sin\theta \\ \sin\theta & \cos\theta \end{pmatrix} \quad \textit{for some } r > 0 \textit{ and } \theta \in \mathbb{R}.$$

(That is, geometrically A acts by a rotation followed by a scaling.)

Proof that (a) $\implies$ (b). Note that both columns of A are nonzero vectors by the assumption that $\det A > 0$. Now applying assumption (1.22) with $w_1 = (1,0)^\top$, $w_2 = (0,1)^\top$ yields that $(a,c) \perp (b,d)$, so that we must have

$$(b,d) = \kappa(-c,a) \tag{1.24}$$

for some $\kappa \in \mathbb{R} \setminus \{0\}$. On the other hand, applying (1.22) with $w_1 = (1,1)^\top$ and $w_2 = (1,-1)^\top$ yields that $(a+b, c+d) \perp (a-b, c-d)$, which is easily seen to be equivalent to $a^2 + c^2 = b^2 + d^2$. When combined with (1.24), this implies that $\kappa = \pm 1$. So A is of one of the two forms $\begin{pmatrix} a & -c \\ c & a \end{pmatrix}$ or $\begin{pmatrix} a & c \\ c & -a \end{pmatrix}$. Finally, the assumption that $\det A > 0$ means that it is the first of those two possibilities that must occur. □

Proof of the implications (b) $\iff$ (c) *and* (b) $\implies$ (a). This is left as an exercise (Exercise 1.7). □

Proof of Theorem 1.6. Assume that f is conformal, continuously differentiable, and orientation-preserving at z_0. Let $\gamma : (a,b) \to \mathbb{C}$ be a differentiable parameterized planar curve with $0 \in (a,b)$, $\gamma(0) = z_0$, and tangent vector $v = \gamma'(0)$ at z_0. By standard properties of differentiable planar maps the tangent vector of $f \circ \gamma$ at $f(w_0)$ is $J_f(z_0)v$ (that is, the Jacobian matrix of f at z_0 acting as a linear map on the vector v, interpreted as a column vector). This means that f is conformal at z_0 if and only if the matrix $J_f(z_0)$ is a linear conformal map in the sense of satisfying condition (1.22) in Lemma 1.7(a). Now adding the knowledge that f is orientation-preserving at z_0, the equivalence stated in the lemma implies that $J_f(z_0)$ must be of the form given on the right-hand side of (1.23). Comparing that form with (1.21), we see that this precisely means that f satisfies the Cauchy–Riemann equations at z_0. This means that the converse part of Proposition 1.5 applies, and we conclude that f is holomorphic at z_0, as claimed. □

Suggested exercises for Section 1.3. 1.3, 1.4, 1.5, 1.6, 1.7, 1.8, 1.9, 1.10, 1.11.

1.4 Additional consequences of the Cauchy–Riemann equations

In the previous section, we saw that the Cauchy–Riemann equations can be used to prove the near-equivalence between holomorphicity with a nonvanishing derivative and conformality. Another curious consequence of the Cauchy–Riemann equations, which gives an alternative geometric picture to that of conformality, is that holomorphicity implies the orthogonality of the level curves of u and of v. That is, if $f = u + iv$ is holomorphic, then

$$\langle \nabla u, \nabla v \rangle = \langle \, (u_x, u_y), (v_x, v_y) \, \rangle = u_x v_x + u_y v_y = v_y v_x - v_x v_y = 0.$$

Since ∇u (resp., ∇v) is orthogonal to the level curve $\{u = c\}$ (resp., the level curve $\{v = d\}$), this proves that the level curves $\{u = c\}$ and $\{v = d\}$ meet at right angles whenever they intersect.

Yet another important and remarkable consequence of the Cauchy–Riemann equations is that, at least under mild smoothness assumptions (which, as we will see later, can be removed) in addition to holomorphicity, u and v are **harmonic functions**. Assume that f is holomorphic at z and is twice continuously differentiable (in the real analysis sense) there. Then

$$\frac{\partial^2 u}{\partial x^2} + \frac{\partial^2 u}{\partial y^2} = \frac{\partial}{\partial x}\left(\frac{\partial u}{\partial x}\right) + \frac{\partial}{\partial y}\left(\frac{\partial u}{\partial y}\right)$$

$$= \frac{\partial}{\partial x}\left(\frac{\partial v}{\partial y}\right) - \frac{\partial}{\partial y}\left(\frac{\partial v}{\partial x}\right) = \frac{\partial^2 v}{\partial x \partial y} - \frac{\partial^2 v}{\partial y \partial x} = 0,$$

i. e., u satisfies Laplace's equation

$$\triangle u = 0,$$

where $\triangle = \frac{\partial^2}{\partial x^2} + \frac{\partial^2}{\partial y^2}$ is the two-dimensional Laplacian operator. A function that satisfies this equation is called a **harmonic function**. Similarly (check), v also satisfies

$$\triangle v = \frac{\partial^2 v}{\partial x^2} + \frac{\partial^2 v}{\partial y^2} = 0.$$

So we have shown that u and v are harmonic functions. This fact is an important connection between complex analysis, real analysis, and the theory of partial differential equations.

We will later see that the assumption of f being twice continuously differentiable is unnecessary, but proving this requires more advanced ideas (see Theorem 1.30 in Section 1.9).

A final remark related to holomorphicity and the Cauchy–Riemann equations is the observation that if $f = u + iv$ is holomorphic, then its Jacobian matrix is given by

$$J_f = \det\begin{pmatrix} u_x & u_y \\ v_x & v_y \end{pmatrix} = u_x v_y - u_y v_x = u_x^2 + v_x^2 = |u_x + iv_x| = |f'(z)|^2. \qquad (1.25)$$

This can also be understood geometrically—spend a moment thinking what the geometric interpretation is.

Suggested exercises for Section 1.4. 1.12.

1.5 Power series

Until now we have not discussed any specific examples of functions of a complex variable. Of course, there are the standard functions that you probably encountered already in your undergraduate studies: polynomials, rational functions, e^z, the trigonometric functions, etc. Aside from these examples, it would be useful to have a general way to construct a large family of functions. Of course, there is such a way: power series, which—nonobviously—turn out to be essentially as general a family of functions as one could hope for.

To make things precise, a **power series** is a function of a complex variable z defined by

$$f(z) = \sum_{n=0}^{\infty} a_n (z - z_0)^n, \tag{1.26}$$

where $z_0 \in \mathbb{C}$, and $(a_n)_{n=0}^{\infty}$ is a sequence of complex numbers. This function is defined wherever the respective series converges.

For which values of z does this formula make sense? Define the number $R \in [0, \infty]$ as

$$R = \left(\limsup_{n \to \infty} |a_n|^{1/n} \right)^{-1},$$

which we refer to as the **radius of convergence** of the power series. Its significance is explained in the following simple result.

Lemma 1.8. 1. *The series* (1.26) *converges absolutely if* $|z - z_0| < R$.
2. *The series* (1.26) *diverges for all z satisfying* $|z - z_0| > R$.

Proof. We assume that $0 < R < \infty$; the edge cases $R = 0$ and $R = \infty$ are left as an exercise (Exercise 1.13). The defining property of R is that for all $\epsilon > 0$, we have that $|a_n| < (\frac{1}{R} + \epsilon)^n$ if n is large enough, and R is the maximal number with that property. Let $z \in D_R(0)$. Since $|z| < R$, we have $|z|(\frac{1}{R} + \epsilon) < 1$ for some fixed $\epsilon > 0$ chosen small enough. This implies that for all $n > N$ (for some large enough N that depends on ϵ),

$$\sum_{n=N}^{\infty} |a_n z^n| < \sum_{n=N}^{\infty} \left[\left(\frac{1}{R} + \epsilon \right) |z| \right]^n,$$

so the series is dominated by a convergent geometric series and hence converges.

Conversely, if $|z| > R$, then $|z|(\frac{1}{R} - \epsilon) > 1$ for some small enough fixed $\epsilon > 0$. Taking a subsequence $(a_{n_k})_{k=1}^{\infty}$ for which $|a_{n_k}| > (\frac{1}{R} - \epsilon)^{n_k}$ for all k (such a subsequence exists by the definition of R), we see that

$$|a_{n_k} z^{n_k}| \geq \left[|z| \left(\frac{1}{R} - \epsilon \right) \right]^{n_k} > 1,$$

that is, the power series (1.26) contains infinitely many terms with modulus > 1 and hence diverges. □

Another important property of power series is given in the following theorem.

Theorem 1.9 (Power series are holomorphic). *Power series are holomorphic functions in the interior of their disc of convergence and can be differentiated termwise there; that is, the derivative of the infinite series is equal to the series of the derivatives.*

Proof. Denote

$$f(z) = \sum_{n=0}^{\infty} a_n z^n = S_N(z) + R_N(z), \quad \text{where}$$

$$S_N(z) = \sum_{n=0}^{N} a_n z^n, \quad R_N(z) = \sum_{n=N+1}^{\infty} a_n z^n,$$

and let

$$g(z) = \sum_{n=1}^{\infty} n a_n z^{n-1}.$$

The claim is that f is differentiable on the disc of convergence and that its derivative is the power series g. Since $n^{1/n} \to 1$ as $n \to \infty$, it is easy to see that $f(z)$ and $g(z)$ have the same radius of convergence. Fix z_0 with $|z_0| < r < R$. We wish to show that $\frac{f(z_0+h)-f(z_0)}{h}$ converges to $g(z_0)$ as $h \to 0$. Observe that

$$\frac{f(z_0 + h) - f(z_0)}{h} - g(z_0) = \left(\frac{S_N(z_0 + h) - S_N(z_0)}{h} - S_N'(z_0) \right)$$
$$+ \frac{R_N(z_0 + h) - R_N(z_0)}{h} + (S_N'(z_0) - g(z_0)). \tag{1.27}$$

In this last expression, the first term converges to 0 as $h \to 0$ for any fixed N. To bound the second term, fix some $\epsilon > 0$, and assume that $|h| < r$, and moreover that $|h|$ is small enough so that $|z_0 + h| < r$. Now make use of the algebraic identity

$$p^n - q^n = (p - q)(p^{n-1} + p^{n-2}q + \cdots + pq^{n-2} + q^{n-1})$$

to get that

$$\left| \frac{R_N(z_0 + h) - R_N(z_0)}{h} \right| \leq \sum_{n=N+1}^{\infty} |a_n| \left| \frac{(z_0 + h)^n - z_0^n}{h} \right|$$
$$= \sum_{n=N+1}^{\infty} |a_n| \left| \frac{h \sum_{k=0}^{n-1} h^k (z_0 + h)^{n-1-k}}{h} \right|$$
$$\leq \sum_{n=N+1}^{\infty} |a_n| n r^{n-1}.$$

The last expression in this chain of inequalities is the tail of an absolutely convergent series, so it can be made $< \epsilon$ be taking N large enough (before taking the limit as $h \to 0$).

Third, we have the limit $S_N'(z_0) \to g(z_0)$ as $N \to \infty$, so we can choose N large enough so that $|S_N'(z_0) - g(z_0)| < \epsilon$. Having thus chosen N, we get finally from (1.27) and the above estimates that

$$\limsup_{h \to 0} \left| \frac{f(z_0 + h) - f(z_0)}{h} - g(z_0) \right| \leq 0 + \epsilon + \epsilon = 2\epsilon.$$

Since ϵ was an arbitrary positive number, this shows that $\frac{f(z_0+h)-f(z_0)}{h} \to g(z_0)$ as $h \to 0$, as claimed. $\qquad\square$

Corollary 1.10. *Holomorphic functions defined as power series are differentiable (in the complex-analytic sense) infinitely many times in the disc of convergence.*

Corollary 1.11. *For a power series $g(z) = \sum_{n=0}^{\infty} a_n(z - z_0)^n$ with positive radius of convergence, we have*

$$a_n = \frac{g^{(n)}(z_0)}{n!}. \tag{1.28}$$

*In other words, $g(z)$ satisfies **Taylor's formula***

$$g(z) = \sum_{n=0}^{\infty} \frac{g^{(n)}(z_0)}{n!}(z - z_0)^n.$$

Suggested exercises for Section 1.5. 1.13, 1.14, 1.15, 1.16.

1.6 Contour integrals

We now introduce **contour integrals**, which are another fundamental building block of the theory.

Contour integrals, like many other types of integrals, take as input a function to be integrated and a "thing" (or "place") over which the function is integrated. In the case of contour integrals, the "thing" is a **contour**, which is (for our current purposes at least) a kind of planar curve. We start by developing some terminology to discuss such objects. A **parameterized curve** is a continuous function $\gamma : [a, b] \to \mathbb{C}$. The value $\gamma(a)$ is called the **starting point**, and $\gamma(b)$ is called the **ending point** (both a, b together are referred to as the **endpoints**). Two curves $\gamma_1 : [a, b] \to \mathbb{C}$, $\gamma_2 : [c, d] \to \mathbb{C}$ are called **equivalent**, denoted $\gamma_1 \sim \gamma_2$, if $\gamma_2(t) = \gamma_1(I(t))$ where $I : [c, d] \to [a, b]$ is a continuous, one-to-one, onto, increasing function. A curve γ is called **simple** if it does not intersect itself, that is, if γ is injective. It is called **closed** if $\gamma(a) = \gamma(b)$.

What we will refer to as a **curve** is, formally speaking, an equivalence class of parameterized curves with respect to the equivalence relation defined above. We also use the word **contour** as a synonym for curve.

In practice, we will usually refer to parameterized curves simply as "curves," which is the usual abuse of terminology that one sees in various places in mathematics, in which one blurs the distinction between equivalence classes and their members, remembering that various definitions, notation, and proof arguments need to "respect the equivalence" in the sense that they do not depend of the choice of a member. (As a meta exercise, try to think of other examples of this phenomenon you might have encountered in your studies.)

For our present context of developing the theory of complex analysis, we will assume that all our curves are piecewise continuously differentiable. More generally, we can assume them to be rectifiable, but we will not bother to develop that theory. There are yet more general contexts in which allowing curves to be merely continuous is beneficial (and indeed some of the ideas we will develop in a complex-analytic context can be carried over to that more general setting), but we will not pursue such distractions either.

You probably encountered curves and parameterized curves in your earlier studies of multivariate calculus, where they were used to define the notion of **line integrals** of vector and scalar fields. Recall that there are two types of line integrals, which are referred to as line integrals of the first and second kind. The line integral of the first kind of a scalar (usually real-valued) function $u(z)$ over a curve γ is defined as

$$\int_\gamma u(z)\, ds = \lim_{\max \Delta s_j \to 0} \sum_{j=1}^n u(z_j)\Delta s_j, \tag{1.29}$$

where the limit is a limit of Riemann sums with respect to a family of tagged partitions of the interval $[a, b]$ over which the curve γ is defined as the norm of the partitions shrinks to 0. Such a partition consists of partition points

$$a = t_0 < t_1 < \cdots < t_n = b,$$

and each partition subinterval $[t_{j-1}, t_j]$ is "tagged" or marked with an arbitrary point τ_j chosen from the subinterval. Given this partition, we denote $z_j = \gamma(\tau_j)$, and the symbols Δs_j refer to finite line elements, namely $\Delta s_j = |z_j - z_{j-1}|$. This notation gives meaning to the right-hand side of (1.29).

The line integral of the second kind is defined for a vector field $\mathbf{F} = (P, Q)$ (using the more traditional notation from calculus; in the complex analysis context, we would regard this object as the complex-valued function $F = P + iQ$) by

$$\int_\gamma \mathbf{F} \cdot d\mathbf{s} = \int_\gamma P\, dx + Q\, dy = \lim_{\max \Delta s_j \to 0} \sum_{j=1}^n P(z_j)\Delta x_j + Q(z_j)\Delta y_j,$$

where the numbers z_j are associated with the tagged partition as above, and

$$x_j = \text{Re}(z_j), \quad y_j = \text{Im}(z_j), \quad \Delta x_j = x_j - x_{j-1}, \quad \Delta y_j = y_j - y_{j-1}.$$

It is well known from calculus that line integrals can be expressed in terms of ordinary (single-variable) Riemann integrals. Take a couple of minutes to remind yourself of why the following formulas are true (assuming that all the functions involved are piecewise continuously differentiable):

$$\int_\gamma u(z)\, ds = \int_a^b u(\gamma(t))|\gamma'(t)|\, dt, \tag{1.30}$$

$$\int_\gamma \mathbf{F} \cdot d\mathbf{s} = \int_a^b \mathbf{F}(\gamma(t)) \cdot \gamma'(t)\, dt. \tag{1.31}$$

(In (1.31), "·" refers to the dot product of vectors in the plane.)

As a further reminder, the basic result known as the **fundamental theorem of calculus for line integrals** states that if $F = \nabla u$, then

$$\int_\gamma \mathbf{F} \cdot d\mathbf{s} = u(\gamma(b)) - u(\gamma(a)).$$

We are now ready to define contour integrals and arc length integrals, which are the complex-analytic analogues of line integrals of the first and second kinds (and are defined in terms of those integrals). For a function $f = u + iv$ of a complex variable z and a curve γ, the contour integral $\int_\gamma f(z)\, dz$ (in words: the integral of f over the curve γ) is defined, loosely speaking, as the line integral of the second kind "$\int_\gamma (u + iv)(dx + i\, dy)$". More precisely, expanding this product of a complex number and a complex differential and separating into real and imaginary components, this definition becomes

$$\int_\gamma f(z)\, dz = \left(\int_\gamma u\, dx - v\, dy \right) + i\left(\int_\gamma v\, dx + u\, dy \right), \tag{1.32}$$

that is, the complex number whose real part is the line integral of $\mathbf{F} \cdot d\mathbf{s}$ and whose imaginary part is the line integral of $\mathbf{G} \cdot d\mathbf{s}$, where $\mathbf{F}$ and $\mathbf{G}$ are the vector fields $\mathbf{F} = (u, -v)$ and $\mathbf{G} = (v, u)$. Appealing to (1.31), you can check easily that the contour integral can be evaluated explicitly as the ordinary Riemann integral

$$\int_\gamma f(z)\, dz = \int_a^b f(\gamma(t))\gamma'(t)\, dt. \tag{1.33}$$

Similarly, the arc length integral is defined as

$$\int_\gamma f(z)\,|dz| = \int_\gamma f(z)\,ds = \int_\gamma u\,ds + i\int_\gamma v\,ds, \tag{1.34}$$

which is simply a line integral of the first kind in which the integrand is complex-valued.

If γ is a closed curve, then we denote the contour integral as $\oint_\gamma f(z)\,dz$, and similarly $\oint_\gamma f(z)\,|dz|$ for the arc length integral.

A particular case of an arc length integral is the length of the curve, denoted $\mathrm{len}(\gamma)$ and defined as the integral of the constant function 1:

$$\mathrm{len}(\gamma) = \int_\gamma |dz| = \int_a^b |\gamma'(t)|\,dt.$$

As mentioned above, our convention of mildly abusing terminology puts on us the burden of having to remember to check that these definitions do not depend on the parameterization of the curve. Indeed, if $\gamma_1 \sim \gamma_2$ are representatives of the same equivalence class of parameterized curves, that is, $\gamma_2(t) = \gamma_1(I(t))$ for some nicely behaved function, then using a standard change of variables in single-variable integrals, we see that

$$\int_{\gamma_2} f(z)\,dz = \int_c^d f(\gamma_2(t))\gamma_2'(t)\,dt = \int_c^d f(\gamma_1(I(t)))(\gamma_1 \circ I)'(t)\,dt$$

$$= \int_c^d f(\gamma_1(I(t)))\gamma_1'(I(t))I'(t)\,dt = \int_a^b f(\gamma_1(\tau))\gamma_1'(\tau)\,d\tau = \int_{\gamma_1} f(z)\,dz. \tag{1.35}$$

The analogous verification in the case of arc length integrals is left as an exercise (Exercise 1.17).

Contour integrals have many surprising properties, but the ones on the following list of basic properties are not of the surprising kind.

Proposition 1.12 (properties of contour integrals). *Contour integrals satisfy the following properties:*

(a) *Linearity as an operator on functions: for functions $f(z)$, $g(z)$ and complex numbers α, β, we have*

$$\int_\gamma (\alpha f(z) + \beta g(z))\,dz = \alpha \int_\gamma f(z)\,dz + \beta \int_\gamma g(z)\,dz.$$

(b) *Linearity as an operator on curves: if a contour Γ is a "composition" of two contours γ_1 and γ_2 (in a sense that is easy to define graphically but tedious to write down precisely), then*

$$\int_{\Gamma} f(z)\, dz = \int_{\gamma_1} f(z)\, dz + \int_{\gamma_2} f(z)\, dz.$$

Similarly, if γ_2 is the "reverse" contour of γ_1, then

$$\int_{\gamma_2} f(z)\, dz = -\int_{\gamma_1} f(z)\, dz.$$

(c) *Triangle inequality:*

$$\left|\int_{\gamma} f(z)\, dz\right| \le \int |f(z)|\, |dz| \le \mathrm{len}(\gamma) \cdot \sup_{z \in \gamma} |f(z)|.$$

Proof. Exercise 1.18. □

Contour integrals have their own version of the fundamental theorem of calculus.

Theorem 1.13 (The fundamental theorem of calculus for contour integrals). *If γ is a curve connecting two points w_1 and w_2 in a region Ω on which a function F is holomorphic, then*

$$\int_{\gamma} F'(z)\, dz = F(w_2) - F(w_1).$$

Equivalently, the theorem says that to compute a general contour integral $\int_{\gamma} f(z)\, dz$, we try to find a **primitive** of f, that is, a holomorphic function F such that $F'(z) = f(z)$ on all of Ω. (A term synonymous with "primitive" is **antiderivative**.) If we found such a primitive, then the contour integral $\int_{\gamma} f(z)\, dz$ is given by $F(w_2) - F(w_1)$.

Proof. For smooth curves, an easy application of the chain rule gives

$$\int_{\gamma} F'(z)\, dz = \int_{a}^{b} F'(\gamma(t))\gamma'(t)\, dt = \int_{a}^{b} (F \circ \gamma)'(t)\, dt = (F \circ \gamma)(t)\big|_{t=a}^{t=b}$$

$$= F(\gamma(b)) - F(\gamma(a)) = F(w_2) - F(w_1).$$

For piecewise smooth curves, this is a trivial extension that is left to the reader. □

Many of our discussions of contour integrals will involve the behavior of integrals over *closed* contours and the interplay between the properties of such integrals and integrals over general contours. As an example of this interplay, the above result has an easy—but important—consequence for integrals over closed contours.

Corollary 1.14. *If $f = F'$ where F is holomorphic on a region Ω—that is, f has a primitive—then for any closed contour γ in Ω, we have*

$$\oint_{\gamma} f(z)\, dz = 0.$$

This last result has the following partial converse.

Proposition 1.15. *If* $f : \Omega \to \mathbb{C}$ *is a continuous function on a region* Ω *such that*

$$\oint_\gamma f(z)\, dz = 0$$

for any closed contour in Ω, *then* f *has a primitive.*

Proof. Fix some $z_0 \in \Omega$. For any $z \in \Omega$, there is some curve $\gamma(z_0, z)$ connecting z_0 and z (since Ω is connected and open, hence pathwise-connected—a standard exercise in topology). Moreover, it is also not hard to see that the curve can be assumed to be piecewise differentiable. Define

$$F(z) = \int_{\gamma(z_0, z)} f(w)\, dw. \tag{1.36}$$

By the assumption this integral does not depend on which curve $\gamma(z_0, z)$ connecting z_0 and z was chosen, so $F(z)$ is well-defined. We now claim that F is holomorphic and its derivative is equal to f. To see this, note that if h is a complex number such that $z + h \in \Omega$, then

$$\frac{F(z+h) - F(z)}{h} - f(z)$$

$$= \frac{1}{h}\left(\int_{\gamma(z_0, z+h)} f(w)\, dw - \int_{\gamma(z_0, z)} f(w)\, dw \right) - f(z)$$

$$= \frac{1}{h} \int_{\gamma(z, z+h)} f(w)\, dw - f(z) = \frac{1}{h} \int_{\gamma(z, z+h)} (f(w) - f(z))\, dw, \tag{1.37}$$

where $\gamma(z, z + h)$ denotes a curve in Ω connecting z and $z + h$. When $|h|$ is sufficiently small so that the disc $D_h(z)$ is contained in Ω, we can take $\gamma(z, z + h)$ as the straight line segment connecting z and $z + h$. For such h, we get that

$$\left| \frac{F(z+h) - F(z)}{h} - f(z) \right| \leq \frac{1}{h} \operatorname{len}(\gamma(z, z+h)) \sup_{w \in D_h(z)} |f(w) - f(z)|$$

$$= \sup_{w \in D_h(z)} |f(w) - f(z)| \xrightarrow[h \to 0]{} 0$$

by the continuity of f. □

Lemma 1.16. *If* f *is holomorphic on* Ω *and* $f' \equiv 0$, *then* f *is a constant.*

Proof. Fix some $z_0 \in \Omega$. For any $z \in \Omega$, as we discussed above, there is a path $\gamma(z_0, z)$ connecting z_0 and z. Then

$$f(z) - f(z_0) = \int\limits_{\gamma(z_0,z)} f'(w)\,dw = 0,$$

and hence $f(z) \equiv f(z_0)$, that is, f is constant. □

Suggested exercises for Section 1.6. 1.17, 1.18.

1.7 The Cauchy, Goursat, and Morera theorems

One of the central results in complex analysis is **Cauchy's theorem**.

Theorem 1.17 (Cauchy's theorem.). *If f is a holomorphic function on a simply connected region Ω, then for any closed curve in Ω, we have*

$$\oint_\gamma f(z)\,dz = 0.$$

The challenges facing us are as follows: first, to prove Cauchy's theorem for curves and regions that are relatively simple (where we do not have to deal with subtle topological considerations); second, to define what "simply connected" means; third, to extend the theorem to the most general setting. This is done in the next section.

Two other theorems closely related to Cauchy's theorem are **Goursat's theorem**, a relatively easy particular case of Cauchy's theorem, and **Morera's theorem**, which is a kind of converse to Cauchy's theorem.

Theorem 1.18 (Goursat's theorem). *If f is holomorphic on a region Ω, T is a triangle contained in Ω, and ∂T is the boundary of T (considered as a curve in the usual sense), then*

$$\oint_{\partial T} f(z)\,dz = 0. \tag{1.38}$$

Theorem 1.19 (Morera's theorem). *If $f : \Omega \to \mathbb{C}$ is a continuous function on a region Ω such that*

$$\oint_\gamma f(z)\,dz = 0$$

for any closed contour in Ω, then f is holomorphic on Ω.

Morera's theorem is proved in Section 1.9.

Proof of Goursat's theorem. The proof can be summarized with a slogan "localize the damage." Namely, try to translate a global statement about the integral around the triangle to a local statement about behavior near a specific point inside the triangle, which would become manageable since we have a good understanding of the local behavior of a holomorphic function near a point. If something goes wrong with the global integral, then something has to go wrong at the local level, and we will show that cannot happen. (Although technically the proof is not a proof by contradiction, conceptually I find this a helpful way to think about it).

The idea can be made more precise using *triangle subdivision*. Specifically, let $T^{(0)} = T$, and define a hierarchy of subdivided triangles:

$$
\begin{aligned}
&\text{order 0 triangle:} && T^{(0)}, \\
&\text{order 1 triangles:} && T_j^{(1)}, 1 \le j \le 4, \\
&\text{order 2 triangles:} && T_{j,k}^{(2)}, 1 \le j, k \le 4, \\
&\text{order 3 triangles:} && T_{j,k,\ell}^{(3)}, 1 \le j, k, \ell \le 4, \\
&\qquad\qquad\vdots \\
&\text{order } n \text{ triangles:} && T_{j_1,\dots,j_n}^{(n)}, 1 \le j_1,\dots,j_n \le 4. \\
&\qquad\qquad\vdots
\end{aligned}
$$

Here the triangles $T_{j_1,\dots,j_n}^{(n)}$ for $j_n = 1, 2, 3, 4$ are obtained by subdividing the order-$(n-1)$ triangle $T_{j_1,\dots,j_{n-1}}^{(n-1)}$ into 4 subtriangles whose vertices are the vertices and/or edge bisectors of $T_{j_1,\dots,j_{n-1}}^{(n-1)}$; see Fig. 1.5.

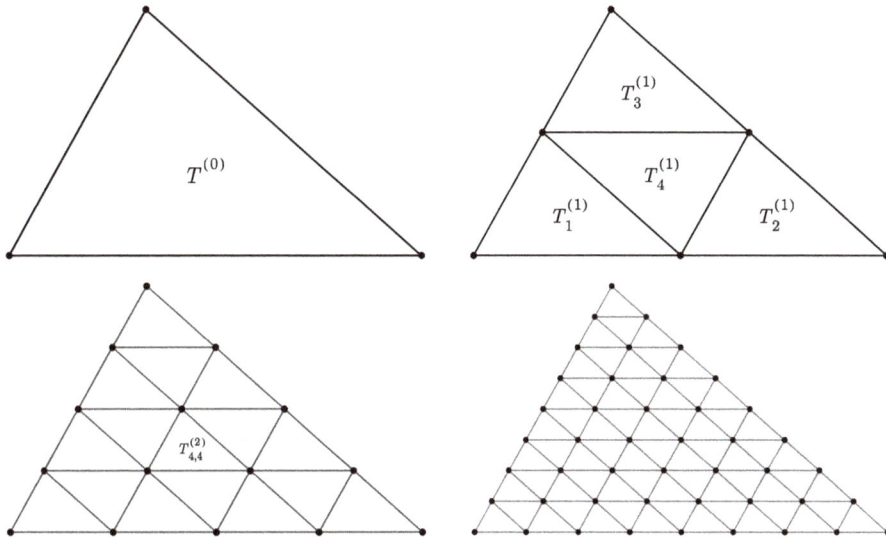

Figure 1.5: The triangle $T = T^{(0)}$ and the first few steps in its hierarchy of subdivided triangles.

Now, given the way this subdivision was done, it is clear that we have the relation

$$\oint_{\partial T^{(n-1)}_{j_1,\dots,j_{n-1}}} f(z)\,dz = \sum_{j_n=1}^{4} \oint_{\partial T^{(n)}_{j_1,\dots,j_n}} f(z)\,dz$$

(where $\partial T^{(n)}_{j_1,\dots,j_n}$ refers as before to the boundary of the triangle $T^{(n)}_{j_1,\dots,j_n}$, considered as a curve oriented in the positive sense) due to cancelation along the internal edges, and hence

$$\oint_{\partial T^{(0)}} f(z)\,dz = \sum_{j_1,\dots,j_n=1}^{4} \oint_{\partial T^{(n)}_{j_1,\dots,j_n}} f(z)\,dz.$$

So the contour integral around the boundary of the original triangle is equal to the sum of the integrals around all 4^n triangles at the nth subdivision level. Now a key observation is that *one of these integrals has to have a modulus that is at least as big as the average*, that is, there exists an n-tuple $\mathbf{j}(n) = (j_1^{(n)},\dots,j_n^{(n)}) \in \{1,2,3,4\}^n$ for which

$$\left|\oint_{\partial T^{(0)}} f(z)\,dz\right| \le \sum_{j_1,\dots,j_n=1}^{4} \left|\oint_{\partial T^{(n)}_{j_1,\dots,j_n}} f(z)\,dz\right| \le 4^n\left|\oint_{\partial T^{(n)}_{\mathbf{j}(n)}} f(z)\,dz\right|. \qquad (1.39)$$

Moreover, we can choose $\mathbf{j}(n)$ inductively in such a way that the triangles $T^{(n)}_{\mathbf{j}(n)}$ are nested, that is, $T^{(n)}_{\mathbf{j}(n)} \subset T^{(n-1)}_{\mathbf{j}(n-1)}$ for $n \ge 1$, or, equivalently, $\mathbf{j}(n) = (j_1^{(n-1)},\dots,j_{n-1}^{(n-1)}, k)$ for some $1 \le k \le 4$. To make this happen, choose a value of k for which $|\oint_{\partial T^{(n)}_{(\mathbf{j}(n-1),k)}} f(z)\,dz|$ is greater than or equal to the average

$$\frac{1}{4}\sum_{d=1}^{4}\left|\oint_{\partial T^{(n)}_{(\mathbf{j}(n-1),d)}} f(z)\,dz\right|,$$

which in turn can be seen (by induction) to be greater than or equal to

$$\frac{1}{4}\left|\sum_{d=1}^{4}\oint_{\partial T^{(n)}_{(\mathbf{j}(n-1),d)}} f(z)\,dz\right| = \frac{1}{4}\left|\oint_{\partial T^{(n-1)}_{\mathbf{j}(n-1)}} f(z)\,dz\right| \ge \frac{1}{4}\cdot 4^{-(n-1)}\left|\oint_{\partial T} f(z)\,dz\right|,$$

thereby justifying (1.39).

We now claim that the sequence of nested triangles $T^{(n)}_{\mathbf{j}(n)}$ shrinks to a single point, that is, we have

$$\bigcap_{n=0}^{\infty} T^{(n)}_{\mathbf{j}(n)} = \{z_0\}.$$

for some point $z_0 \in T$. Indeed, the diameter of the triangles goes to 0 as $n \to \infty$, so certainly there cannot be two distinct points in the intersection. On the other hand, the triangles $T_{j(n)}^{(n)}$ are all compact, and the finite intersections $\bigcap_{n=0}^{N} T_{j(n)}^{(n)}$ are nonempty, so by the standard finite intersection property of compact sets the full intersection $\bigcap_{n=0}^{\infty} T_{j(n)}^{(n)}$ is also nonempty.

Having defined z_0, write $f(z)$ for z near z_0 as

$$f(z) = f(z_0) + f'(z_0)(z - z_0) + \psi(z)(z - z_0),$$

where

$$\psi(z) = \frac{f(z) - f(z_0)}{z - z_0} - f'(z_0).$$

The holomorphicity of f at z_0 implies that $\psi(z) \to 0$ as $z \to z_0$. Denote by $d^{(n)}$ the diameter of $T_{j(n)}^{(n)}$ and by $p^{(n)}$ its perimeter. Each subdivision shrinks both the diameter and perimeter by a factor of 2, so we have

$$d^{(n)} = 2^{-n} d^{(0)}, \quad p^{(n)} = 2^{-n} p^{(0)}.$$

It follows that

$$\left| \int_{\partial T_{j(n)}^{(n)}} f(z)\, dz \right| = \left| \int_{\partial T_{j(n)}^{(n)}} \left(f(z_0) + f'(z_0)(z - z_0) + \psi(z)(z - z_0) \right) dz \right|$$

$$= \left| \int_{\partial T_{j(n)}^{(n)}} \psi(z)(z - z_0)\, dz \right| \le p^{(n)} d^{(n)} \sup_{z \in T_{j(n)}^{(n)}} |\psi(z)|$$

$$= 4^{-n} p^{(0)} d^{(0)} \sup_{z \in T_{j(n)}^{(n)}} |\psi(z)|.$$

This estimate allows us to finish, since combining it with (1.39), we get that

$$\left| \int_{\partial T^{(0)}} f(z)\, dz \right| \le p^{(0)} d^{(0)} \sup_{z \in T_{j(n)}^{(n)}} |\psi(z)| \xrightarrow[n \to \infty]{} 0,$$

which establishes (1.38). □

The next few results illustrate how Goursat's theorem, for all its apparent simplicity, can be used to quickly derive even stronger versions of Cauchy's theorem, gradually building up our knowledge toward the general version that will be proved in the next section.

Corollary 1.20 (Goursat's theorem for rectangles). *Theorem 1.18 is also true when we replace the word "triangle" with "rectangle."*

Proof. Obviously, a rectangle can be decomposed as the union of two triangles, with the contour integral around the rectangle being the sum of the integrals around the two triangles due to cancelation of the integrals going in both directions along the diagonal.

□

Corollary 1.21 (existence of a primitive for a holomorphic function on a disc). *If f is holomorphic on a disc D, then $f = F'$ for some holomorphic function F on D.*

Proof. The claim is identical to Proposition 1.15, but with a different set of assumptions. In fact, the proof of that proposition can be easily adapted to prove the existence of a primitive in the current setting. Specifically, we again define the purported primitive F for f using (1.36), but this time using a particular choice of path $\gamma(z_0, z)$ connecting z_0 and z, namely, we take $\gamma(z_0, z)$ to be the straight line segment from z_0 to z.

We now claim that with this definition, for h small in magnitude (so that $z + h$ is still in the disc D), the chain of equalities (1.37) still holds, where in this chain, we also interpret $\gamma(z, z + h)$ as the straight line segment connecting z and $z + h$. If we can show this, then the rest of the proof carries through as before. Now, upon inspection of (1.37), we see that the first and third equalities still hold trivially; it is only the middle equality that needs to be explained. This equality can be rewritten as

$$\int_{\gamma(z_0,z)} f(w)\, dw + \int_{\gamma(z,z+h)} f(w)\, dw - \int_{\gamma(z_0,z+h)} f(w)\, dw = 0,$$

a relationship between the contour integrals of f along the three straight line segments $\gamma(z_0, z)$, $\gamma(z_0, z + h)$, and $\gamma(z, z + h)$. This is simply the statement that the contour integral along the boundary of the triangle with vertices z_0, z, and $z + h$ is 0, which follows from Goursat's theorem.

□

Theorem 1.22 (Cauchy's theorem for a disc). *If f is holomorphic on a disc, then $\oint_\gamma f\, dz = 0$ for any closed contour γ in the disc.*

Proof. By Corollary 1.21, f has a primitive, so Corollary 1.14 implies the claimed consequence.

□

1.8 Simply connected regions and the general version of Cauchy's theorem

We now develop the additional concepts required to formulate and prove the general version of Cauchy's theorem. A key notion is that of **homotopy of curves**. Given a region $\Omega \subset \mathbb{C}$, two parameterized curves $\gamma_1, \gamma_2 : [0, 1] \to \Omega$ (assumed for simplicity of notation to be defined on $[0, 1]$) are said to be **homotopic (with fixed endpoints)** if $\gamma_1(0) = \gamma_2(0)$, $\gamma_1(1) = \gamma_2(1)$, and there exists a function $F : [0, 1] \times [0, 1] \to \Omega$ such that
i) F is continuous.

ii) $F(0, t) = \gamma_1(t)$ for all $t \in [0, 1]$.
iii) $F(1, t) = \gamma_2(t)$ for all $t \in [0, 1]$.
iv) $F(s, 0) = \gamma_1(0)$ for all $s \in [0, 1]$.
v) $F(s, 1) = \gamma_1(1)$ for all $s \in [0, 1]$.

The map F is called a **homotopy** between γ_1 and γ_2. Intuitively, for each $s \in [0, 1]$, the function $F_s : t \mapsto F(s, t)$ defines a curve connecting the two endpoints $\gamma_1(0)$ and $\gamma_1(1)$. As s grows from 0 to 1, this family of curves transitions in a continuous way between the curve γ_1 and γ_2 with the endpoints being fixed in place; see Fig. 1.6.

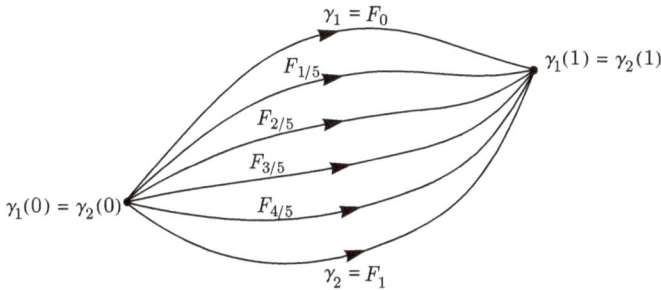

Figure 1.6: A homotopy between two curves γ_1 and γ_2, visualized as a one-parameter family of curves $t \mapsto F_s(t)$ that interpolate continuously between γ_1 and γ_2, with the endpoints staying fixed.

A common alternative way to define the notion of homotopy of curves is for closed curves, where the endpoints are not fixed, but the homotopy must keep the curves closed as it is deforming them. The definition of a simply connected region then becomes a region in which any two closed curves are homotopic. It is not hard to show that those two definitions are equivalent.

It is easy (but recommended!) to check that the relation of being homotopic is an equivalence relation; see Exercise 1.19.

Next, we define the notion of a **simply connected** region. A region Ω is called simply connected if any two curves γ_1, γ_2 in Ω with the same endpoints are homotopic. Note that this is a *topological* property (in the sense that it is preserved under homeomorphism). The complex plane, the unit disc, and any region homeomorphic to the unit disc are simply connected regions (Exercise 1.20).

Theorem 1.23. *If f is a holomorphic function on a region Ω, and γ_0, γ_1 are two curves on Ω with the same endpoints that are homotopic, then*

$$\int_{\gamma_0} f(z)\, dz = \int_{\gamma_1} f(z)\, dz.$$

Proof. As with the proof of Goursat's theorem in the previous section, this proof is based on the idea of reducing the global statement about the equality of the two contour integrals into a local statement. Denote by $F : [0,1] \times [0,1] \to \Omega$ the homotopy between γ_0 and γ_1, and for any $s \in [0,1]$, denote by $\gamma_s : [0,1] \to \mathbb{C}$ the curve $\gamma_s(t) = F(s,t)$. The strategy of the proof is to show that there are values $0 = s_0 < s_1 < s_2 < \cdots < s_n = 1$ such that

$$\int_{\gamma_{s_0}} f(z)\, dz = \int_{\gamma_{s_1}} f(z)\, dz = \cdots = \int_{\gamma_{s_{n-1}}} f(z)\, dz = \int_{\gamma_{s_n}} f(z)\, dz.$$

In fact, we can take $s_k = k/n$ for $0 \le k \le n$ with large n; we will define n more precisely below. Fix $1 \le k \le n$. To prove the equality between the two integrals $\int_{\gamma_{s_{k-1}}} f(z)\, dz$ and $\int_{\gamma_{s_k}} f(z)\, dz$, we decompose each of the two integrals into a sum of integrals over small pieces of the contours $\gamma_{s_{k-1}}$ and γ_{s_k} by writing them as

$$\int_{\gamma_{s_{k-1}}} f(z)\, dz = \sum_{j=1}^{m} \int_{\gamma_{s_{k-1}|[t_{j-1},t_j])}} f(z)\, dz, \tag{1.40}$$

$$\int_{\gamma_{s_k}} f(z)\, dz = \sum_{j=1}^{m} \int_{\gamma_{s_k|[t_{j-1},t_j]}} f(z)\, dz. \tag{1.41}$$

Here $\gamma_{s_k|[t_{j-1},t_j]}$ denotes the restriction of the contour γ to the interval $[t_{j-1}, t_j]$, where t_j denotes some sequence of points $0 = t_0 < t_1 < \cdots < t_n = 1$ partitioning $[0,1]$ into subintervals $[t_{j-1}, t_j]$. We will show at the end of the proof that the partition $t_j = j/n$ for $0 \le j \le n - 1$, where n is large (and is the same n that was used for the definition of s_k above), works well for our purposes. Specifically, we will show that with the way we defined s_k and t_j above and with n taken sufficiently large, the following assumption is satisfied: for all $1 \le k, j \le n$, there exists an open disc $D_{k,j} \subset \Omega$ containing the two curve segments $\gamma_{s_{k-1}|[t_{j-1},t_j]}$ and $\gamma_{s_k|[t_{j-1},t_j]}$.

Under this assumption, to prove that the two integrals (1.40)–(1.41) are equal, it suffices to prove that for any $1 \le j \le n$, we have the equality

$$\int_{\gamma_{s_{k-1}|[t_{j-1},t_j])}} f(z)\, dz = \int_{\gamma_{s_k|[t_{j-1},t_j]}} f(z)\, dz \tag{1.42}$$

between the integrals over the small subcontours.

For each $0 \le j \le n$, let $\eta_{k,j}$ denote a straight line segment (considered as a parameterized curve) from $\gamma_{s_{k-1}}(t_j)$ to $\gamma_{s_k}(t_j)$, and for each $1 \le j \le m$, let $\Gamma_{k,j}$ denote the closed curve $\gamma_{s_{k-1}}([t_{j-1}, t_j]) + \eta_{k,j} - \gamma_{s_k}([t_{j-1}, t_j]) - \eta_{k,j-1}$ (in words: the concatenation of the four curves $\gamma_{s_{k-1}}([t_{j-1}, t_j])$, $\eta_{k,j}$, "the reverse of $\gamma_{s_k}([t_{j-1}, t_j])$," and "the reverse of $\eta_{k,j-1}$"). By the assumption on the disc $D_{k,j}$ the curve $\Gamma_{k,j}$ is contained in $D_{k,j}$. Therefore by Cauchy's theorem for discs (Theorem 1.22) we have

$$\oint_{\Gamma_{k,j}} f(z)\, dz = 0,$$

or, more explicitly,

$$\int_{\gamma_{s_{k-1}}|[t_{j-1},t_j]} f(z)\, dz - \int_{\gamma_{s_k}|[t_{j-1},t_j]} f(z)\, dz = \int_{\eta_{k,j-1}} f(z)\, dz - \int_{\eta_{k,j}} f(z)\, dz.$$

Summing this relation over j and recalling (1.40)–(1.41), we get that

$$\int_{\gamma_{s_{k-1}}} f(z)\, dz - \int_{\gamma_{s_k}} f(z)\, dz = \sum_{j=1}^{m}\left(\int_{\eta_{k,j-1}} f(z)\, dz - \int_{\eta_{k,j}} f(z)\, dz \right)$$

$$= \int_{\eta_{k,0}} f(z)\, dz - \int_{\eta_{k,m}} f(z)\, dz = 0.$$

(Here, in the next-to-last step the sum is telescoping, and in the last step, we note that $\eta_{k,0}$ and $\eta_{k,m}$ are both degenerate curves, each of which simply stays at a single point.) This is precisely equality (1.42) we wanted.

It remains to justify the assumption about the discs $D_{k,j}$. This is done as follows. First, since the set $A = F([0,1] \times [0,1])$ is compact, it is easy to see (for example, using the Heine–Borel property) that there exists a number $\epsilon > 0$ such that the discs $D_\epsilon(z)$ are contained in Ω for all $z \in A$. Second, since F is continuous, and hence also uniformly continuous, on $[0,1] \times [0,1]$, there exists a number $\delta > 0$ such that for any $0 \le s, t \le 1$ with $|s - s'| + |t - t'| < \delta$, we have

$$\left| \gamma_{s'}(t') - \gamma_s(t) \right| = \left| F(s',t') - F(s,t) \right| < \epsilon.$$

Let n be an integer larger than $2/\delta$, and let $s_k = k/n$ and $t_j = j/n$ as before. We define the discs $D_{k,j}$ by $D_{k,j} = D_\epsilon(\gamma_{s_{k-1}}(t_{j-1}))$ and claim that they satisfy our assumption. Indeed, if $t \in [t_{j-1}, t_j]$, then $|t - t_{j-1}| \le 1/n < \delta/2$, so $|\gamma_{s_{k-1}}(t) - \gamma_{s_{k-1}}(t_{j-1})| < \epsilon$. This shows that the curve segment $\gamma_{s_{k-1}|[t_{j-1},t_j]}$ is contained in $D_{k,j}$. Similarly, $|t - t_{j-1}| + |s_k - s_{k-1}| \le 1/n + 1/n < \delta$, so $|\gamma_{s_k}(t) - \gamma_{s_{k-1}}(t_{j-1})| < \epsilon$, that is, the curve segment $\gamma_{s_k|[t_{j-1},t_j]}$ is also contained in $D_{k,j}$. This proves that our assumption about the discs $D_{k,j}$ is satisfied and finishes the proof. □

Theorem 1.24 (Cauchy's theorem, general version). *If f is holomorphic on a simply connected region Ω, then for any closed curve in Ω, we have*

$$\oint_{\gamma} f(z)\, dz = 0.$$

Proof. Assume without loss of generality that γ is parameterized as a curve on $[0,1]$. Then it can be thought of as the concatenation of two curves γ_1 and $-\gamma_2$, where $\gamma_1 =$

$\gamma_{|[0,1/2]}$, and γ_2 is the "reverse" of the curve $\gamma_{|[1/2,1]}$. Note that γ_1 and γ_2 have the same endpoints. By Theorem 1.23 we have

$$\int_\gamma f(z)\,dz = \int_{\gamma_1-\gamma_2} f(z)\,dz = \int_{\gamma_1} f(z)\,dz - \int_{\gamma_2} f(z)\,dz = 0. \qquad \square$$

Combining Theorem 1.24 with Proposition 1.15, we get the following result.

Corollary 1.25. *Any holomorphic function on a simply connected region has a primitive.*

One subtle issue that is glossed over in many complex analysis textbooks is the question of how to recognize when a region is simply connected. In many practical situations, it is easy to recognize or at least accept as intuitively plausible, that the region under discussion is homeomorphic to a disc, which of course implies the property of being simply connected. This informal style of reasoning will be sufficient for our needs in this book. For those readers who prefer a higher level of rigor, we cite without proof the following result from topology.

Theorem 1.26. *Given any simple closed curve γ in the plane, there is a region Ω such that:*
1. *Ω is bounded;*
2. *Ω is the unique connected component of $\mathbb{C} \setminus \gamma$ that is bounded;*
3. *Ω is homeomorphic to a disc.*

Because of the second property of Ω given in the theorem, Ω is usually referred to as "the region enclosed by γ."

Theorem 1.26 is a version of the **Jordan–Schoenflies theorem**, which in turn is a strengthened version of the **Jordan curve theorem**. These results have elementary proofs that do not require complex analysis; see [9, 69] and [W6] for additional discussion and references. A planar curve that is simple and closed is often referred to as a **Jordan curve**.

Suggested exercises for Section 1.8. 1.19, 1.20, 1.21, 1.22.

1.9 Consequences of Cauchy's theorem

Theorem 1.27 (Cauchy's integral formula). *If f is holomorphic on a region Ω containing the closed disc $D_{\leq R}(z_0)$, then*

$$\frac{1}{2\pi i} \oint_{C_R(z_0)} \frac{f(w)}{w-z}\,dw = \begin{cases} f(z) & \text{if } z \in D_R(z_0), \\ 0 & \text{if } z \in \Omega \setminus D_{\leq R}(D), \\ \text{undefined} & \text{if } z \in C_R(z_0). \end{cases} \qquad (1.43)$$

Proof. The case where $z \in \Omega \setminus D_{\leq R}(D)$ is covered by Cauchy's theorem in a disc, since in that case the function $w \mapsto f(w)/(w-z)$ is holomorphic in an open set containing $D_{\leq R}(D)$.

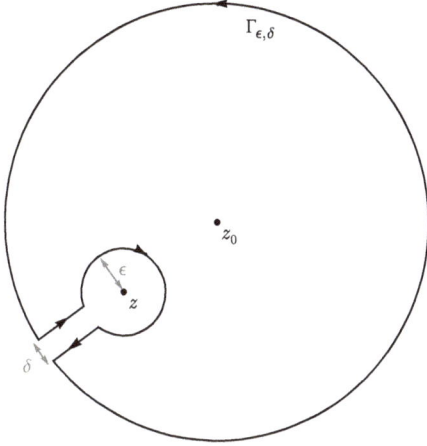

Figure 1.7: The keyhole contour $\Gamma_{\epsilon,\delta}$.

It remains to deal with the case $z \in D_R(z_0)$. In this case, denote $F_z(w) = f(w)/(w-z)$. The idea is now to consider instead the integral

$$\oint_{\Gamma_{\epsilon,\delta}} F_z(w)\, dw = \oint_{\Gamma_{\epsilon,\delta}} \frac{f(w)}{w-z}\, dw,$$

where $\Gamma_{\epsilon,\delta}$ is a so-called **keyhole contour**, namely a contour comprising a large circular arc around z_0 that is a subset of the circle $C_R(z_0)$, and another smaller circular arc of radius ϵ centered at z, with two straight line segments connecting the two circular arcs to form a closed curve, such that the width of the "neck" of the keyhole is δ. (Here ϵ and δ are two small positive parameters; think of ϵ as being small and of δ as being much smaller than ϵ.) See Fig. 1.7. Note that the function $F_z(w)$ is holomorphic inside the region enclosed by $\Gamma_{\epsilon,\delta}$. Moreover, this region is clearly homeomorphic to a disc and so is simply connected. Therefore Cauchy's theorem gives that

$$\oint_{\Gamma_{\epsilon,\delta}} F_z(w)\, dw = 0.$$

We now take the limit of this equation as $\delta \to 0$. The two parts of the integral along the "neck" of the contour $\Gamma_{\epsilon,\delta}$ cancel out in the limit because F_z is continuous, and hence uniformly continuous, on the compact set $D_{\leq R}(z_0) \setminus D_\epsilon(z)$. So we can conclude that

$$\oint_{C_R(z_0)} F_z(w)\, dw = \oint_{C_\epsilon(z)} F_z(w)\, dw. \tag{1.44}$$

The next and final step is to take the limit as $\epsilon \to 0$ of the right-hand side of this equation. Write

$$F_z(w) = \frac{f(w) - f(z)}{w - z} + f(z) \cdot \frac{1}{w - z}. \qquad (1.45)$$

Integrating each of these two terms separately, for the first term, we have

$$\left| \oint_{C_\epsilon(z)} \frac{f(w) - f(z)}{w - z} \, dw \right| \leq 2\pi\epsilon \cdot \sup_{|w - z| = \epsilon} \frac{|f(w) - f(z)|}{\epsilon}$$

$$= 2\pi \sup_{|w - z| = \epsilon} |f(w) - f(z)| \xrightarrow{\epsilon \to 0} 0 \qquad (1.46)$$

by the continuity of f; and for the second term,

$$\oint_{C_\epsilon(z)} f(z) \cdot \frac{1}{w - z} \, dw = f(z) \oint_{C_\epsilon(z)} \frac{1}{w - z} \, dw = 2\pi i f(z) \qquad (1.47)$$

(by a standard calculation; see Exercise 1.21). Combining (1.44) and (1.47) gives that $\oint_{C_R(z_0)} \frac{1}{2\pi i} F_z(w) \, dw = f(z)$, which was the formula to be proved. $\square$

An important particular case of (1.43) is the one in which $z = z_0$. Cauchy's integral formula gives in this case that

$$f(z) = \frac{1}{2\pi} \oint_{C_R(z_0)} f(w) \frac{dw}{i(w - z)} = \frac{1}{2\pi} \int_0^{2\pi} f(z + Re^{it}) \, dt.$$

In other words, we have proved the following result.

Theorem 1.28 (Mean value property for holomorphic functions). *If f is holomorphic on a region Ω containing the closed disc $D_{\leq R}(z_0)$, then the value $f(z_0)$ is equal to the average of the values of f around the circle $C_R(z_0)$.*

Considering what the mean value property means for the real and imaginary parts of $f = u + iv$, which are harmonic functions, we see that they in turn also satisfy a similar mean value property:

$$u(x, y) = \frac{1}{2\pi} \int_0^{2\pi} u(x + R\cos t, y + R\sin t) \, dt. \qquad (1.48)$$

In fact, (1.48) holds for all harmonic functions and is a result known as the **mean value property for harmonic functions**. This result is proved in many textbooks using methods from real analysis or partial differential equations. Alternatively, it can be derived from the above considerations by proving that every harmonic function in a disc is the real part of a holomorphic function.

Theorem 1.29 (Cauchy's integral formula, extended version). *Under the same assumptions as in Theorem 1.27, f is differentiable infinitely many times, and for $z \in D_R(z_0)$, its derivatives $f^{(n)}(z)$ are given by*

$$f^{(n)}(z) = \frac{n!}{2\pi i} \oint_{C_R(z_0)} \frac{f(w)}{(w-z)^{n+1}} \, dw. \tag{1.49}$$

Proof. We prove by induction that for all $n \geq 0$, $f^{(n)}(z)$ exists, is differentiable, and is given by the expression on the right-hand side of (1.49). For $n = 0$, this is the statement of (1.43) in the case $z \in D_R(z_0)$. For the inductive step, assuming that we have proved the claim for a given value of n, the idea is now to show that the expression on the right-hand side of (1.49) can be differentiated under the integral sign. More precisely, observe that, by the inductive hypothesis, if $z + h \in D_R(z_0)$ (which is the case where h is close enough to 0), then

$$\frac{f^{(n)}(z+h) - f^{(n)}(z)}{h} = \frac{n!}{2\pi i} \oint_C f(w) \cdot \frac{1}{h}\left(\frac{1}{(w-z-h)^{n+1}} - \frac{1}{(w-z)^{n+1}}\right) dw.$$

It is easily seen that as $h \to 0$, the divided difference $\frac{(w-z-h)^{-n-1} - (w-z)^{-n-1}}{h}$ converges to $(n+1)(w-z)^{-n-2}$, *uniformly over $w \in C$*. (The same claim without the uniformity is just the rule for differentiation of a power function; to get the uniformity, we need to "go back to basics" and repeat the elementary algebraic calculation that was originally used to derive this power rule; we leave this as an exercise.) It follows that

$$\lim_{h \to 0} \frac{f^{(n)}(z+h) - f^{(n)}(z)}{h} = \frac{n!}{2\pi i} \oint_{C_R(z_0)} (n+1)\frac{f(w)}{(w-z)^{n+2}} \, dz$$

$$= \frac{(n+1)!}{2\pi i} \oint_{C_R(z_0)} \frac{f(w)}{(w-z)^{n+2}} \, dz. \tag{1.50}$$

This implies that $f^{n+1}(z)$ exists and is equal to the last expression in (1.50), which was precisely the claim in the $(n+1)$th case. The induction is complete. $\qquad\square$

In Theorem 1.29, we have stated one of the most remarkable facts about holomorphic functions but hid it inside a technical-looking claim in a way that makes it seem almost like an afterthought. Let us state it more explicitly to pay it proper respect.

Theorem 1.30 (Infinite differentiability of holomorphic functions). *If a function f of a complex variable is holomorphic in a region Ω, then it is differentiable infinitely many times there.*

The real-analysis analogue of Theorem 1.30 is, of course, (very) false. As another illustration of how remarkable this result is, recall that in Section 1.4, we proved that the real and imaginary parts of a holomorphic function are harmonic functions subject to the extra assumption that those functions are twice continuously differentiable. We now see that this assumption is not needed and the conclusion that u, v are harmonic already follows just from the holomorphicity assumption. Moreover, as an added bonus, we also get "for free" the statement that u and v are themselves *infinitely many times* differentiable; that is, they are C^∞ functions. (The fact that harmonic functions are C^∞ can also be proved just using real analysis techniques, but it is nonetheless pleasing to see it emerging out of the theory we are developing.)

Proof of Morera's theorem. We already proved that if f is a function all of whose contour integrals over closed curves vanish, then f has a primitive F. By Theorem 1.29, $F' = f$ is also holomorphic. □

As another immediate corollary to the (extended) Cauchy integral formula, we now get an extremely useful family of inequalities that bounds a function $f(z)$ and its derivatives at some specific point $z \in C$ in terms of the values of the function on the boundary of a circle centered at z.

Theorem 1.31 (Cauchy inequalities). *For f holomorphic in a region Ω that contains the closed disc $D_{\leq R}(z)$, we have*

$$\left|f^{(n)}(z)\right| \leq n! R^{-n} \sup_{|w-z|=R} \left|f(w)\right|. \tag{1.51}$$

Yet another remarkable fact we can now prove is the equivalence between the class of holomorphic functions and the class of functions that are locally expressible as power series. One direction in this equivalence—the easy one—was already proved in Theorem 1.9. The other is given in the following result.

Theorem 1.32 (Holomorphic functions have convergent power series). *If f is holomorphic in a region Ω that contains a closed disc $D_{\leq R}(z_0)$, then f has a power series expansion at z_0*

$$f(z) = \sum_{n=0}^{\infty} a_n (z - z_0)^n,$$

which is convergent for all $z \in D_R(z_0)$. The coefficients a_n in this expansion are given (in accordance with (1.28)) by $a_n = f^{(n)}(z_0)/n!$.

Proof. The basic idea here is that Cauchy's integral formula gives us a representation of $f(z)$ as a weighted "sum" (in fact, an integral, which is a limit of sums) of functions of the form $z \mapsto (w - z)^{-1}$. Each of the functions in the weighted sum has a power series expansion since it is, essentially, a geometric series, so the sum also has a power series expansion.

To make this precise, write

$$\frac{1}{w-z} = \frac{1}{(w-z_0)-(z-z_0)} = \frac{1}{w-z_0} \cdot \frac{1}{1-\left(\frac{z-z_0}{w-z_0}\right)}$$

$$= \frac{1}{w-z_0} \sum_{n=0}^{\infty} \left(\frac{z-z_0}{w-z_0}\right)^n = \sum_{n=0}^{\infty} (w-z_0)^{-n-1}(z-z_0)^n.$$

This is a power series in $z-z_0$, which, for any fixed $w \in C_R(z_0)$, converges absolutely for all z such that $|z-z_0| < R$ (that is, for all $z \in D_R(z_0)$). Moreover, the convergence is clearly uniform in $w \in C_R(z_0)$. Since infinite summations that are absolutely and uniformly convergent can be interchanged with integration operations, we then get, appealing to both the regular and extended versions of Cauchy's integral formula, that

$$f(z) = \frac{1}{2\pi i} \oint_{C_R(z_0)} \frac{f(w)}{w-z} \, dw$$

$$= \frac{1}{2\pi i} \oint_{C_R(z_0)} f(w) \sum_{n=0}^{\infty} (w-z_0)^{-n-1}(z-z_0)^n \, dw$$

$$= \sum_{n=0}^{\infty} \left(\frac{1}{2\pi i} \oint_{C_R(z_0)} f(w)(w-z_0)^{-n-1} \, dw \right)(z-z_0)^n$$

$$= \sum_{n=0}^{\infty} \frac{f^{(n)}(z_0)}{n!} (z-z_0)^n,$$

which is precisely the expansion we were after. ☐

Theorem 1.33 (Liouville's theorem). *A bounded entire function is constant.*

Proof. Let f be bounded and entire, and let $M = \sup_{z \in \mathbb{C}} |f(z)| < \infty$. By the case $n = 1$ of the Cauchy inequalities (1.51), for any $z \in \mathbb{C}$ and $R > 0$, we have

$$|f'(z)| \le \frac{M}{R}.$$

Taking the limit as $R \to \infty$ gives that $f'(z) = 0$. Since f' is identically 0, f is constant by Lemma 1.16. ☐

Exercises 1.23, 1.24, and 1.25 explore some additional ideas related to Liouville's theorem and additional results that can be proved using a similar technique.

Proposition 1.34. *If f is holomorphic on a region Ω, and $f(z) = 0$ for z in a set containing a limit point in Ω, then f is identically zero on Ω.*

The condition that the limit point z_0 is in Ω in this result is needed. For example, the function $e^{1/z} - 1$ has zeros in every neighborhood of $z_0 = 0$ but is not identically zero.

Proof of Proposition 1.34. Let $z_0 \in \Omega$ be a limit point of zeros of 0. This means that there is a sequence $(w_k)_{k=1}^{\infty}$ of points in Ω such that $f(w_k) = 0$ for all n, $w_k \to z_0$ as $k \to \infty$, and $w_k \neq z_0$ for all k. We know that in a neighborhood of z_0, f has a convergent power series expansion. If we assume that f is not identically zero in a neighborhood of z_0, then we can write the power series expansion as

$$f(z) = \sum_{n=0}^{\infty} a_n(z - z_0)^n = \sum_{n=m}^{\infty} a_n(z - z_0)^n$$

$$= a_m(z - z_0)^m \sum_{n=0}^{\infty} \frac{a_{n+m}}{a_m}(z - z_0)^n = a_m(z - z_0)^m(1 + g(z)),$$

where we define m to be the smallest index such that $a_m \neq 0$, and define $g(z) = \sum_{n=1}^{\infty} \frac{a_{n+m}}{a_m}(z - z_0)^n$. Note that g is a holomorphic function in a neighborhood of z_0 that satisfies $g(z_0) = 0$. It follows that for all k,

$$a_m(w_k - z_0)^m(1 + g(w_k)) = f(w_k) = 0,$$

but for large enough k, this is impossible, since $w_k - z_0 \neq 0$ for all k and $g(w_k) \to g(z_0) = 0$ as $k \to \infty$.

The conclusion is that f is identically zero at least in a neighborhood of z_0. Now we claim that this also implies that f is identically zero on all of Ω, because Ω is a region (open and connected). More precisely, denote by U the set of points $z \in \Omega$ such that f is equal to 0 in a neighborhood of z. It is obvious that U is open; U is also closed by the argument above, which shows that any point that is a limit of points in U must be in U; and U is nonempty (it contains z_0, again by what we showed above). It follows that $U = \Omega$ by the well-known characterization of a connected set in the plane as a set E that has no "clopen" (closed and open) sets other than the empty set and E itself. □

Proposition 1.34 has an equivalent form that is more memorable, given in the next result.

Theorem 1.35 (Zeros of holomorphic functions are isolated). *If f is holomorphic on Ω, is not identically zero on Ω, and $f(z_0) = 0$ for $z_0 \in \Omega$, then for some $\epsilon > 0$, the punctured neighborhood $D_\epsilon(z_0) \setminus \{z_0\}$ of z_0 contains no zeros of f. In other words, the set of zeros of f contains only isolated points.*

Corollary 1.36. *If f, g are holomorphic on a region Ω, and $f(z) = g(z)$ for z in a set with limit point in Ω (e. g., an open disc or even a sequence of points z_n converging to some $z \in \Omega$), then $f \equiv g$ everywhere in Ω.*

Proof. Apply the previous result to $f - g$. □

The previous result is usually reformulated slightly as the following conceptually important result.

Theorem 1.37 (Principle of analytic continuation). *If f is holomorphic on a region Ω, and f_+ is holomorphic on a bigger region $\Omega_+ \supset \Omega$ and satisfies $f_+(z) = f(z)$ for all $z \in \Omega$, then f_+ is the* unique *such extension, in the sense that if $\tilde{f}_+$ is another function with the same properties, then $\tilde{f}_+(z) = f_+(z)$ for all $z \in \Omega_+$.*

The function f_+ in Theorem 1.37, if it exists, is usually referred to as the **analytic continuation** of f.

The principle of analytic continuation is of fundamental importance in complex analysis. One of the common ways in which it is used is as a tool for justifying the construction of interesting holomorphic functions in several stages, where one starts by defining the function on a small region and then shows how to extend the definition to a larger region (see Chapter 2 for two of the most famous examples of this idea). There are often several ways of performing the extension, with no single one of them being necessarily more natural or canonical than the others, so we typically appeal to the principle of analytic continuation to explain why we end up with the same extended function regardless of which particular construction is used. In that sense, the principle of analytic continuation gives a philosophical justification for regarding naturally occurring holomorphic functions, such as the Euler gamma function and Riemann zeta function discussed in Chapter 2, as having a kind of idealized Platonic existence that transcends any particular formula used to represent them.

This philosophical point of view can be illustrated in an amusing way in a more elementary setting. In real analysis, we learn that "formulas" such as

$$1 - 1 + 1 - 1 + 1 - 1 + \cdots = \frac{1}{2}, \tag{1.52}$$

$$1 + 2 + 4 + 8 + 16 + 32 + \cdots = -1 \tag{1.53}$$

do not have any meaning, despite the fact that they can be easily "proved" using algebraic manipulations of a somewhat dubious nature. However, in the context of complex analysis, we can in fact make perfect sense of such identities using the principle of analytic continuation! Do you see how? (Exercise 1.27.) Additional seemingly meaningless formulas of this type, beloved by complex analysts and recreational mathematicians alike, are

$$1 + 2 + 3 + 4 + \cdots = -\frac{1}{12}, \tag{1.54}$$

$$1 - 2 + 3 - 4 + \cdots = \frac{1}{4}. \tag{1.55}$$

These formulas have attracted considerable attention in recent years, being the subject of a popular online video [W7], newspaper articles [W8], discussions on mathematics blogs and forums [W9], [W10], [W11], a Wikipedia article [W12], and more. We will learn in Chapter 2 that they, too, can be given a formal meaning that is no less precise or rigorous than the formulas involving convergent series that you are more familiar with from real analysis; see Exercise 2.11.

We now discuss a particular case of analytic continuation that constitutes the most minimalistic kind of continuation we can imagine, namely, a scenario in which a holomorphic function is extended to a region that is larger by a *single point* relative to the original domain on which it is defined. This is usually described in terms of the so-called **removable singularity**. A point $z_0 \in \Omega$ is called a removable singularity of a function $f : \Omega \to \mathbb{C} \cup \{\text{undefined}\}$ if f is holomorphic in a punctured neighborhood of z_0, is not holomorphic at z_0, but its value at z_0 can be redefined so as to make it holomorphic at z_0, that is, if we can perform an analytic continuation of f from $\Omega \setminus \{z_0\}$ to Ω. Of course, in this case the fact that the analytic continuation is unique is trivial; the issue here is to understand when the continuation *exists*, and the next result gives a useful condition.

Theorem 1.38 (Riemann's removable singularities theorem). *If f is holomorphic in Ω except at a point $z_0 \in \Omega$ (where it may be undefined or be defined but not known to be holomorphic or even continuous). Assume that f is bounded in a punctured neighborhood $D_r(z_0) \setminus \{z_0\}$ of z_0. Then z_0 is a removable singularity of f.*

Proof. Fix some disc $D = D_R(z_0)$ around z_0 whose closure is contained in Ω. Define the function

$$\widetilde{f}(z) = \frac{1}{2\pi i} \oint_{C_R(z_0)} \frac{f(w)}{w - z} \, dw \quad (z \in D). \tag{1.56}$$

We claim that $\widetilde{f}$ extends f to a holomorphic function on D, which requires showing that $\widetilde{f}(z) = f(z)$ for all $z \in D\setminus\{z_0\}$ and that $\widetilde{f}$ is holomorphic at z_0. For the first part of the claim, let $z \in D\setminus\{z_0\}$. Consider a "double keyhole" contour $K_{\epsilon,\delta}$ that surrounds most of the disc D but makes diversions to avoid the points z_0 and z, circling them in the negative direction around most of a circle of radius ϵ (Fig. 1.8). We assume that $0 < \delta < \epsilon < \frac{1}{4}|z - z_0|$. Now the region enclosed by $K_{\epsilon,\delta}$ is simply connected, so, after applying Cauchy's theorem and a limiting argument similar to that used in the proof of Theorem 1.27 (taking the limit as $\delta \to 0$ with ϵ fixed), we get that

$$\widetilde{f}(z) = \frac{1}{2\pi i} \oint_{C_\epsilon(z)} \frac{f(w)}{w - z} \, dw + \frac{1}{2\pi i} \oint_{C_\epsilon(z_0)} \frac{f(w)}{w - z} \, dw. \tag{1.57}$$

On the right-hand side, the first term is equal to $f(z)$ by a straightforward application of Cauchy's integral formula. The second term can be bounded in magnitude using the assumption that f is bounded in a neighborhood of z_0; more precisely, denote $M = \sup_{w \in D_r(z_0)\setminus\{z_0\}} |f(w)| < \infty$. We have

$$\left| \oint_{C_\epsilon(z_0)} \frac{f(w)}{w - z} \, dw \right| \leq 2\pi\epsilon \sup_{w \in C_\epsilon(z_0)} |f(w)| \cdot \frac{1}{|z - z_0| - \epsilon} \leq \frac{\pi M \epsilon}{|z - z_0|}.$$

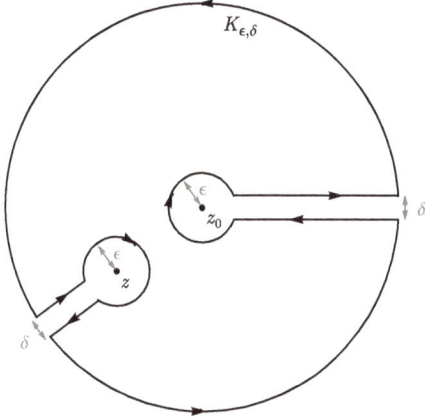

Figure 1.8: The double keyhole contour $K_{\epsilon,\delta}$.

Thus the claim that $\widetilde{f}(z) = f(z)$ follows by taking the limit of (1.57) as $\epsilon \to 0$.

It remains to prove that $\widetilde{f}$ defined in (1.56) is holomorphic at z_0. This is easy to see and is something we already knew implicitly. For example, the relevant argument (involving a direct manipulation of the divided differences $\frac{1}{h}(\widetilde{f}(z+h) - \widetilde{f}(z))$) appeared in the proof of Theorem 1.29. Another approach is to show that integrating $\widetilde{f}$ over closed contours gives 0 (which requires interchanging the order of two integration operations, which will not be hard to justify) and then use Morera's theorem. The details are left as an exercise. □

We now introduce the concept of **uniform convergence on compact subsets**. If f and $(f_n)_{n=1}^{\infty}$ are holomorphic functions on a region Ω, we say that the sequence f_n converges to f **uniformly on compact subsets** if for any compact set $K \subset \Omega, f_n(z) \to f(z)$ uniformly on K. This mode of convergence is preserved under differentiation, as the following result makes precise.

Theorem 1.39. *If $f_n \to f$ uniformly on compact subsets in Ω and f_n are holomorphic, then f is holomorphic, and $f_n' \to f'$ uniformly on compact subsets in Ω.*

Proof. The fact that f is holomorphic can be shown through a combination of Cauchy's and Morera's theorems. More precisely, note that for each closed disc $D_{\leq r}(z_0) \subset \Omega$, we have $f_n(z) \to f$ uniformly on $D_{\leq r}(z_0)$. In particular, for each curve γ whose image is contained in the open disc $D_r(z_0)$,

$$\int_\gamma f_n(z)\,dz \xrightarrow[n \to \infty]{} \int_\gamma f(z)\,dz.$$

By Cauchy's theorem the integrals in this sequence are all zero, so $\int_\gamma f(z)\,dz = 0$. Since this is true for all curves γ in the disc $D_r(z_0)$, by Morera's theorem, f is holomorphic on $D_r(z_0)$. This holds for any disc whose closure is in Ω, and holomorphicity is a local property, so we have shown that f is holomorphic on all of Ω, as claimed.

Next, to show that $f_n' \to f'$ uniformly on compact sets, we start by proving that uniform convergence holds on a certain family of discs. Let $D_r(z_0)$ be a disc whose closure is contained in Ω. For $z \in D_r(z_0)$, we have by Cauchy's integral formula that

$$f_n'(z) - f'(z) = \frac{1}{2\pi i} \oint_{C_r(z_0)} \frac{f_n(w)}{(w-z)^2} \, dw - \frac{1}{2\pi i} \oint_{C_r(z_0)} \frac{f(w)}{(w-z)^2} \, dw$$

$$= \frac{1}{2\pi i} \oint_{C_r(z_0)} \frac{f_n(w) - f(w)}{(w-z)^2} \, dw.$$

This implies that $f_n'(z) \to f'(z)$ as $n \to \infty$, *uniformly as z ranges on the disc $D_{r/2}(z_0)$,* since $f_n(w) \to f(w)$ uniformly for $w \in C_r(z_0) \subset D_{\leq r}(z_0)$, and since the bound $|w-z|^{-2} \leq (r/2)^{-2}$ holds for $z \in D_{r/2}(z_0)$ and $w \in C_r(z_0)$.

Now let $K \subset \Omega$ be compact. For each $z \in K$, let $r(z)$ be the radius of a closed disc $D_{\leq r(z)}(z)$ around z that is contained in Ω. The family of discs $\{B_z := D_{r(z)/2}(z) : z \in \Omega\}$ is an open covering of K, so by the Heine–Borel property of compact sets it has a finite subcovering $B_{z_1}, \ldots, B_{z_n}$. We showed that $f_n'(z) \to f'(z)$ uniformly on every B_{z_j}, so we also have uniform convergence on their union, which contains K, so we get that $f_n' \to f'$ uniformly on K, as claimed. $\qquad\square$

Suggested exercises for Section 1.9. 1.23, 1.24, 1.25, 1.26, 1.27.

1.10 Zeros, poles, and the residue theorem

We say that a complex number z_0 is a **zero** of a holomorphic function f if $f(z_0) = 0$. Zeros in complex analysis behave rather like zeros of polynomial, in the sense that a zero must have an integer multiplicity, known as its **order**. More precisely, we say that z_0 is a **zero of order** $m \geq 1$ of a nonconstant holomorphic function f if it can be represented in the form

$$f(z) = (z - z_0)^m g(z) \tag{1.58}$$

in some neighborhood of z_0, where $m \geq 1$, and g is a holomorphic function in that neighborhood such that $g(z_0) \neq 0$. A zero of order 1 is called a **simple zero**.

Lemma 1.40. *The order of a zero is a well-defined concept. That is, if f is a nonconstant holomorphic function and $f(z_0) = 0$, then representation (1.58) with the properties of g as given above exists for a unique integer $m \geq 1$.*

Proof. We make use of power series expansions and a calculation similar to that used in the proof of Proposition 1.34. Write the power series expansion (known to converge in a neighborhood of z_0)

$$f(z) = \sum_{n=0}^{\infty} a_n(z - z_0)^n = \sum_{n=m}^{\infty} a_n(z - z_0)^n,$$

where m is the smallest index ≥ 0 such that $a_m \neq 0$. Since $a_0 = f(z_0) = 0$, it must be the case that $m \geq 1$. If we now define

$$g(z) = \sum_{k=0}^{\infty} a_{m+k}(z - z_0)^k,$$

then clearly $f(z) = (z - z_0)^m g(z)$, and $g(z_0) = a_m \neq 0$; this proves the existence of representation (1.58). On the other hand, given a representation of this form, expanding $g(z)$ as a power series around z_0 shows that m has to be the smallest index of a nonzero coefficient in the power series expansion of $f(z)$ around z_0. This proves the uniqueness claim. □

In the definition above, in the case where z_0 is *not* a zero of f, the same representation (1.58) holds with $m = 0$ (and $g = f$), so in certain contexts, we may occasionally describe this situation by saying that z_0 is a zero of order 0.

If f is holomorphic in a punctured neighborhood of a point z_0, then we say that it has a **pole of order** m at z_0 if the function $h(z) = 1/f(z)$ (defined to be 0 at z_0) has a zero of order m at z_0. A pole of order 1 is called a **simple pole**. As with the case of zeros, we can extend this definition in an obvious way by saying that f has a pole of order 0 if f is holomorphic at z_0 or has a removable singularity there, and the value $f(z_0)$ (or the redefined value $\lim_{z \to z_0} f(z)$ that makes f holomorphic at z_0 in the case of a removable singularity) is nonzero.

Lemma 1.41. *A function f has a pole of order m at z_0 if and only if it can be represented in the form*

$$f(z) = (z - z_0)^{-m} g(z)$$

in a punctured neighborhood of z_0, where g is holomorphic in a neighborhood of z_0 and satisfies $g(z_0) \neq 0$.

Proof. Apply the previous lemma to $1/f(z)$. □

Theorem 1.42. *If f has a pole of order m at z_0, then it can be represented in a unique way as*

$$f(z) = \frac{a_{-m}}{(z - z_0)^m} + \frac{a_{-m+1}}{(z - z_0)^{m-1}} + \cdots + \frac{a_{-1}}{z - z_0} + G(z), \tag{1.59}$$

where G is holomorphic in a neighborhood of z_0, and $a_{-1}, \ldots, a_{-m}$ are arbitrary complex numbers with $a_{-m} \neq 0$.

Proof. The function $g(z) = (z - z_0)^m f(z)$ is holomorphic in a neighborhood of z_0 and satisfies $g(z_0) \neq 0$. Write its power series expansion as

$$g(z) = \sum_{n=0}^{\infty} b_n (z - z_0)^n \tag{1.60}$$

$$= b_0 + b_1(z - z_0) + \cdots + b_{m-1}(z - z_0)^{m-1} + \sum_{n=m}^{\infty} b_m(z - z_0)^n. \tag{1.61}$$

Here $b_0 = g(z_0) \neq 0$. Now defining $G(z) = \sum_{n=m}^{\infty} b_m(z - z_0)^{n-m}$ and converting (1.61) to an expression for f, we get that

$$f(z) = \frac{b_0}{(z - z_0)^m} + \frac{b_1}{(z - z_0)^{m-1}} + \cdots + \frac{b_{m-1}}{z - z_0} + G(z),$$

which is of the correct form (1.59) if we further define $a_{-j} = b_{m-j}$ for $1 \leq j \leq m$. This proves the existence part of the claim; the uniqueness part is left as an easy exercise. $\square$

In representation (1.59) the expression

$$f(z) - G(z) = \frac{a_{-m}}{(z - z_0)^m} + \frac{a_{-m+1}}{(z - z_0)^{m-1}} + \cdots + \frac{a_{-1}}{z - z_0}$$

is called the **principal part** of f at the pole z_0. The coefficient a_{-1} is called the **residue of** f at z_0 and denoted $\mathrm{Res}_{z_0}(f)$.

The definitions of the order of a zero and a pole can be unified into a single consistent definition of the **(generalized) order of a zero**, where if f has a pole of order m at z_0, then we say instead that f has a zero of order $-m$. Denote the order of a zero of f at z_0—an integer, which may be positive, negative, or zero—by $\mathrm{ord}_{z_0}(f)$. With these definitions, it is easy to check (Exercise 1.28) that

$$\mathrm{ord}_{z_0}(f + g) \geq \min(\mathrm{ord}_{z_0}(f), \mathrm{ord}_{z_0}(g)), \tag{1.62}$$

$$\mathrm{ord}_{z_0}(fg) = \mathrm{ord}_{z_0}(f) + \mathrm{ord}_{z_0}(g). \tag{1.63}$$

The **residue theorem** is a famous formula for evaluating integrals around closed contours of functions holomorphic inside the region enclosed by the contour, except for a discrete set of points. This theorem, like Cauchy's theorem, has several different formulations addressing different levels of generality. We further give three versions of the theorem, which are sufficient for our needs.

Theorem 1.43 (The residue theorem; simple version). *Assume that f is holomorphic in a region containing a closed disc $D_{\leq R}(z_0)$, except for a pole at $z_0 \in D$. Then*

$$\oint_{C_R(z_0)} f(z)\, dz = 2\pi i \operatorname{Res}_{z_0}(f).$$

Proof. By the standard argument involving a keyhole contour we see that the circle $C_R(z_0)$ in the integral can be replaced with a circle $C_\epsilon(z_0)$ of a small radius $\epsilon > 0$ around z_0, that is, we have

$$\oint_{C_R(z_0)} f(z)\, dz = \oint_{C_\epsilon(z_0)} f(z)\, dz.$$

When ϵ is small enough, to evaluate the integral over $C_\epsilon(z_0)$, we can use decomposition (1.59) of f into its principal part and the remaining holomorphic part. Integrating the right-hand side of (1.59) termwise over the contour $C_\epsilon(z_0)$ gives 0 for the integral of $G(z)$ by Cauchy's theorem; 0 for the integral powers $(z - z_0)^k$ with $-m \leq k \leq -2$ by a standard computation (Exercise 1.21); and $2\pi i\, a_{-1} = 2\pi i \operatorname{Res}_{z_0}(f)$ for the integral of $(z - z_0)^{-1}$ by the same standard computation. This gives the result. □

Theorem 1.44 (The residue theorem for discs). *Assume that f is holomorphic in a region containing a closed disc $D_{\leq R}(z_0)$, except for a finite number of poles at $z_1, \dots, z_N \in D_R(z_0)$. Then*

$$\oint_{C_R(z_0)} f(z)\, dz = 2\pi i \sum_{k=1}^{N} \operatorname{Res}_{z_k}(f).$$

Proof. The idea is the same as in the proof of Theorem 1.43, except that now we use a contour with multiple keyholes (one for each z_j) to deduce after a limiting argument that

$$\oint_{C_R(z_0)} f(z)\, dz = \sum_{k=1}^{N} \oint_{C_\epsilon(z_k)} f(z)\, dz$$

for a small enough ϵ, and then proceeds as before. □

Theorem 1.45 (The residue theorem for simple closed contours). *Let f be a function defined in a region Ω containing a simple closed curve γ (oriented in the positive direction). Denote by R_γ the region enclosed by γ. Assume that f is holomorphic everywhere in Ω except for the finite set of points $z_1, \dots, z_N \in R_\gamma$, where it has poles. Then*

$$\oint_\gamma f(z)\, dz = 2\pi i \sum_{k=1}^{N} \operatorname{Res}_{z_k}(f).$$

Sketch of proof. Again, construct a multiple keyhole version of the original contour γ and then use a limiting argument to conclude that

$$\oint_\gamma f(z)\,dz = \sum_{k=1}^{N} \oint_{C_\epsilon(z_k)} f(z)\,dz$$

for a small enough ϵ. Then proceed as before. □

Suggested exercises for Section 1.10. 1.28.

1.11 Meromorphic functions, holomorphicity at ∞, and the Riemann sphere

We extend the notion of holomorphicity in two directions by introducing the notions of **meromorphicity** and **holomorphicity at ∞**. First, a function $f : \Omega \to \mathbb{C} \cup \{\text{undefined}\}$ on a region Ω is called **meromorphic** if f is holomorphic except for a discrete set of points, all of which are poles of f.

Second, let $U \subset \mathbb{C}$ be an open set containing the complement $\mathbb{C} \setminus D_{\leq R}(0)$ of a closed disc around 0. A function $f : U \to \mathbb{C}$ is **holomorphic at ∞** if $g(z) = f(1/z)$ (defined on a neighborhood $D_{1/R}(0)$ of 0) has a removable singularity at 0. In that case, we define $f(\infty) = g(0)$ (more precisely, the value that makes g holomorphic at 0).

Conceptually, the above definitions can be thought of as extending the notion of what a complex number is to include an additional "point at infinity." Formally, we define the set of **extended complex numbers**, also known as the **Riemann sphere**, as the set $\widehat{\mathbb{C}} = \mathbb{C} \cup \{\infty\}$ equipped with several layers of additional structure:

- **Topological structure.** We think of $\widehat{\mathbb{C}}$ as the **one-point compactification** of $\mathbb{C}$; that is, we add to $\mathbb{C}$ an additional element ∞ and say that the open neighborhoods of ∞ are the complements of compact sets in $\mathbb{C}$. This turns $\widehat{\mathbb{C}}$ into a topological space in a simple way.

- **Geometric structure.** We can identify $\widehat{\mathbb{C}}$ with an actual sphere embedded in $\mathbb{R}^3$, namely

$$S^2 = \left\{ (X, Y, Z) \in \mathbb{R}^3 \ : \ X^2 + Y^2 + \left(Z - \frac{1}{2} \right)^2 = \frac{1}{4} \right\}$$

(the sphere of radius 1/2 centered at $(0, 0, 1/2)$). The identification works as follows: the point at ∞ is identified with the north pole $(0, 0, 1)$ of the sphere; for other points, the identification $(X, Y, Z) \in S^2 \longleftrightarrow a + ib \in \mathbb{C}$ is given by two reciprocal relations

$$a + ib = \frac{X}{1 - Z} + i\frac{Y}{1 - Z},$$

$$(X, Y, Z) = \left(\frac{a}{1 + a^2 + b^2}, \frac{b}{1 + a^2 + b^2}, \frac{a^2 + b^2}{1 + a^2 + b^2} \right).$$

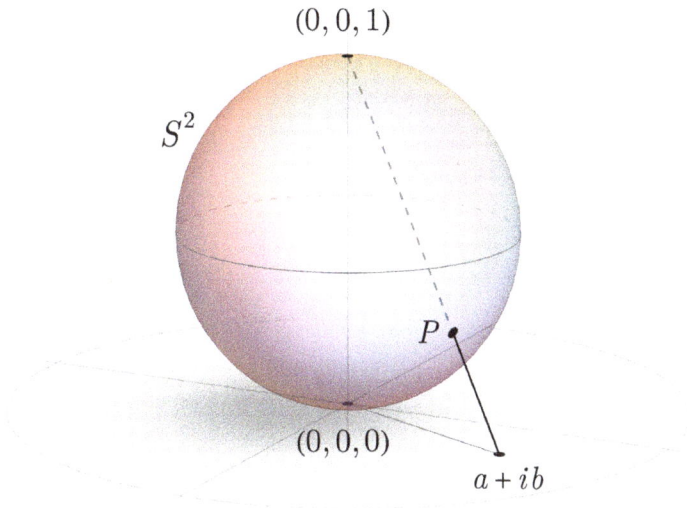

Figure 1.9: The Riemann sphere $\widehat{\mathbb{C}} = S^2$ and the translation between points $a + ib$ on the complex plane and points $P = (X, Y, Z)$ on the sphere via stereographic projection. The equator on the sphere is mapped to the unit circle in $\mathbb{C}$.

Geometrically, this identification corresponds to **stereographic projection**, where the point $a + bi$ is calculated from $P = (X, Y, Z)$ by projecting the straight line segment from the north pole $(0, 0, 1)$ to P further out onto its unique intersection point with the x–y plane, identified with the complex plane $\mathbb{C}$ in the obvious way; see Fig. 1.9. We can check without difficulty that this geometric identification is a homeomorphism between S^2, equipped with the obvious topology inherited from $\mathbb{R}^3$, and $\widehat{\mathbb{C}}$ with the one-point compactification topology defined above.

– **Holomorphic structure.** The above definition of what it means for a function on a neighborhood of ∞ to be holomorphic at ∞ provides a way of giving $\widehat{\mathbb{C}}$ the structure of a **Riemann surface** (the simplest nontrivial case of a manifold with a complex-analytic structure). We will not discuss the topic of Riemann surfaces here; for more details on this point of view, see, e. g., [23, 60].

From this new point of view of the Riemann sphere, the concept of a meromorphic function $f : \Omega \to \mathbb{C} \cup \{\text{undefined}\}$ can be seen to coincide with the notion of a holomorphic function $f : \Omega \to \widehat{\mathbb{C}}$; that is, the underlying concept of the definition is still holomorphicity, but it applies to functions taking values in $\widehat{\mathbb{C}}$, a different Riemann surface, instead of $\mathbb{C}$. Similarly, the idea of a function $f : \Omega \to \mathbb{C}$ being holomorphic at ∞ corresponds exactly to the notion of a function whose "true" domain of definition is actually $\Omega \cup \{\infty\}$ in the sense that it can be extended to a holomorphic function on this larger domain.

To conclude this section, we also generalize the notion of the order of a zero or pole at a point to include the behavior at the point at ∞. Let $U \subset \mathbb{C}$ be an open set containing the complement $\mathbb{C} \setminus D_{\leq R}(0)$ of a closed disc around 0. We say that a function $f : U \to \mathbb{C}$ has a zero (resp., pole) of order m at ∞ if $g(z) = f(1/z)$ has a zero (resp., pole) at $z = 0$ after appropriately defining the value of g at 0.

1.12 Classification of singularities and the Casorati–Weierstrass theorem

If a function $f : \Omega \to \widehat{\mathbb{C}} \cup \{\text{undefined}\}$ is holomorphic in a punctured neighborhood $D_r(z_0) \setminus \{z_0\}$ of z_0, then we say that f has a **singularity** at z_0 if f is not holomorphic at z_0. We classify singularities into three types, two of which we already defined:
- **Removable singularities**: when f can be made holomorphic at z_0 by defining or redefining its value at z_0;
- **poles**;
- any singularity that is not removable or a pole is called an **essential singularity**.

For a function defined on a neighborhood of ∞ that is not holomorphic at ∞, we say that f has a singularity at ∞ and classify the singularity as a removable singularity, a pole, or an essential singularity according to the type of singularity that $z \mapsto f(1/z)$ has at $z = 0$.

The function $z \mapsto e^{1/z}$ is an example of a function with an essential singularity at the point $z = 0$. Its behavior near that singularity is rather difficult to visualize. Indeed, the next result shows that this is the case more generally.

Theorem 1.46 (Casorati–Weierstrass theorem). *If f is holomorphic in a punctured neighborhood $D_r(z_0) \setminus \{z_0\}$ of z_0 and has an essential singularity at z_0, then the image $f(D_r(z_0) \setminus \{z_0\})$ of the punctured neighborhood under f is dense in $\mathbb{C}$.*

Proof. We prove the contrapositive of the claim: assume that for some $r > 0$, the image $f(D_r(z_0) \setminus \{z_0\})$ is not dense. Then the closure $\mathrm{cl}(f(D_r(z_0) \setminus \{z_0\}))$ of this image does not contain some point $w \in \mathbb{C}$. It follows that the function g defined by $g(z) = \frac{1}{f(z)-w}$ is holomorphic and bounded in $D_r(z_0) \setminus \{z_0\}$. By Theorem 1.38 its singularity at z_0 is removable, so we can assume that it is holomorphic at z_0 after defining its value there. It then follows that

$$f(z) = w + \frac{1}{g(z)}$$

has either a pole or a removable singularity at z_0, that is, the singularity at z_0 is not essential. $\qquad\square$

1.13 The argument principle and Rouché's theorem

We define the **logarithmic derivative** of a holomorphic function $f(z)$ as the function $f'(z)/f(z)$. Intuitively, this can be thought of as "the derivative of the logarithm of f." A word of caution is in order however: we have not actually defined what "the logarithm of f" means, and when we actually define it a bit later (in Section 1.15), we will see that "the logarithm of f" does not always exist. The logarithmic derivative on the other hand clearly exists, so it is best to get used to thinking about it as a separate concept from that of a logarithm rather than being derived from it.

Lemma 1.47. *The logarithmic derivative of a product of holomorphic functions is the sum of their logarithmic derivatives, that is,*

$$\frac{(\prod_{k=1}^n f_k)'}{\prod_{k=1}^n f_k} = \sum_{k=1}^n \frac{f_k'(z)}{f_k(z)}.$$

Proof. Show this for $n = 2$ and proceed by induction. $\quad\square$

Theorem 1.48 (The argument principle). *Assume that f is meromorphic in a region Ω and that γ is a simple closed contour in Ω enclosing a region R_γ such that f has no zeros or poles on the circle γ. Denote its zeros and poles inside R_γ by $z_1, \ldots, z_n$, where z_k is a zero of generalized order $m_k = \mathrm{ord}_{z_k}(f)$ (in the sense discussed in Section 1.10, where m_k is a positive integer if z_k is a zero and a negative integer if z_k is a pole). Then*

$$\frac{1}{2\pi i} \oint_\gamma \frac{f'(z)}{f(z)} \, dz = \sum_{k=1}^n m_k$$

$$= \textit{[total number of zeros of f inside R_γ, counting multiplicities]}$$
$$- \textit{[total number of poles of f inside R_γ, counting multiplicities]}.$$

Proof. Define

$$g(z) = \prod_{k=1}^n (z - z_k)^{-m_k} f(z).$$

Then $g(z)$ is meromorphic on Ω, has no singularities or zeros on γ, and has no poles or zeros inside R_γ, only removable singularities at $z_1, \ldots, z_n$ (so after redefining its values at these points, we can assume that it is holomorphic on R_γ). It follows that

$$f(z) = \prod_{k=1}^n (z - z_k)^{m_k} g(z).$$

Taking the logarithmic derivative of this equation gives that

$$\frac{f'(z)}{f(z)} = \sum_{k=1}^{n} \frac{m_k}{z - z_k} + \frac{g'(z)}{g(z)}.$$

The result now follows by integrating this equation and using the residue theorem (the term $g'(z)/g(z)$ is holomorphic on an open set containing $\mathrm{cl}(R_\gamma)$, so by Cauchy's theorem its contribution to the integral is 0). □

There is another way to look at the integral $\frac{1}{2\pi i} \oint_\gamma \frac{f'(z)}{f(z)} \, dz$, which gives an alternative explanation for why it is an integer, as well as an alternative geometric interpretation of its value. To see this, start by rewriting the integral (using the chain rule (1.84) from Exercise 1.6) as

$$\frac{1}{2\pi i} \oint_\gamma \frac{f'(z)}{f(z)} \, dz = \frac{1}{2\pi i} \int_a^b \frac{f'(\gamma(t))\gamma'(t)}{f(\gamma(t))} \, dt = \frac{1}{2\pi i} \int_a^b \frac{(f \circ \gamma)'(t)}{(f \circ \gamma)(t)} \, dt = \frac{1}{2\pi i} \int_{f \circ \gamma} \frac{1}{w} \, dw,$$

that is, an integral of dw/w over the contour $f \circ \gamma$, the image of γ under f. Now note that the differential form dw/w has a special geometric meaning in complex analysis; namely, we have

$$\frac{dw}{w} = \text{``}d(\log w)\text{''} = \text{``}d(\log |w| + i \arg w)\text{''}.$$

We put these expressions in quotes since the logarithm and argument are not single-valued functions (see Section 1.15), so it needs to be explained what such formulas mean. However, at least $\log |w|$ is well-defined for a curve that does not cross 0, so when integrating over the closed curve $f \circ \gamma$, the real part is zero by the fundamental theorem of calculus. The imaginary part (which becomes real after the division by $2\pi i$) can be interpreted intuitively as the *change in the argument over the curve*. That is, initially at time parameter $t = a$, we fix a specific value of $\arg w = \arg \gamma(a)$; then as t increases from $t = a$ to $t = b$, we track the increase or decrease in the argument as we travel along the curve $\gamma(t)$; if this is done correctly (i.e., in a continuous fashion), at the end the argument must have a well-defined value. Since the curve is closed, the total change in the argument must be an integer multiple of 2π, so the division by 2π turns it into an integer. The value of the integer has the intuitive meaning of "the total number of times the curve $f \circ \gamma$ goes around the origin."

This discussion leads us to another important concept, that of **winding numbers**. Given a closed curve Γ that does not cross 0, the above reasoning involving the differential form dw/w, applied to the curve Γ instead of $f \circ \gamma$, shows that an integral of the form

$$\frac{1}{2\pi i} \oint_\Gamma \frac{dz}{z}$$

carries the meaning of "the total number of times the curve γ goes around the origin," with the number being positive if the curve goes in the positive direction around the origin; negative if the curve goes in the negative direction around the origin; or zero if there is no net change in the argument. This number is more properly called the **winding number** of Γ around 0 (also sometimes referred to as the **index** of the curve around 0) and denoted $\mathrm{Ind}_\Gamma(0)$:

$$\mathrm{Ind}_\Gamma(0) = \frac{1}{2\pi i} \oint_\Gamma \frac{dz}{z}.$$

More generally, we define the winding number of Γ around z_0, denoted $\mathrm{Ind}_\Gamma(z_0)$, as

$$\mathrm{Ind}_\Gamma(z_0) = \frac{1}{2\pi i} \oint_\Gamma \frac{dz}{z - z_0},$$

assuming that Γ does not cross z_0. This can be interpreted as the number of times the curve Γ "winds around" an arbitrary point z_0.

To summarize the discussion above, we defined the notion of winding numbers and explained why the quantity $\frac{1}{2\pi i} \oint_\gamma \frac{f'(z)}{f(z)} dz$ that is the subject of the argument principle has the additional interpretation as the winding number of the curve $f \circ \gamma$ around 0. Note that the winding number is a *topological* concept of planar geometry that can be considered and studied without any reference to complex analysis. It is not very difficult to define it in purely topological terms without mentioning contour integrals and then show that the complex analytic and topological definitions coincide, but we will not pursue this here. Try to think what such a definition might look like.

Theorem 1.49 (Rouché's theorem). *Assume that f, g are holomorphic on a region Ω containing a circle $\gamma = C$ and the disc U enclosed by it (or, more generally, a simple closed contour γ enclosing a region U). If $|f(z)| > |g(z)|$ for all $z \in \gamma$, then f and $f + g$ have the same number of zeros in U.*

Proof. Define $f_t(z) = f(z) + tg(z)$ for $t \in [0,1]$, and note that $f_0 = f$ and $f_1 = f + g$, and that the condition $|f(z)| > |g(z)|$ on γ implies that f_t has no zeros on γ for any $t \in [0,1]$. Denote

$$n_t = \frac{1}{2\pi i} \oint_\gamma \frac{f_t'(z)}{f_t(z)} dz,$$

which by the argument principle is the number of "generalized zeros" (zeros or poles, counting multiplicities) of f_t in U. In particular, the function $t \mapsto n_t$ is integer-valued. If we also knew that it was continuous, then it would have to be constant (by the easy exercise: any integer-valued continuous function on an interval $[a, b]$ is constant), so in particular we would get the desired conclusion that $n_1 = n_0$.

To prove continuity of n_t, fix a number $\epsilon > 0$. Note that the function $g(t,z) = f_t'(z)/f_t(z)$ is continuous, hence also uniformly continuous, on the compact set $[0,1] \times y$. Therefore there exists $\delta > 0$ such that if $0 \leq t, s \leq 1$ satisfy $|t - s| < \delta$, then $|g(t,z) - g(s,z)| < 2\pi\epsilon/\text{len}(y)$ (recall that $\text{len}(y)$ denotes the length of the curve y). It follows that for such t, s, we have

$$|n_t - n_s| \leq \frac{1}{2\pi} \oint_y |g(t,z) - g(s,z)| \cdot |dz| \leq \frac{1}{2\pi} \oint_y \frac{2\pi\epsilon}{\text{len}(y)} |dz| = \epsilon.$$

This is exactly what is needed to show that $t \mapsto n_t$ is continuous. □

Rouché's theorem has a rather amusing intuitive explanation (which I learned from the book [48]). The slogan to remember is "walking the dog." Imagine that you are walking in a large empty park containing at some "origin" point 0 a large lamppost. You start at some point X and go for a walk along some curve, ending back at the same starting point X. Let N denote your winding number around the lamppost at the origin—that is, the total number of times you went around the lamppost with appropriate sign.

Now imagine that you also have a dog that is walking alongside you in some erratic path that is sometimes close to you, sometimes less close. As you traverse your curve C_1, the dog walks along on its own curve C_2, which also begins and ends in the same place. Let M denote the *dog's* winding number around the lamppost at the origin. Can we say that $N = M$? The answer is *yes, we can, provided that we know the dog's distance to you was always less than your distance to the lamppost*. To see this, imagine that you had the dog on a retractable leash; if the distance condition was not satisfied, it would be possible for the dog to reach the lamppost and go in a short tour around it while you were still far away and not turning around the lamppost, causing an entanglement of the leash with the pole.

The above scenario maps in a precise way to Rouché's theorem using the following dictionary: the curve $f \circ y$ represents your path; the curve $(f + g) \circ y$ represents the dog's path; $g \circ y$ represents the vector pointing from you to the dog; the condition $|f| > |g|$ along y is precisely the condition that the dog stays closer to you than your distance to the pole; and the conclusion that the two winding numbers are the same is precisely the statement of the theorem that f and $f + g$ have the same number of generalized zeros in the region U enclosed by y (see the discussion above regarding the connection between the integral $(2\pi i)^{-1} \oint_y f'/f \, dz$ and the winding number of $f \circ y$ around 0).

I recommend spending a few minutes thinking about the above correspondence and making sure you understand it. You may forget the technical details of the proof of Rouché's theorem in a few weeks or months, but I hope you will remember this intuitive explanation for a long time.

Rouché's theorem is an important tool both for numerically estimating the numbers of roots of polynomials and other functions in regions of interests and for theoretical applications. One illustration of the power of Rouché's theorem is given in Exercise 1.30.

In the next section, we also use Rouché's theorem to prove two more well-known properties of holomorphic functions, the **open mapping theorem** and the **maximum modulus principle**.

Suggested exercises for Section 1.13. 1.29, 1.30.

1.14 The open mapping theorem and maximum modulus principle

Theorem 1.50 (Open mapping theorem). *Any holomorphic function that is not constant is an **open mapping**, that is, it maps open sets to open sets.*

Proof. Let f be holomorphic and nonconstant in a region Ω. Fix an arbitrary $z_0 \in \Omega$, and denote $w_0 = f(z_0)$. We need to show that $f(\Omega)$ contains a neighborhood of w_0, that is, that there exists some $\delta > 0$ for which $f(\Omega) \supset D_\delta(w_0)$. The reason Rouché's theorem can be brought into play is that the inclusion $f(\Omega) \supset D_\delta(w_0)$ amounts to the statement that for $w \in D_\delta(w_0)$, the function $f(z) - w$ has at least one zero; and we know that this is true for the function $f(z) - w_0$, so we are precisely in a situation in which we want to compare the number of zeros of two functions, where (if we restrict our point of view to what is happening in a small neighborhood of z_0) one function can be regarded as a perturbation of the other.

To make this idea precise, define

$$F(z) = f(z) - w_0,$$
$$G_w(z) = w_0 - w,$$
$$h_w(z) = F(z) + G_w(z) = f(z) - w.$$

Let $\epsilon > 0$ be a number small enough so that the closed disc $D_{\leq \epsilon}(z_0)$ is contained in Ω and such that the point $z = z_0$ is the only zero of $F(z)$ in the disc $D_\epsilon(z_0)$. (Such ϵ exists by the property that zeros of holomorphic functions are isolated.) Now define

$$\delta = \inf\{|f(z) - w_0| \: : \: z \in D_{\leq \epsilon}(z_0)\}. \tag{1.64}$$

By construction we have that $\delta > 0$ and $|f(z) - w_0| \geq \delta$ for z on the circle $|z - z_0| = \epsilon$. This means that for any $w \in D_\delta(w_0)$, the condition $|F(z)| > |G_w(z)|$ in Rouché's theorem will be satisfied for $z \in \partial D_\epsilon(z_0)$. The conclusion is that the equation $h_w(z) = 0$ (or, equivalently, $f(z) = w$) has the same number in solutions as the equation $f(z) = w_0$ in the disc $D_\epsilon(z_0)$. The latter equation has precisely one solution, the point $z = z_0$. Thus we have shown that for $w \in D_\delta(w_0)$ with δ defined in (1.64), there exists $z \in D_\epsilon(z_0)$ such that $f(z) = w$. This was precisely what we needed to establish that f is an open mapping. ☐

Theorem 1.51 (Maximum modulus principle). *If f is a nonconstant holomorphic function on a region Ω, then $|f|$ cannot attain a maximum on Ω.*

Proof. This follows immediately from the open mapping theorem. ☐

For an interesting application of the maximum modulus principle, see Section 3.5.

1.15 The logarithm function

The logarithm function can be defined as

$$\log z = \log |z| + i \arg z$$

on any region Ω that does not contain 0 and where we can make a consistent, smoothly varying choice of $\arg z$ as z ranges over Ω. It is easy to see that this formula gives an inverse to the exponential function.[2]

For example, if

$$\Omega = \mathbb{C} \setminus (-\infty, 0]$$

(the "slit complex plane" with the negative real axis removed), then we can set

$$\mathrm{Log}\, z = \log |z| + i \,\mathrm{Arg}\, z,$$

where $\mathrm{Arg}\, z$ is defined as a choice of $\arg z$ that takes values in $(-\pi, \pi)$. The function $\mathrm{Log}\, z$ is called the **principal branch of the logarithm**, a kind of standard version of the log function that complex analysts have agreed to use whenever this is reasonably convenient. However, sometimes we may want to consider the logarithm function on stranger or more complicated regions. When can this be made to work? The answer is: when Ω is simply connected. We further give two results making this notion precise, the first involving a situation where the logarithm exists and can be made unique in a relatively canonical way, and the second in a more general setting that forces us to accept a (mild) lack of uniqueness.

Theorem 1.52 (Existence of the logarithm: first version). *Assume that Ω is a simply connected region with $0 \notin \Omega$, $1 \in \Omega$. There exists a unique function $F : \Omega \rightarrow \mathbb{C}$ with the following properties:*
i) *F is holomorphic in Ω.*
ii) *$e^{F(z)} = z$ for all $z \in \Omega$.*
iii) *$F(r) = \log r$ (the usual logarithm for real numbers) for all real numbers $r \in \Omega$ sufficiently close to 1.*

---

2 Logarithms in complex analysis are a subtle concept. One common source of confusion is that the language used to refer to them is inconsistent with their properties: it is common to speak of "*the* logarithm function" when the use of the definite article is potentially at odds with the fact that a function satisfying the properties of a logarithm is not unique.

Proof. Uniqueness: if F and G are two functions satisfying the properties listed in the theorem, then since $F(r) = G(r)$ for real r in a neighborhood of 1, we must have $F \equiv G$ by Corollary 1.36.

Existence: we define F as a primitive function of the function $z \mapsto 1/z$, guaranteed to exist by Corollary 1.25. We can assume without loss of generality that $F(1) = 0$. We then have that

$$\frac{d}{dz}(ze^{-F(z)}) = e^{-F(z)} - zF'(z)e^{-F(z)} = e^{-F(z)}(1 - z/z) = 0,$$

so $ze^{-F(z)}$ is a constant function. Since its value at $z = 1$ is 1, we see that $e^{F(z)} = z$, as required. Finally, let ϵ be chosen small enough so that the interval $(1-\epsilon, 1+\epsilon)$ is contained in Ω. Then for $r \in (1 - \epsilon, 1 + \epsilon)$, the fundamental theorem of calculus gives that

$$F(r) = F(1) + \int_1^r F'(x)\, dx = 0 + \int_1^r \frac{dx}{x} = \log r. \qquad \square$$

Note that, a bit counterintuitively, the conclusion that $F(r) = \log r$ in the theorem may not be satisfied for all positive real $r \in \Omega$; see Exercise 1.33.

Theorem 1.53 (Existence of the logarithm: second version). *Assume that Ω is a simply connected region with $0 \notin \Omega$. There exists a function $F : \Omega \to \mathbb{C}$ with the following properties:*
i)   *F is holomorphic in Ω.*
ii) *$e^{F(z)} = z$ for all $z \in \Omega$.*

The function F is unique up to an additive integer multiple of $2\pi i$ in the following sense: if G is another function satisfying the same properties, then we have

$$G(z) \equiv F(z) + 2\pi i k \qquad\qquad (1.65)$$

for some integer k; conversely, any function G of the form (1.65) for some $k \in \mathbb{Z}$ satisfies the same properties.

Proof. Exercise 1.34. $\qquad \square$

A function F with the properties given in Theorem 1.53 is called a **branch of the logarithm function on Ω**.

Next, we generalize the concept of a logarithm further by considering the following question: given a region Ω and a holomorphic function $f : \Omega \to \mathbb{C}$, when can we "take the logarithm of f"? That is, does there exist a holomorphic function g for which $e^{g(z)} \equiv f(z)$? An obvious necessary condition is that f must not have any zeros; this generalizes the requirement that $0 \notin \Omega$ from Theorems 1.52 and 1.53. If Ω is simply connected, then this is also a sufficient condition. The precise result, including the extent to which the choice of logarithm is unique, is as follows.

Theorem 1.54 (Existence of the logarithm of a function). *If f is a holomorphic function on a simply connected region Ω and $f \neq 0$ on Ω, then there exists a holomorphic function g on Ω satisfying*

$$e^{g(z)} = f(z).$$

The function g is unique up to an additive constant of the form $2\pi ik$ with integer k.

Proof. The idea is to define g as a primitive function of the function $z \mapsto f'(z)/f(z)$, then the reasoning is similar to the proof of Theorem 1.52. The details are left as an exercise (Exercise 1.35). □

On a simply connected region Ω, we can now define the **power function** $z \mapsto z^{\alpha}$ for an arbitrary $\alpha \in \mathbb{C}$ by setting

$$z^{\alpha} = e^{\alpha F(z)},$$

where F is some branch of the logarithm on Ω.[3] In the particular case $\alpha = 1/n$ with positive integer n, this has the meaning of the nth root function $z \mapsto z^{1/n}$, which satisfies

$$\left(z^{1/n}\right)^{n} = \left(e^{\frac{1}{n}F(z)}\right)^{n} = e^{n\frac{1}{n}F(z)} = e^{F(z)} = z.$$

If $f(z) = z^{1/n}$ is an nth root function associated with some branch of the logarithm, then for any $0 \leq k \leq n - 1$, the function $g(z) = e^{2\pi ik/n}f(z)$ will be another function satisfying $g(z)^{n} = z$. Conversely, it is easy to see that those are precisely the possible choices for an nth root function. That is, nth root functions are unique up to multiplication by an arbitrary nth root of unity.

Generalizing power functions further in a similar way as we did for the logarithm, if Ω is a simply connected region, f is a holomorphic function on Ω that has no zeros, g is a branch of the logarithm of f, and $\alpha \in \mathbb{C}$ is an arbitrary complex number, then the function $h(z) = e^{\alpha g(z)}$ can be interpreted as the power function "f raised to the power α." In particular, for $\alpha = 1/n$ (n a positive integer), this function is usually referred to as (a branch of) the nth root of f and has the property that $h(z)^{n} = f(z)$.

Suggested exercises for Section 1.15. 1.31, 1.32, 1.33, 1.34, 1.35.

1.16 The local behavior of holomorphic functions

In Section 1.3, we considered what the property of being holomorphic at a point z_0 says about the local behavior of the function near the point, focusing on the case when the

---

3 As with the phrase "the logarithm function," saying "the power function" is somewhat misleading; it is more correct to say "a branch of the power function." However, mathematicians are human and prone to employing mental shortcuts just like everyone else, so in practice, you will rarely encounter mathematicians in the real world employing such precise terminology.

derivative $f'(z_0)$ does not vanish. We now give a more complete analysis that covers a more general situation. As we will show in Theorem 1.57, for a function f holomorphic in the neighborhood of a point z_0, we can canonically express the function as $c + w^k$, where w is a new variable associated with z near z_0, which takes values in the unit disc. Thus, loosely speaking, f behaves locally "like a power function $w \mapsto w^k$."

We say that a holomorphic function $f : \Omega \to \mathbb{C}$ is **locally injective** near a point $z_0 \in \Omega$ if there is a neighborhood U of z_0 such that the restriction of f to U is injective.

Lemma 1.55. *Let* $f : \Omega \to \mathbb{C}$ *be a holomorphic function, and let* $z_0 \in \Omega$. *If* $f'(z_0) \neq 0$, *then* f *is locally injective near* z_0.

Proof. Denote $\lambda = f'(z_0)$ and $\epsilon = |\lambda|^2$. Denoting $\langle z, w \rangle = \text{Re}(z\overline{w})$ (the standard inner product in the plane), we have $\langle f'(z_0), \lambda \rangle = |\lambda|^2 > \epsilon/2$, and therefore by continuity also $\langle f'(z), \lambda \rangle > \epsilon/2$ for all z in some disc $D_\delta(z_0)$. Let z_1 and z_2 be distinct points in $D_\delta(z_0)$. Then we have

$$f(z_2) - f(z_1) = \int_{z_1}^{z_2} f'(z)\,dz = (z_2 - z_1) \int_0^1 f'(z_1 + t(z_2 - z_1))\,dt.$$

This implies that

$$\overline{(z_2 - z_1)}\langle f(z_2) - f(z_1), \lambda \rangle = \overline{(z_2 - z_1)}\left\langle (z_2 - z_1) \int_0^1 f'(z_1 + t(z_2 - z_1))\,dt, \lambda \right\rangle$$

$$= |z_2 - z_1|^2 \int_0^1 \langle f'(z_1 + t(z_2 - z_1)), \lambda \rangle\,dt \geq \frac{\epsilon}{2}|z_2 - z_1|^2 > 0.$$

In particular, $f(z_2) - f(z_1) \neq 0$. $\square$

The next two classic results are both important consequences.

Theorem 1.56 (Inverse function theorem). *Let* $f : \Omega \to \mathbb{C}$ *be a holomorphic function. Let* $z_0 \in \Omega$, *and denote* $w_0 = f(z_0)$. *Assume that* $f'(z_0) \neq 0$. *Then* f *has a local holomorphic inverse. More precisely, there exist an open neighborhood* U *of* z_0, *an open neighborhood* V *of* w_0, *and a holomorphic function* $g : V \to U$ *such that:*
1. *f maps U bijectively onto V;*
2. *g maps V bijectively onto U'*
3. *$g = f^{-1}$ (in the set-theoretic sense of an inverse function);*
4. *The derivative of the inverse function g is given by*

$$g'(w_0) = \frac{1}{f'(z_0)}. \tag{1.66}$$

Proof. By Lemma 1.55, f maps an open neighborhood $U = D_\delta(z_0)$ bijectively into $V = f(U)$. By the open mapping theorem, V is an open neighborhood V of w_0. Since the restriction $f_{|U} : U \to V$ of f to U is continuous and open, it is a homeomorphism. Denote its inverse by $g : V \to U$. To see that g is holomorphic at z_0, observe that

$$\lim_{w \to w_0} \frac{g(w) - g(w_0)}{w - w_0} = \lim_{z \to z_0} \frac{g(f(z)) - g(f(z_0))}{f(z) - f(z_0)} = \lim_{z \to z_0} \frac{z - z_0}{f(z) - f(z_0)}$$

$$= \left(\lim_{z \to z_0} \frac{f(z) - f(z_0)}{z - z_0} \right)^{-1} = \frac{1}{f'(z_0)}, \tag{1.67}$$

which also gives formula (1.66). Similarly, replacing z_0 and w_0 in (1.67) by an arbitrary pair of points $z_1 \in U$ and $w_1 = f(z_1)$ proves that g is holomorphic on all of V. □

Theorem 1.57 (Local behavior of holomorphic functions). *Let f be holomorphic in a region Ω. Let $z_0 \in \Omega$, and let $k \geq 1$ denote the order of the zero of $f(z) - f(z_0)$ at z_0. Then there exist an open neighborhood U of z_0, a number $r > 0$, and a function $\varphi : U \to D_r(0)$ such that:*

1. *φ is holomorphic and bijective, and the inverse function φ is also holomorphic;[4]*
2. *$\varphi(z_0) = 0$;*
3. *We have*

$$f(z) = f(z_0) + \varphi(z)^k \quad (z \in U). \tag{1.68}$$

In other words, under the change of variables $w = \varphi(z)$, the function $z \mapsto f(z)$, $z \in U$, is represented as $w \mapsto f(z_0) + w^k$, $w \in D_r(0)$, in terms of the new variable w.

Proof. By the definition of k we can represent f as

$$f(z) = f(z_0) + (z - z_0)^k g(z)$$

with g holomorphic and $g(z_0) \neq 0$. Since zeros of holomorphic functions are isolated, g is also nonzero in some disc $D_\epsilon(z_0)$, so by the discussion about nth roots at the end of the previous section we can express g as

$$g(z) = h(z)^k \quad (z \in D_\epsilon(z_0)) \tag{1.69}$$

for some function h that is holomorphic (and also nonzero by (1.69)) in $D_\epsilon(z_0)$. If we now define

$$H(z) = (z - z_0)h(z),$$

then we see that $f(z)$ can be expressed as

---

4 A function with these properties is called a **biholomorphism**; see Chapter 3.

$$f(z) = f(z_0) + ((z - z_0)h(z))^k = f(z_0) + H(z)^k, \tag{1.70}$$

a representation that is similar to (1.68), but not yet with the correct domain and range claimed in the theorem. Note that $H(z_0) = 0$ and $H'(z_0) = h(z_0) \neq 0$. By Lemma 1.55, H is locally injective near z_0, that is, its restriction $H_{|D_\delta(z_0)}$ to a smaller disc $D_\delta(z_0)$ for some $\delta \in (0, \epsilon)$ is injective. The restricted function, being holomorphic, is also an open continuous mapping, so it maps the disc $D_\delta(z_0)$ homeomorphically to some open set V containing $H(z_0) = 0$. Let $r > 0$ be such that $D_r(0) \subset U$, and denote $U = (H_{|D_\delta(z_0)})^{-1}(D_r(0))$. Then $\varphi := H_{|U}$ (the further restriction of H to U) maps U bijectively and homeomorphically onto $D_r(0)$, and its inverse is holomorphic. The above remarks together with (1.70) show that it satisfies the properties claimed in the theorem. The proof is complete. □

Corollary 1.58. *A holomorphic function $f : \Omega \to \mathbb{C}$ is locally injective near $z_0 \in \Omega$ if and only if $f'(z_0) \neq 0$.*

Suggested exercises for Section 1.16. 1.36.

1.17 Infinite products and the product representation of the sine function

Complex analysis abounds in esthetically appealing identities involving integrals and infinite sums. We will also encounter a variety of beautiful identities involving **infinite products**. In this section, we develop the basic theory of such products and illustrate it in one particularly elegant example, the infinite product identity for the sine function.

1.17.1 Infinite products of complex numbers

Let $(c_n)_{n=1}^\infty$ be a sequence of complex numbers. The infinite product $\prod_{n=1}^\infty c_n$ is defined as the limit of finite (partial) products $\lim_{N \to \infty} \prod_{n=1}^N c_k$ if the limit exists. In that case, we say that the product $\prod_{n=1}^\infty c_n$ **converges**.

Proposition 1.59. *For a sequence of complex numbers $(a_n)_{n=1}^\infty$, if $\sum_{n=1}^\infty |a_n| < \infty$, then the infinite product $\prod_{n=1}^\infty (1 + a_n)$ converges, and its value is 0 if and only if one of the factors $1 + a_n$ is equal to 0.*

Proof. Under the assumption, there exists some large enough $N_0 \geq 1$ such that $|a_n| < 1/2$ for all $n \geq N_0$. This implies that $1 + a_n = \exp(\text{Log}(1 + a_n))$, where $\text{Log}(z)$ is the principal branch of the logarithm function. Now by the Taylor expansion of the function $z \mapsto \text{Log}(z)$ (Exercise 1.31) there is some constant $C > 0$ such that

$$|\text{Log}(1 + w)| \leq C|w| \quad \text{if } |w| < 1/2.$$

It follows that

$$\sum_{n=N_0}^{\infty} \left| \mathrm{Log}(1 + a_n) \right| \le C \sum_{n=N_0}^{\infty} |a_n| < \infty,$$

so in particular, the series $\sum_{n=N_0}^{\infty} \mathrm{Log}(1 + a_n)$ converges. We can now write

$$\prod_{n=N_0}^{\infty} (1 + a_n) = \lim_{N \to \infty} \prod_{n=N_0}^{N} (1 + a_n) = \lim_{N \to \infty} \prod_{n=N_0}^{N} \exp(\mathrm{Log}(1 + a_n))$$

$$= \lim_{N \to \infty} \exp\left(\sum_{n=N_0}^{N} \mathrm{Log}(1 + a_k) \right) = \exp\left(\lim_{N \to \infty} \sum_{n=N_0}^{N} \mathrm{Log}(1 + a_k) \right)$$

$$= \exp\left(\sum_{n=N_0}^{\infty} \mathrm{Log}(1 + a_k) \right).$$

Thus we have proved that the infinite product $\prod_{n=N_0}^{\infty} (1 + a_n)$ converges, and, moreover, it converges to a nonzero value. Therefore, trivially, the product $\prod_{n=N_0}^{\infty} (1 + a_n)$ also converges and is equal to zero if and only if one of the factors $1 + a_n$ for $1 \le n < N_0$ is zero. $\qquad\square$

1.17.2 Infinite products of holomorphic functions

Proposition 1.60. *Let $(f_n)_{n=1}^{\infty}$ be a sequence of holomorphic functions on a region Ω. If the series $\sum_{n=1}^{\infty} |f_n|$ converges uniformly on compacts in Ω, then the infinite product $F(z) = \prod_{n=1}^{\infty}(1 + f_n(z))$ also converges uniformly on compacts. The limiting function $F(z)$ is holomorphic and is nonzero everywhere except at the points z for which $1 + f_n(z) = 0$ for some n.*

Proof. Proposition 1.59 implies that the infinite product $\prod_{n=1}^{\infty}(1 + f_n(z))$ converges to a nonzero limit for any $z \in \Omega$. By repeating the same estimates in the proof of that proposition in the context of z being allowed to range on a compact subset $K \subset \Omega$, we see that the sequence of partial products $\prod_{k=1}^{n}(1 + f_n)$ actually converges uniformly on compacts, so the limiting function is holomorphic. The claim about the set of points z for which $F(z) = 0$ is an immediate consequence of the corresponding condition in Proposition 1.59. $\qquad\square$

Proposition 1.61. *Under the assumptions of Proposition 1.60, the logarithmic derivative of the infinite product $\prod_{n=1}^{\infty}(1+f_n)$ is the sum of the logarithmic derivatives of the individual factors, that is,*

$$\frac{\left(\prod_{n=1}^{\infty}(1+f_n) \right)'}{\prod_{n=1}^{\infty}(1+f_n)} = \sum_{n=1}^{\infty} \frac{f_n'}{1+f_n}. \tag{1.71}$$

Moreover, the infinite series in (1.71) *converges uniformly on compacts in the set* $\{z \in \Omega :$ $\prod_{n=1}^{\infty}(1 + f_n(z)) \neq 0\}.$

Proof. Exercise 1.37 □

1.17.3 The sine function

As an illustration of the theory of infinite products, we prove the following classic result.

Theorem 1.62 (Infinite product formula for the sine function).

$$\sin(\pi z) = \pi z \prod_{n=1}^{\infty}\left(1 - \frac{z^2}{n^2}\right) \quad (z \in \mathbb{C}). \tag{1.72}$$

Theorem 1.62 often comes up in an equivalent form of an infinite series identity, obtained by taking the logarithmic derivative of both sides of (1.72). This result is known as the **partial fraction expansion of the cotangent function**.

Theorem 1.63 (Partial fraction expansion of the cotangent function). *The rescaled cotangent function* $\pi \cot(\pi z)$ *has three representations*

$$\pi \cot(\pi z) = \lim_{N \to \infty} \sum_{n=-N}^{N} \frac{1}{z + n} = \frac{1}{z} + \sum_{n \in \mathbb{Z} \setminus \{0\}}\left(\frac{1}{z + n} - \frac{1}{n}\right) = \frac{1}{z} + \sum_{n=1}^{\infty}\frac{2z}{z^2 - n^2}, \tag{1.73}$$

valid for all $z \in \mathbb{C} \setminus \mathbb{Z}.$

The equivalence of the three sums in (1.73) and the convergence of the respective expressions are easy to verify (Exercise 1.38). The first of the three formulas is sometimes written in the form of the infinite series

$$\text{P.V.} \sum_{n=-\infty}^{\infty}\frac{1}{z + n}, \tag{1.74}$$

with the caveat that this is to be interpreted in the "principal value" sense, where the summation is performed symmetrically on positive and negative indices. This also gives a bit of intuition of why we expect an identity such as (1.74) to hold: the series (1.74), assuming that we can make sense of it as defining a genuine function, is periodic with period 1, and its local behavior around $z = n$ for each integer n is the correct principal value of the function $\pi \cot(\pi z)$ around that point, namely the simple pole $\frac{1}{z-n}$. This intuition is not quite a proof, but can be turned into one with some additional arguments (see [3, Ch. 26]). Here we give a more complex-analytic proof based on contour integration.

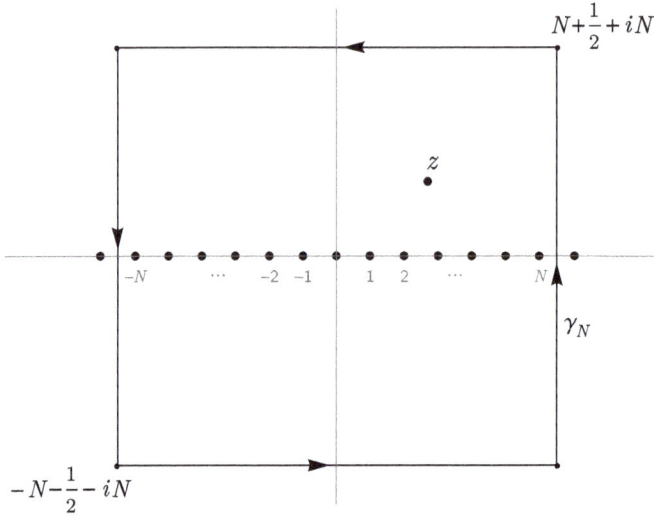

Figure 1.10: The integration contour γ_N and the poles (as a function of w with z fixed) of the integrand $f_z(w) = \frac{\pi \cot(\pi w)}{(w+z)^2}$.

Proof of Theorem 1.63. Let $z \in \mathbb{C} \setminus \mathbb{Z}$. Fix a large positive integer N. We use the residue theorem to evaluate the contour integral

$$I_N(z) := \oint_{\gamma_N} \frac{\pi \cot(\pi w)}{(w + z)^2} \, dw$$

over the contour γ_N going in the positive direction around the rectangle with vertices $(\pm(N + 1/2), \pm N)$; see Fig. 1.10. The integrand $f_z(w) = \frac{\pi \cot(\pi w)}{(w+z)^2}$ has at its poles enclosed by the contour the points $w = -z$ (assuming that N is chosen large enough) and $w = k \in \mathbb{Z}$, $-N \leq k \leq N$. The residues are evaluated without much difficulty as

$$\text{Res}_{-z}(f_z) = -\frac{\pi^2}{\sin^2(\pi z)},$$

$$\text{Res}_k(f_z) = \frac{1}{(z + k)^2} \quad (-N \leq k \leq N),$$

so the residue theorem gives that

$$I_N(z) = 2\pi i \left[-\frac{\pi^2}{\sin^2(\pi z)} + \sum_{k=-N}^{N} \frac{1}{(z + k)^2} \right]. \tag{1.75}$$

Now consider what this means in the limit as $N \to \infty$. We claim that

$$I_N(z) \xrightarrow[N \to \infty]{} 0,$$

which, together with (1.75), would imply the identity

$$\frac{\pi^2}{\sin^2(\pi z)} = \sum_{n=-\infty}^{\infty} \frac{1}{(z+n)^2} \quad (z \in \mathbb{C} \setminus \mathbb{Z}). \tag{1.76}$$

To prove this, first note the auxiliary identities

$$|\sin(x+iy)|^2 = \sin^2 x + \sinh^2 y,$$
$$\quad\quad\quad\quad\quad\quad\quad\quad\quad\quad (x, y \in \mathbb{R}),$$
$$|\cos(x+iy)|^2 = \cos^2 x + \sinh^2 y,$$

which we leave to the reader to verify. Taking $x = \pm\pi(N + 1/2)$ and y arbitrary, these identities imply the bound

$$\left|\cot\left(\pm\pi\left(N + \frac{1}{2}\right) + \pi i y\right)\right| = \frac{\sinh^2(\pi y)}{1 + \sinh^2(\pi y)} \leq 1, \tag{1.77}$$

and similarly, for $y = \pm N$ and x arbitrary, we have

$$|\cot(\pi x \pm \pi i N)| \leq \frac{1 + \sinh^2(\pi N)}{\sinh^2(\pi N)} \leq 2. \tag{1.78}$$

The bounds (1.77)–(1.78) together show that on the contour γ_N, the integrand $f_z(w)$ is bounded in magnitude by $\frac{2\pi}{(N-|z|)^2}$, which implies that

$$|I_N(z)| \leq \frac{10\pi N}{(N-|z|)^2} \xrightarrow[N\to\infty]{} 0,$$

as claimed, proving (1.76).

Finally, to derive (1.63), let

$$F(z) = \pi \cot(\pi z), \quad G(z) = \frac{1}{z} + \sum_{n=1}^{\infty} \frac{2z}{z^2 - n^2}.$$

Note that

$$F'(z) = -\frac{\pi^2}{\sin^2(\pi z)},$$

$$G'(z) = -\frac{1}{z^2} - 2\sum_{n=1}^{\infty} \frac{z^2 + n^2}{(z^2 - n^2)^2} = -\frac{1}{z^2} - \sum_{n=1}^{\infty}\left(\frac{1}{(z+n)^2} + \frac{1}{(z-n)^2}\right)$$

$$= -\sum_{n=-\infty}^{\infty} \frac{1}{(z+n)^2},$$

so that, by (1.76), $F'(z) \equiv G'(z)$. It follows that $F(z) \equiv G(z) + c$ for some constant c. However, F and G are both odd functions, so we must have $c = 0$, i. e., $F \equiv G$. □

Proof of Theorem 1.62. Define the holomorphic functions

$$S(z) = \sin(\pi z), \quad T(z) = \pi z \prod_{n=1}^{\infty}\left(1 - \frac{z^2}{n^2}\right),$$

noting that the convergence of the infinite product to a holomorphic function is justified by Proposition 1.60. Taking logarithmic derivatives, we see (using (1.71)) that

$$\frac{S'(z)}{S(z)} = \pi\cot(\pi z), \quad \frac{T'(z)}{T(z)} = \frac{1}{z} + \sum_{n=1}^{\infty}\frac{2z}{z^2 - n^2} \quad (z \in \mathbb{C} \setminus \mathbb{Z}).$$

By (1.73) we therefore see that $S'/S \equiv T'/T$ or, equivalently, $(S/T)' \equiv 0$. It follows that $S \equiv c_0 T$ for some constant c_0. Rewriting this in the form

$$\frac{\sin(\pi z)}{\pi z} \equiv c_0 \prod_{n=1}^{\infty}\left(1 - \frac{z^2}{n^2}\right)$$

and taking the limit as $z \to 0$ show that $c_0 = 1$ and finish the proof. □

Aside from being a remarkable result in its own right, Theorem 1.62 has a number of interesting consequences, discussed in Exercise 1.39. We will also use this result (in the equivalent form (1.73)) several times in our studies of modular forms in Chapter 5.

Corollary 1.64. *We have the infinite product formulas*

$$\cos(\pi z) = \prod_{n=1}^{\infty}\left(1 - \frac{z^2}{(n-1/2)^2}\right), \tag{1.79}$$

$$e^z - 1 = ze^{z/2}\prod_{n=1}^{\infty}\left(1 + \frac{z^2}{4\pi^2 n^2}\right). \tag{1.80}$$

Proof. Exercise 1.40. □

Suggested exercises for Section 1.17. 1.37, 1.38, 1.39, 1.40, 1.41, 1.42.

1.18 Laurent series

A **Laurent series** is a generalization of a power series expansion and takes the form of a two-sided infinite series

$$f(z) = \sum_{n=-\infty}^{\infty} a_n(z - z_0)^n \tag{1.81}$$

for some $z_0 \in \mathbb{C}$ and complex coefficients $(a_n)_{n=-\infty}^{\infty}$. Given such a series, it is easy to see that it converges absolutely and uniformly on compacts in the annulus-shaped region

$$A_{r,R}(z_0) = \{z \: : \: r < |z - z_0| < R\},$$

where R is the radius of convergence of the power series $\sum_{n=0}^{\infty} a_n z^n$, and r is the *reciprocal* of the radius of convergence of the power series $\sum_{n=1}^{\infty} a_{-n} w^n$. Note that this region can be empty, e. g., if $r = \infty$, $R = 0$, or $r > R$.

The basic question about when a function can be expressed as a Laurent series is answered by the following theorem.

Theorem 1.65. *Let $0 \le r < R \le \infty$. Let f be holomorphic in a region Ω containing the annulus $A_{r,R}(z_0)$. Then $f(z)$ has a unique representation as a Laurent series (1.81), which is absolutely convergent uniformly on compacts on $A_{r,R}(z_0)$. The coefficients a_n are given by*

$$a_n = \frac{1}{2\pi i} \oint_{C_\rho(z_0)} \frac{f(z)}{(z - z_0)^{n+1}} \, dz \quad (n \in \mathbb{Z}) \tag{1.82}$$

with arbitrary $\rho \in (r, R)$.

Proof. Uniqueness: Given an expansion of the form (1.81) known to converge absolutely uniformly on compacts on $A_{r,R}(z_0)$, let $\rho \in (r, R)$, and observe that

$$\frac{1}{2\pi i} \oint_{C_\rho(z_0)} \frac{f(z)}{(z - z_0)^{n+1}} \, dz = \frac{1}{2\pi i} \oint_{C_\rho(z_0)} \frac{1}{(z - z_0)^{n+1}} \left(\sum_{m=-\infty}^{\infty} a_m (z - z_0)^m \right) dz$$

$$= \sum_{m=-\infty}^{\infty} a_m \left(\frac{1}{2\pi i} \oint_{C_\rho(z_0)} (z - z_0)^{m-n-1} \, dz \right) = a_n.$$

(The last step uses the standard formula (1.92) from Exercise 1.21.) Thus the a_n are determined uniquely and are given by (1.82).

Existence: Fix $z \in A_{r,R}(z_0)$. Take numbers ρ_- and ρ_+, with $r < \rho_- < |z - z_0| < \rho_+ < R$. Then, by a standard limiting argument involving a keyhole contour we can show that

$$f(z) = \frac{1}{2\pi i} \oint_{C_{\rho_+}(z_0)} \frac{f(w)}{w - z} \, dw - \frac{1}{2\pi i} \oint_{C_{\rho_-}(z_0)} \frac{f(w)}{w - z} \, dw.$$

In this representation the factors $\frac{1}{w-z}$ inside the two integrals can be expanded as geometric series (in two different ways, one being valid for $w \in C_{\rho_-}(z_0)$ and the other for w on the circle $C_{\rho_-}(z_0)$). This leads to

$$f(z) = \frac{1}{2\pi i} \oint_{C_{\rho_+}(z_0)} \frac{1}{(w - z_0)} \frac{1}{1 - \frac{z-z_0}{w-z_0}} f(w) \, dw$$

$$+ \frac{1}{2\pi i} \oint_{C_{\rho_-}(z_0)} \frac{1}{z - z_0} \frac{1}{1 - \frac{w - z_0}{z - z_0}} f(w) \, dw$$

$$= \frac{1}{2\pi i} \oint_{C_{\rho_+}(z_0)} \sum_{n=0}^{\infty} \frac{(z - z_0)^n}{(w - z_0)^{n+1}} f(w) \, dw$$

$$+ \frac{1}{2\pi i} \oint_{C_{\rho_-}(z_0)} \sum_{m=0}^{\infty} \frac{(w - z_0)^m}{(z - z_0)^{m+1}} f(w) \, dw$$

$$= \sum_{n=0}^{\infty} \frac{1}{2\pi i} \left(\oint_{C_{\rho_+}(z_0)} \frac{f(w)}{(w - z_0)^{n+1}} \, dw \right)(z - z_0)^n$$

$$+ \sum_{m=0}^{\infty} \frac{1}{2\pi i} \left(\oint_{C_{\rho_-}(z_0)} f(w)(w - z_0)^m \, dw \right)(z - z_0)^{-m-1},$$

where in the last step, we interchanged the summation and integration; this is easy to justify, since the geometric series converge uniformly on the integration contours. We have therefore obtained a representation for $f(z)$, which we see is of the form (1.81) with the coefficients a_n given by

$$a_n = \begin{cases} \dfrac{1}{2\pi i} \displaystyle\oint_{C_{\rho_+}(z_0)} \dfrac{f(w)}{(w - z_0)^{n+1}} \, dw & \text{if } n \geq 0, \\[1.5em] \dfrac{1}{2\pi i} \displaystyle\oint_{C_{\rho_-}(z_0)} \dfrac{f(w)}{(w - z_0)^{n+1}} \, dw & \text{if } n < 0. \end{cases} \qquad (1.83)$$

Finally, observe that (1.83) is equivalent to (1.82), since, by another application of Cauchy's formula on an appropriate keyhole contour, it can easily be shown that the integral on the right-hand side of (1.82) is independent of the radius ρ of the integration contour. □

Suggested exercises for Section 1.18. 1.43, 1.44.

Exercises for Chapter 1

1.1 An immediate corollary of the fundamental theorem of algebra is that any complex polynomial

$$p(z) = a_n z^n + a_{n-1} z^{n-1} + \cdots + a_0$$

(where $a_0, \ldots, a_n \in \mathbb{C}$ and $a_n \neq 0$) can be factored as

$$p(z) = a_n \prod_{k=1}^{n} (z - z_k)$$

for some $z_1, \ldots, z_n \in \mathbb{C}$; these are the roots of $p(z)$ counted with multiplicities. Use this to prove that any such polynomial where the coefficients $a_0, \ldots, a_n$ are *real* has a factorization

$$p(z) = a_n Q_1(z) Q_2(z) \ldots Q_m(z),$$

where each $Q_k(z)$ is a linear or quadratic monic polynomial (i. e., is of one of the forms $z - c$ or $z^2 + bz + c$) with real coefficients.

1.2 If $p(z) = a_n z^n + a_{n-1} z^{n-1} + \cdots + \cdots + a_0$ is a polynomial of degree n such that

$$|a_n| > \sum_{j=0}^{n-1} |a_j|,$$

then prove that $p(z)$ has exactly n zeros (counting multiplicities) in the unit disc $\mathbb{D}$.
Guidance. Use the fundamental theorem of algebra.
Note. This is a particular case of a less elementary fact, which can be proved using Rouché's theorem; see Exercise 1.29.

1.3 For each of the following functions, determine where it is holomorphic.
a. $f(z) = z$ c. $f(z) = |z|$ e. $f(z) = \bar{z}$
b. $f(z) = \mathrm{Re}(z)$ d. $f(z) = |z|^2$ f. $f(z) = 1/z$

1.4 For each of relations (1.11)–(1.15) in Lemma 1.4, explain precisely what holomorphicity assumptions are needed for the relation to hold and prove its correctness under those assumptions.

1.5 Draw (approximately, or with as much precision as you can) the image in the w-plane of the following figure in the z-plane

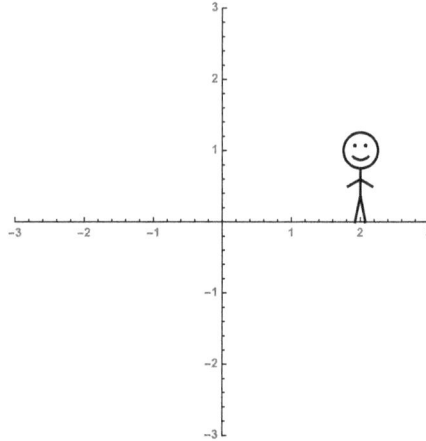

under each of the following maps $w = f(z)$:

a. $w = \frac{1}{2}z$ c. $w = \bar{z}$ e. $w = 1/z$

b. $w = iz$ d. $w = (2 + i)z - 3$ f. $w = z^2 - 1$

1.6 Let f be a holomorphic function in a region Ω, and let $\gamma : (a, b) \to \mathbb{C}$ be a differentiable parameterized curve in Ω. Prove that

$$\frac{d}{dt}(f(\gamma(t))) = f'(\gamma(t))\gamma'(t). \tag{1.84}$$

1.7 Complete the proof of Lemma 1.7 by proving the remaining implications (b) $\Longleftrightarrow$ (c) and (b) $\Longrightarrow$ (a), which were not proved in the text.

1.8 For each of the following functions $u(x,y)$, determine if there exists a function $v(x,y)$ such that $f(x + iy) = u(x,y) + iv(x,y)$ is an entire function, and if so, then find it and try to find a formula for $f(z)$ directly in terms of z rather than in terms of its real and imaginary parts.

a. $u(x,y) = x^2 - y^2$ c. $u(x,y) = x^4 - 6x^2y^2 + 3x + y^4 - 2$

b. $u(x,y) = y^3$ d. $u(x,y) = \cos x \cosh y$

1.9 **Alternative form of the Cauchy–Riemann equations.** A function $f = u + iv$ of a complex variable $z = x + iy$ is traditionally thought of as a function of the two coordinates x and y. However, if we think of the equations

$$z = x + iy, \quad \bar{z} = x - iy$$

as representing a formal change of variables from the "real coordinates" (x,y) to the "complex conjugate coordinates" $(z, \bar{z})$, then it may make sense to think of f as a function of the two variables z and $\bar{z}$ (pretending that those are two independent variables). Thus we may suggestively write $u = u(z, \bar{z})$ and $v = v(z, \bar{z})$ and consider operations such as taking the partial derivatives of f, u, v with respect to z and $\bar{z}$.

Show that from this point of view, the Cauchy–Riemann equations

$$\frac{\partial u}{\partial x} = \frac{\partial v}{\partial y}, \quad \frac{\partial u}{\partial y} = -\frac{\partial v}{\partial x}$$

can be rewritten in the more concise equivalent form

$$\frac{\partial f}{\partial \bar{z}} = 0,$$

assuming that it is okay to apply the chain rule from multivariable calculus; and, moreover, that in this notation, we also have the identity

$$f'(z) = \frac{\partial f}{\partial z}.$$

1.10 Let $f : \Omega \to \mathbb{C}$ be a function defined on a region Ω such that both functions $f(z)$ and $zf(z)$ have real and imaginary parts that are harmonic functions. Prove that $f(z)$ is holomorphic on Ω.

1.11 Let $p(z) = a_n z^n + a_{n-1} z^{n-1} + \cdots + a_0$ be a complex polynomial of degree $n \ge 2$ (that is, $a_0, \ldots, a_n \in \mathbb{C}$ and $a_n \ne 0$), and let $z_1, \ldots, z_n$ be its roots counted with multiplicities. Let $w_1, \ldots, w_{n-1}$ denote the roots of $p'(z)$. Prove the following claim, known as the **Gauss–Lucas theorem**.

Theorem 1.66 (Gauss–Lucas theorem). *The points $w_1, \ldots, w_{n-1}$ all lie in the convex hull of $z_1, \ldots, z_n$, that is, each w_k can be expressed as a convex combination*

$$w_k = a_1^{(k)} z_1 + a_2^{(k)} z_2 + \cdots + a_n^{(k)} z_n$$

for some coefficients $a_1^{(k)}, \ldots, a_n^{(k)} \ge 0$ satisfying $\sum_{j=1}^{n} a_j^{(k)} = 1$.

See Fig. 1.11 for an illustration of this phenomenon.

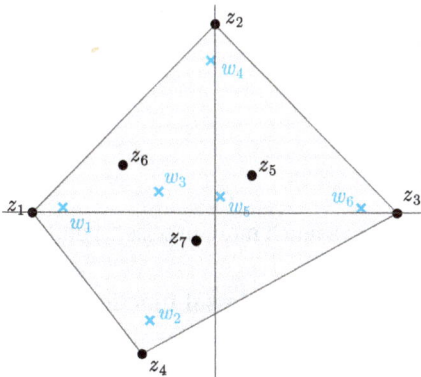

Figure 1.11: An illustration of the Gauss–Lucas theorem discussed in Exercise 1.11, showing the roots $z_1, \ldots, z_7$ of a polynomial $p(z)$ of degree 7, their convex hull, and the roots $w_1, \ldots, w_6$ of $p'(z)$.

1.12 Illustrate the claim from p. 19 regarding the orthogonality of the level curves of the real and imaginary parts of holomorphic functions by plotting some of the level curves of $\text{Re}(f)$ and $\text{Im}(f)$ for each of the following functions:

    a. $f(z) = z^2$ b. $f(z) = 1/z$ c. $f(z) = e^z$

1.13 Complete the argument of the proof of Lemma 1.8 in the extreme cases $R = 0, \infty$.

1.14 Using the formula $e^z = \sum_{n=0}^{\infty} \frac{z^n}{n!}$ as the definition of the exponential function, prove that

$$e^{w+z} = e^w e^z \quad (w, z \in \mathbb{C}).$$

1.15 **The Bernoulli numbers.** The Bernoulli numbers are the numbers $(B_n)_{n=0}^{\infty}$ defined by the power series expansion

$$\frac{z}{e^z - 1} = \sum_{n=0}^{\infty} \frac{B_n}{n!} z^n. \tag{1.85}$$

For example, the first three Bernoulli numbers are $B_0 = 1$, $B_1 = -1/2$, and $B_2 = 1/6$.

(a) Find the radius of convergence of the series (1.85).

(b) Prove that the Bernoulli numbers satisfy the following identities:

$$B_{2k+1} = 0 \quad (k = 1, 2, \ldots), \tag{1.86}$$

$$(n + 1)B_n = -\sum_{k=0}^{n-1} \binom{n+1}{k} B_k \quad (n \geq 1), \tag{1.87}$$

$$(2n + 1)B_{2n} = -\sum_{k=1}^{n-1} \binom{2n}{2k} B_{2k} B_{2n-2k} \quad (n \geq 2), \tag{1.88}$$

$$\frac{z}{2} \coth\left(\frac{z}{2}\right) = \sum_{n=0}^{\infty} \frac{B_{2n}}{(2n)!} z^{2n}. \tag{1.89}$$

    Hint for (1.88). Show that the function $g(z) = f(z) + z/2$ satisfies the ordinary differential equation $g(z) - zg'(z) = g(z)^2 - z^2/4$.

(c) Prove that

$$\limsup_{n \to \infty} \left| \frac{B_{2n}}{2n!} \right|^{1/(2n)} = \frac{1}{2\pi}. \tag{1.90}$$

    (See also Exercise 1.39, where we will derive a much more precise formula for the asymptotic behavior of B_{2n} for large n.)

1.16 **Bessel functions.** The **Bessel functions** (also known as **Bessel functions of the first kind**) are a family of functions $(J_n)_{n=-\infty}^{\infty}$ of a complex variable, defined by

$$J_n(z) = \sum_{k=0}^{\infty} \frac{(-1)^k}{k!(k+n)!} \left(\frac{z}{2}\right)^{2k+n}. \tag{1.91}$$

(For example, note that $J_0(-2\sqrt{x}) = \sum_{k=0}^{\infty} \frac{x^k}{(k!)^2}$, which is reminiscent of the exponential function and already seems like a fairly natural function to study.)

(a) For which $z \in \mathbb{C}$ does the series (1.91) converge?

(b) Prove that the Bessel functions satisfy the following properties:

$$J_{-n}(z) = (-1)^n J_n(z),$$

$$J_{n+1}(z) = \frac{2n}{z} J_n(z) - J_{n-1}(z),$$

$$J_n''(z) = -\frac{1}{z} J_n'(z) - \frac{z^2 - n^2}{z^2} J_n(z).$$

(c) Prove the following additional identities:

$$\exp\left[\frac{z}{2}\left(t - \frac{1}{t}\right)\right] = \sum_{n=-\infty}^{\infty} J_n(z) t^n,$$

$$\cos(z \sin t) = J_0(z) + 2 \sum_{n=1}^{\infty} J_{2n}(z) \cos(2nt),$$

$$\sin(z \sin t) = 2 \sum_{n=0}^{\infty} J_{2n+1}(z) \sin((2n+1)t),$$

$$\cos(z \cos t) = J_0(z) + 2 \sum_{n=1}^{\infty} (-1)^n J_{2n}(z) \cos(2nt),$$

$$\sin(z \cos t) = 2 \sum_{n=0}^{\infty} (-1)^n J_{2n+1}(z) \cos((2n+1)t),$$

$$J_n(z) = \frac{1}{\pi} \int_0^\pi \cos(z \sin t - nt)\, dt.$$

1.17 Show that, analogously to the calculation in (1.35), the arc length integral (1.34) does not depend on the particular parameterization chosen for the curve γ.

1.18 Prove Proposition 1.12. (Part of the exercise is to define precisely the notions of "composition of curves" and "reverse curve").

1.19 Prove that homotopy of curves defined at the beginning of Section 1.8 is an equivalence relation.

1.20 Prove that $\mathbb{C}$ is simply connected.

1.21 (a) For $r > 0$ and $n \in \mathbb{Z}$, show that

$$\oint_{|z|=r} z^n\, dz = \begin{cases} 2\pi i & \text{if } n = -1, \\ 0 & \text{otherwise.} \end{cases} \tag{1.92}$$

(b) For which $n \in \mathbb{Z}$ does the function $f(z) = z^n$ have a primitive in $\mathbb{C}\backslash\{0\}$? Explain.

(c) Is the "punctured complex plane" $\mathbb{C} \backslash \{0\}$ simply connected? Explain.

1.22 **Cauchy's theorem and irrotational vector fields.** Recall from vector calculus that a planar vector field $\mathbf{F} = (P, Q)$ defined on some region $\Omega \subset \mathbb{C} = \mathbb{R}^2$ is called **conservative** if it is of the form $\mathbf{F} = \nabla g = (\frac{\partial g}{\partial x}, \frac{\partial g}{\partial y})$ (the gradient of g) for some scalar function $g : \Omega \to \mathbb{R}$. By the fundamental theorem of calculus for line integrals, for such a vector field, we have

$$\oint_\gamma \mathbf{F} \cdot \mathbf{ds} = 0$$

for any closed curve γ. Recall also that (as is easy to check) any conservative vector field is **irrotational**, that is, it satisfies

$$\text{curl}\, \mathbf{F} = 0,$$

where, in the context of two-dimensional vector fields, the curl operator is defined by

$$\text{curl}\, \mathbf{F} = \frac{\partial Q}{\partial x} - \frac{\partial P}{\partial y}.$$

The following converse to this result can be shown: if the region Ω is simply connected, then a theorem in vector calculus says that an irrotational vector field is also conservative.

Use these background results to show that if $f = u + iv$ is holomorphic on a simply connected region Ω, then

$$\oint_\gamma f(z)\, dz = 0$$

for any closed curve γ in Ω. (This is, of course, Cauchy's theorem.)

1.23 Show that Liouville's theorem (Theorem 1.33) can be proved directly using the "simple" ($n = 0$) case of Cauchy's integral formula, instead of using the case $n = 1$ of the extended formula as we did in the lecture.

1.24 Show that Liouville's theorem can in fact be deduced even just from the mean value property of holomorphic functions (Theorem 1.28), which, as you may recall, is the particular case of Cauchy's integral formula in which z is taken as the center of the circle around which the integration is performed.

Guidance. Here it makes sense to consider a modified version of the mean value property (that follows easily from the original version) that says that $f(z)$ is the average value of $f(w)$ over a *disc* $D_R(z)$ (instead of a circle $C_R(z)$), that is,

$$f(z) = \frac{1}{\pi R^2} \iint_{D_R(z)} f(x + iy)\, dx\, dy,$$

where the integral is an ordinary two-dimensional Riemann integral. Explain why this formula holds, then use it to bound $|f(z_1) - f(z_2)|$ (for arbitrary complex numbers z_1, z_2) from above by a quantity that goes to 0 as $R \to \infty$.

1.25 Prove the following generalization of Liouville's theorem: let f be an entire function that for all $z \in \mathbb{C}$ satisfies the inequality

$$|f(z)| \leq A + B|z|^n$$

for some constants $A, B > 0$ and integer $n \geq 0$. Then f is a polynomial of degree at most n.

1.26 **Integration of a family of holomorphic functions with respect to a parameter.** Let $I \subset \mathbb{R}$ be an interval, and let Ω be a complex region. Let $F(t, z)$ be a function of a real parameter $t \in I$ and a complex variable $z \in \Omega$. Assume that $F(t, z)$ is continuous on $I \times \Omega$, holomorphic in z for any fixed $t \in I$, and that for any compact set $K \subset \Omega$, $\sup_{z \in K} \int_I |F(t, z)| \, dt < \infty$. Prove that the function $f : \Omega \to \mathbb{C}$ defined by

$$f(z) = \int_I F(t, z) \, dt$$

is holomorphic on Ω.

1.27 (a) Explain how to derive the formulas (1.52)–(1.53) through purely formal algebraic manipulations. Are these manipulations valid in any sense you are familiar with from real analysis?

(b) Explain how the principle of analytic continuation can breathe new life into the two formulas by providing a context within which the formulas can be interpreted as having a precise, well-defined (and correct) meaning.

1.28 Prove properties (1.62)–(1.63) of the generalized order of a zero of a holomorphic function at a point z_0. Can you give a useful condition for when equality holds in (1.62)?

1.29 If $p(z) = a_n z^n + a_{n-1} z^{n-1} + \cdots + \cdots + a_0$ is a polynomial of degree n such that for some $0 \leq k \leq n$, we have

$$|a_k| > \sum_{\substack{0 \leq j \leq n \\ j \neq k}} |a_j|,$$

then prove that $p(z)$ has exactly k zeros (counting multiplicities) in the unit disc $\mathbb{D}$.

Guidance. Use Rouché's theorem.

Note. This result generalizes the result of Exercise 1.2.

1.30 Show how Rouché's theorem can be used to give yet a proof of the fundamental theorem of algebra. This proof is one way to make precise the intuitively compelling "topological" proof idea discussed in Section 1.2.

1.31 Prove that the principal branch of the logarithm has the Taylor series expansion

$$\operatorname{Log} z = \sum_{n=1}^{\infty} \frac{(-1)^{n-1}}{n} (z-1)^n \quad (|z-1| < 1) \tag{1.93}$$

around $z = 1$.

1.32 What is (or are) the complex number (or numbers) represented by i^i?

1.33 (a) Draw a simply connected region $\Omega \subset \mathbb{C}$ such that $0 \notin \Omega, 1, 2 \in \Omega$, and such that there exists a branch $F(z)$ of the logarithm function on Ω satisfying

$$F(1) = 0, \quad F(2) = \log 2 + 2\pi i$$

(where $\log 2$ is the ordinary natural logarithm of 2 in the usual sense of real analysis).

(b) More generally, let $k \in \mathbb{Z}$. If we were to replace the above condition $F(2) = \log 2 + 2\pi i$ with the more general condition $F(2) = \log 2 + 2\pi i k$ but keep all the other conditions, would an appropriate simply connected region $\Omega = \Omega(k)$ exist to make that possible? If so, then what would this region look like, roughly, as a function of k?

1.34 Prove Theorem 1.53.

1.35 Prove Theorem 1.54.

1.36 Prove Theorem 1.56.

1.37 Prove Proposition 1.61.

1.38 Prove that the three infinite series in (1.73) all converge for $z \in \mathbb{C} \setminus \mathbb{Z}$ and represent the same function.

1.39 **Consequences of the infinite product formula for the sine function.**

(a) By specializing the value of z in (1.72) to an appropriate specific value obtain the following infinite product formula for π, known as **Wallis' product** (first proved by John Wallis in 1655):

$$\frac{\pi}{2} = \frac{2}{1} \cdot \frac{2}{3} \cdot \frac{4}{3} \cdot \frac{4}{5} \cdot \frac{6}{5} \cdot \frac{6}{7} \cdot \frac{8}{7} \cdot \frac{8}{9} \cdot \ldots$$

(b) By comparing the first terms in the Taylor expansion around $z = 0$ of both sides of (1.72), derive the well-known identities

$$\sum_{n=1}^{\infty} \frac{1}{n^2} = \frac{\pi^2}{6}, \quad \sum_{n=1}^{\infty} \frac{1}{n^4} = \frac{\pi^4}{90}. \tag{1.94}$$

(c) More generally, we can use (1.72) or, more conveniently, its equivalent cousin (1.73) to obtain closed formulas for all the series

$$\zeta(2k) = \sum_{n=1}^{\infty} \frac{1}{n^{2k}} = 1 + \frac{1}{2^{2k}} + \frac{1}{3^{2k}} + \frac{1}{4^{2k}} + \cdots \quad (k = 1, 2, \ldots).$$

(The notation $\zeta(2k)$ for these infinite sums is standard and has to do with the fact that these are the special values of the Riemann zeta function $\zeta(s) = \sum_{n=1}^{\infty} \frac{1}{n^s}$ at the positive even integers; see Chapter 2.) To see this, expand both sides of the relation

$$\pi \cot(\pi z) = \frac{1}{z} + \sum_{\substack{n=-\infty \\ n \neq 0}}^{\infty} \left(\frac{1}{z+n} - \frac{1}{n} \right) \quad (z \in \mathbb{C} \setminus \mathbb{Z})$$

in a Taylor series around $z = 0$, making use of identity (1.89) from Exercise 1.15. Compare the coefficients and simplify to get the famous formula

$$\zeta(2k) = \frac{(-1)^{k-1}(2\pi)^{2k}}{2(2k)!} B_{2k} \quad (k \geq 1). \tag{1.95}$$

For example, using the first few values $B_2 = \frac{1}{6}, B_4 = -\frac{1}{30}, B_6 = \frac{1}{42}$, and $B_8 = -\frac{1}{30}$, we get

$$\zeta(2) = \sum_{n=1}^{\infty} \frac{1}{n^2} = \frac{\pi^2}{6},$$

$$\zeta(4) = \sum_{n=1}^{\infty} \frac{1}{n^4} = \frac{\pi^4}{90},$$

$$\zeta(6) = \sum_{n=1}^{\infty} \frac{1}{n^6} = \frac{\pi^6}{945},$$

$$\zeta(8) = \sum_{n=1}^{\infty} \frac{1}{n^8} = \frac{\pi^8}{9\,450},$$

where of course the first two values coincide with those from (1.94).

(d) Show that $\zeta(2k) = 1 + O(2^{-2k})$ as $k \to \infty$ and deduce that the asymptotic behavior of the Bernoulli numbers is given by

$$B_{2k} = (1 + O(2^{-2k}))(-1)^{k-1} \frac{2(2k)!}{(2\pi)^{2k}}, \quad k \to \infty.$$

Note that this is consistent with the earlier weaker estimate (1.90).

1.40 Prove identities (1.79)–(1.80).

1.41 (a) Prove the infinite product formula

$$\sin(z) = z \prod_{n=1}^{\infty} \cos\left(\frac{z}{2^n} \right) \quad (z \in \mathbb{C}). \tag{1.96}$$

Hint. $\sin(z) = 2 \sin(z/2) \cos(z/2)$.

(b) By substituting an appropriate value of z into (1.96) prove the formula

$$\frac{2}{\pi} = \frac{\sqrt{2}}{2} \cdot \frac{\sqrt{2+\sqrt{2}}}{2} \cdot \frac{\sqrt{2+\sqrt{2+\sqrt{2}}}}{2} \cdot \frac{\sqrt{2+\sqrt{2+\sqrt{2+\sqrt{2}}}}}{2} \cdot \dots,$$

first proved by François Viète in 1593.

1.42 Evaluate the following infinite products:

(a) $\prod_{n=2}^{\infty} \frac{n^2-1}{n^2} = \frac{3}{4} \cdot \frac{8}{9} \cdot \frac{15}{16} \cdot \frac{24}{25} \cdot \frac{48}{49} \cdot \dots = ?$

(b) $\prod_{n=1}^{\infty} \frac{n^2+1}{n^2} = \frac{2}{1} \cdot \frac{5}{4} \cdot \frac{10}{9} \cdot \frac{17}{16} \cdot \frac{26}{25} \cdot \dots = ?$

Later, in Chapter 2, we will encounter an interesting variation on these infinite products; see Exercise 2.17.

1.43 Let f be holomorphic in a punctured neighborhood of a point $z_0 \in \mathbb{C}$. Assume that f has a pole of order k at z_0. Show that the Laurent series (1.81) in this case takes the form

$$f(z) = \sum_{n=-k}^{\infty} a_n(z-z_0)^n.$$

1.44 Let $f(z) = \frac{1}{z(2-z)}$. By Theorem 1.65, $f(z)$ has a Laurent series (1.81) that converges in the punctured disc $\{0 < |z| < 2\}$, and separately from that, $f(z)$ has a Laurent series that converges in $\{2 < |z| < \infty\}$. Find the coefficients a_n explicitly for both those Laurent series.

1.45 Let $f(z) = p(z)/q(z)$ be a rational function such that $\deg q \geq \deg p + 2$ (where $\deg p$ denotes the degree of a polynomial p). Prove that the sum of the residues of $f(z)$ over all its poles is equal to 0.

1.46 **Sendov's conjecture**, an elementary statement in complex analysis proposed by the mathematician Blagovest Sendov in 1959 and still open today, is the claim that if $p(z) = (z-z_1)\dots(z-z_n)$ is a complex polynomial whose roots $z_j, j = 1, \dots, n$, all lie in the closed unit disc $|z| \leq 1$, then for each root z_j, there is a root α of the derivative $p'(z)$ for which $|z_j - \alpha| \leq 1$.

(a) Prove the conjecture for the case $n = 2$ of quadratic polynomials.

(b) Prove that if in the inequality $|z_j - \alpha| \leq 1$, the number 1 is replaced by any smaller number, then the claim is false.

(c) Prove the conjecture for the case $n = 3$ of cubic polynomials. (This is not a completely trivial result; for one possible proof, see [11].)

1.47 Use Cauchy's theorem and the residue theorem to calculate the following definite integrals:

(a) $\displaystyle\int_{-\infty}^{\infty} e^{-x^2}\, dx = \sqrt{\pi}.$

(b) $\displaystyle\int_{-\infty}^{\infty} e^{-\pi x^2} e^{2\pi i u x}\, dx = e^{-\pi u^2}$ $(u \in \mathbb{R})$.

(c) $\displaystyle\int_{0}^{\infty} \sin(t^2)\, dt = \int_{0}^{\infty} \cos(t^2)\, dt = \frac{\sqrt{\pi}}{2\sqrt{2}}$.

(d) $\displaystyle\int_{-\infty}^{\infty} \frac{1}{\cosh(x)}\, dx = \pi$.

(e) $\displaystyle\int_{-\infty}^{\infty} \frac{1}{\cosh(\pi x)} e^{2\pi i u x}\, dx = \frac{1}{\cosh(\pi u)}$ $(u \in \mathbb{R})$.

(f) $\displaystyle\int_{-\infty}^{\infty} \frac{1}{x^2 + 1} e^{i u x}\, dx = \pi e^{-|u|}$ $(u \in \mathbb{R})$.

(g) $\displaystyle\int_{-\infty}^{\infty} \frac{e^{ux}}{1 + e^x}\, dx = \frac{\pi}{\sin(\pi u)}$ $(0 < u < 1)$.

2 The prime number theorem

My distinguished friend:

Your remarks concerning the frequency of primes were of interest to me in more ways than one. You have reminded me of my own endeavors in the field which began in the very distant past, in 1792 or 1793, after I had acquired the Lambert supplements to the logarithmic tables. Even before I had begun my more detailed investigations into higher arithmetic, one of my first projects was to turn my attention to the decreasing frequency of primes, to which end I counted the primes in several chiliads and recorded the results on the attached white pages. I soon recognized that behind all of its fluctuations, this frequency is on the average inversely proportional to the logarithm, so that the number of primes below a given bound n is approximately equal to

$$\int \frac{dn}{\log n},$$

where the logarithm is understood to be hyperbolic.

Carl Friedrich Gauss, letter to Johann Encke dated December 24, 1847

2.1 Motivation: analytic number theory and the distribution of prime numbers

Humans have been fascinated by the prime numbers since antiquity. Euclid famously proved that there exist infinitely many prime numbers; his ingenious proof still delights us today. Erathostenes developed his eponymous sieve algorithm for finding all primes up to some prescribed upper limit. They and the mathematicians who came after them continued to puzzle over the apparent erraticism with which prime numbers seem to be spread out among the natural numbers. For a long time, the only empirical observation anyone dared to make concerning the primes was that as we look at higher and higher numbers, primes seem to occur with a diminishing frequency.

It was only in the late eighteenth century that mathematicians started making more quantitative guesses. Gauss observed privately in 1792 or 1793 (when he was around 16 years old!) that the density of primes found around a certain integer n falls like the inverse of the logarithm of n; see the epigraphic quote above, and the historical survey [30]. This is easily seen to be equivalent to the statement that the number $\pi(x)$ of prime numbers up to a given upper bound x behaves like $\frac{x}{\log x}$ as $x \to \infty$. Legendre, who was unaware of Gauss's unpublished investigations, published an equivalent formula in 1808. This statement is now known as the **prime number theorem**.

Theorem 2.1 (Prime number theorem). *The prime-counting function $\pi(x)$ behaves asymptotically as*

$$\pi(x) \sim \frac{x}{\log x} \quad \textit{as } x \to \infty. \tag{2.1}$$

Open Access. © 2023 the author(s), published by De Gruyter. [CC BY-NC-ND] This work is licensed under the Creative Commons Attribution-NonCommercial-NoDerivatives 4.0 International License.
https://doi.org/10.1515/9783110796810-003

What is so striking about this result is that it takes a set of objects that appear to be the epitome of disorder, at least when inspected on a small scale, and in one clean, simple statement decrees that they nonetheless obey a very rigid law on the *large scale*. Moreover, the connection to calculus in the form of the appearance of the natural logarithm function seems surprising in view of the existence of prime numbers as fundamentally discrete objects that do not appear to have any connection to the types of continuous phenomena calculus was developed to understand.

Gauss' conjecture, though bold and (as it turned out) correct, was ahead of its time; he and his contemporaries lacked the tools to make any significant progress on the problem until several decades later. In fact, the entire field of complex analysis had yet to be invented, and it would turn out that that branch of mathematics is rather crucial to the methods involved in an eventual proof. Even when Riemann came up with some of the ideas that would turn out to be the most significant, in a famous paper he published in 1859—one of the most famous papers in the entire history of mathematics, titled *On the Number of Primes Less Than a Given Magnitude*—significant work still needed to be done, and several more decades would pass before the result became a proper theorem. This happened in 1896, when two proofs were published independently by Hadamard and de la Vallée Poussin.

The work that led to the proof has become a cornerstone of what is now its own rich area of mathematics, known as **analytic number theory**. At the heart of this field is one of the greatest mathematical questions of all time, the still unsolved **Riemann hypothesis**, which can be thought of as being, in a rather precise sense, the "ultimate" version of the prime number theorem [46].

In this chapter, our ostensible goal is a proof of the prime number theorem, which in my opinion is the quintessential application of complex analysis.[1] However, this is a case where the journey is no less interesting than the destination and will take us through a study of two special functions that play a crucial role in the proof: the **Euler gamma function** and the **Riemann zeta function**. These functions are well worth learning about for their own sake, independently of the prime number theorem, and because of their applicability to many other problems in pure and applied mathematics.

2.2 The Euler gamma function

The **Euler gamma function** (often referred to simply as the gamma function) is one of the most important special functions in mathematics. It has applications to many areas, such as combinatorics, number theory, differential equations, probability, and

---

1 To be fair, so-called "elementary" proofs of the prime number theorem that avoid the use of complex analysis have been found, but this development came much later, required great effort and ingenuity, and many mathematicians seem to agree that these proofs are conceptually less appealing and fruitful for understanding the behavior of the prime numbers than the complex-analytic proofs.

more, and is probably the most ubiquitous transcendental function after the "elementary" transcendental functions (the exponential function, logarithms, trigonometric functions, and their inverses) that we learn about in calculus. The gamma function is a natural meromorphic function of a complex variable that extends the factorial function to noninteger values. In complex analysis, it is particularly important in connection with the theory of the **Mellin transform** (a version of the Fourier transform associated with the multiplicative group of positive real numbers in the same way that the ordinary Fourier transform is associated with the additive group of the real numbers).

Most textbooks define the gamma function in one way and proceed to prove several other equivalent representations of it. I have always found that approach to be slightly misleading; the truth is that none of the representations of the gamma function is more fundamental or "natural" than the others. It seems more logical to me to present the topic by listing the various formulas and properties associated with the gamma function and then proving that that list adds up to a consistent whole, that is, that there exists a unique mathematical object satisfying them.

Theorem 2.2 (Euler gamma function). *There exists a unique function Γ of a complex variable s that has the following properties:*
1. $\Gamma(s)$ *is a meromorphic function on $\mathbb{C}$.*
2. **Connection to factorials:** $\Gamma(n + 1) = n!$ *for $n = 0, 1, 2, \ldots$.*
3. **Important special value:** $\Gamma(1/2) = \sqrt{\pi}$.
4. **Integral representation:**

$$\Gamma(s) = \int_0^\infty e^{-x} x^{s-1}\, dx \quad (\operatorname{Re} s > 0). \tag{2.2}$$

5. **Infinite product representation:**

$$\Gamma(s) = s^{-1} e^{-\gamma s} \prod_{n=1}^\infty \left(1 + \frac{s}{n}\right)^{-1} e^{s/n} \quad (s \in \mathbb{C}), \tag{2.3}$$

where $\gamma = \lim_{n \to \infty}(1 + \frac{1}{2} + \frac{1}{3} + \cdots + \frac{1}{n} - \log n) \doteq 0.577215$ is the Euler–Mascheroni constant.
6. **Limit of finite products representation:**

$$\Gamma(s) = \lim_{n \to \infty} \frac{n!\, n^s}{s(s+1)\cdots(s+n)} \quad (s \in \mathbb{C}). \tag{2.4}$$

7. **Zeros:** *the gamma function has no zeros (so $\Gamma(s)^{-1}$ is an entire function).*
8. **Poles:** *the gamma function has poles precisely at the nonpositive integers $s = 0, -1, -2, \ldots$ and is holomorphic everywhere else. The pole at $s = -n$ is a simple pole with residue*

$$\operatorname{Res}_{s=-n}(\Gamma) = \frac{(-1)^n}{n!} \quad (n = 0, 1, 2, \ldots).$$

9. **Functional equation:**

$$\Gamma(s+1) = s\,\Gamma(s) \quad (s \in \mathbb{C}). \tag{2.5}$$

10. **Reflection formula:**

$$\Gamma(s)\Gamma(1-s) = \frac{\pi}{\sin(\pi s)} \quad (s \in \mathbb{C}). \tag{2.6}$$

To begin the proofs, we do have to define the function we are claiming exists *somehow*, so we take formula (2.2) as our working definition of $\Gamma(s)$. Fix $\alpha > 0$. If s is in the half-plane $\{\mathrm{Re}(s) \geq \alpha > 0\}$, then

$$\left| \int_0^\infty e^{-x} x^{s-1}\,dx \right| \leq \int_0^\infty e^{-x} |x^{s-1}|\,dx = \int_0^\infty e^{-x} x^{\mathrm{Re}(s)-1}\,dx < \int_0^\infty e^{-x} x^{\alpha-1}\,dx < \infty.$$

Thus the improper integral (2.2) converges in the region $\mathrm{Re}(s) > 0$ (uniformly on any half-plane $\mathrm{Re}(s) \geq \alpha > 0$) and therefore defines a function $\Gamma(s)$ which, by the result of Exercise 1.26, is holomorphic in that region.

Next, perform an integration by parts, to get that for $\mathrm{Re}(s) > 0$, we have

$$\Gamma(s+1) = \int_0^\infty e^{-x} x^s\,dx = -e^{-x} x^s \big|_{x=0}^{x=\infty} + \int_0^\infty e^{-x} s x^{s-1}\,dx = s\,\Gamma(s),$$

which is the functional equation (2.5).

Combining the trivial evaluation $\Gamma(1) = \int_0^\infty e^{-x}\,dx = 1$ with the functional equation shows by induction that $\Gamma(n+1) = n!$.

Why is the gamma function shifted from the factorial by 1?

The titular question above is a standard one that gets asked by many students introduced to the gamma function but is rarely discussed in print. If you assume that that the gamma function is a well-behaved extension of the factorial function to noninteger values is one of its most important properties, then the shifting of the value of the argument by 1 seems to make little sense, and the competing definition of a "factorial function"

$$\Pi(s) = \Gamma(s+1)$$

would appear to be the more logical and natural one. In fact, historically, both definitions coexisted for some time, and the reasons why the notation $\Gamma(s)$ won the day and became established as the standard one are not entirely clear; this may be more of an accident of history than anything else.

Nonetheless, there are indeed some good reasons to accept $\Gamma(s)$ as the more natural and sensible notational convention, at least in the context of complex-analytic applications (as opposed to, say, uses of the gamma function in combinatorics). See [W13] for further discussion of this issue.

The special value $\Gamma(1/2) = \sqrt{\pi}$ follows immediately by a change of variable $x = u^2$ in the integral (2.2) and an appeal to the standard Gaussian integral $\int_{-\infty}^{\infty} e^{-u^2}\,du = \sqrt{\pi}$:

$$\Gamma(1/2) = \int_0^{\infty} e^{-x} x^{-1/2}\,dx = \int_0^{\infty} e^{-u^2} 2\,du = \int_{-\infty}^{\infty} e^{-u^2}\,du = \sqrt{\pi}.$$

The functional equation (2.5), which so far we have only established in the region $\operatorname{Re}(s) > 0$, where our working definition (2.2) is valid, can now be used to perform an analytic continuation of $\Gamma(s)$ to a meromorphic function on $\mathbb{C}$. This is done in a series of steps: as the first step, define

$$\Gamma_1(s) = \frac{\Gamma(s+1)}{s},$$

which is a function that is holomorphic on $\operatorname{Re}(s) > -1$, $s \neq 0$, and coincides with $\Gamma(s)$ for $\operatorname{Re}(s) > 0$. By the principle of analytic continuation this provides a unique extension of $\Gamma(s)$ to a meromorphic function in the region $\operatorname{Re}(s) > -1$. Because of the factor $1/s$ and the fact that $\Gamma(1) = 1$, we also see that $\Gamma_1(s)$ has a simple pole at $s = 0$ with residue 1.

Next, for $\operatorname{Re}(s) > -2$, we define

$$\Gamma_2(s) = \frac{\Gamma_1(s+1)}{s} = \frac{\Gamma(s+2)}{s(s+1)},$$

a function that is holomorphic on $\operatorname{Re}(s) > -2$, $s \neq 0, -1$, and coincides with $\Gamma_1(s)$ for $\operatorname{Re}(s) > -1$, $s \neq 0$. Again, this provides an analytic continuation of $\Gamma(s)$ to that region. The factors $1/s(s+1)$ show that $\Gamma_2(s)$ has a simple pole at $s = -1$ with residue -1.

Continuing by induction, having defined an analytic continuation $\Gamma_{n-1}(s)$ of $\Gamma(s)$ to the region $\operatorname{Re}(s) > -n+1$, $s \neq 0, -1, -2, \ldots, -n+2$, we now define

$$\Gamma_n(s) = \frac{\Gamma_{n-1}(s+1)}{s} = \cdots = \frac{\Gamma(s+n)}{s(s+1)\cdots(s+n-1)}.$$

By inspection we see that this gives a meromorphic function in $\operatorname{Re}(s) > -n$ whose poles are precisely at $s = -n+1, \ldots, 0$ and have the claimed residues.

We constructed a sequence of meromorphic functions $\Gamma_n(s)$ that are analytic continuations of the original function $\Gamma(s)$ defined in (2.2) to a growing sequence of regions whose union is the entire complex plane. By packaging all these continuations into a single object we see that we have proved the existence of a unique meromorphic function on all of $\mathbb{C}$ that is an analytic continuation of the original $\Gamma(s)$ and whose restriction to each of the half-planes $\operatorname{Re}(s) > -n$ coincides with the nth function $\Gamma_n(s)$ in the sequence. By a standard abuse of notation, we continue to denote this global analytically continued version of $\Gamma(s)$ by $\Gamma(s)$.

As a partial summary, we established the existence and uniqueness of $\Gamma(s)$ as a function of a complex variable satisfying properties 1, 2, 3, 4, 8, and 9 in Theorem 2.2. We now proceed with the proof of the remaining properties.

Lemma 2.3. *For* $\mathrm{Re}(s) > 0$, *we have*

$$\Gamma(s) = \lim_{n\to\infty} \int_0^n \left(1 - \frac{x}{n}\right)^n x^{s-1}\, dx. \tag{2.7}$$

Proof. The right-hand side of (2.7) can be rewritten as $\int_0^\infty (1 - \frac{x}{n})^n \chi_{[0,n]}(x) x^{s-1}\, dx$ (where χ_A denotes the characteristic function of a set). The integrand in this expression converges to $e^{-x} x^{s-1}$ pointwise as $n \to \infty$. By the elementary inequality $1 - t \le e^{-t}$ $(t \in \mathbb{R})$ we have

$$\left| \left(1 - \frac{x}{n}\right)^n \chi_{[0,n]}(x) x^{s-1} \right| \le e^{-x} x^{\mathrm{Re}(s)-1} \quad (x > 0).$$

The claim therefore follows from the dominated convergence theorem. □

Lemma 2.4. *For* $\mathrm{Re}(s) > 0$, *we have*

$$\int_0^n \left(1 - \frac{x}{n}\right)^n x^{s-1}\, dx = \frac{n!\, n^s}{s(s+1)\cdots(s+n)}.$$

Proof. For $n = 1$, the claim is that

$$\int_0^1 (1 - x) x^{s-1}\, dx = \frac{1}{s(s+1)},$$

which is easy to verify directly. For the general claim, using a linear change of variables and integration by parts, we see that

$$\int_0^n \left(1 - \frac{x}{n}\right)^n x^{s-1}\, dx = n^s \int_0^1 (1 - t)^n t^{s-1}\, dt$$

$$= n^s \left[(1 - t)^n \frac{t^s}{s} \Big|_{t=0}^{t=1} + \int_0^1 n(1 - t)^{n-1} \frac{t^s}{s}\, dt \right]$$

$$= n^s \cdot \frac{n}{s} \int_0^1 (1 - t)^{n-1} t^{(s+1)-1}\, dt,$$

so the claim follows by induction on n. □

Combining the results of Lemmas 2.3 and 2.4, we obtain the "limit of finite products" representation (2.4), except that we only proved it for $\mathrm{Re}(s) > 0$. To establish it for general s, note first that

$$\frac{n!\,n^s}{s(s+1)\cdots(s+n)} = s^{-1}e^{s\log n}\left[\left(1+\frac{s}{1}\right)\left(1+\frac{s}{2}\right)\cdots\left(1+\frac{s}{n}\right)\right]^{-1}$$

$$= s^{-1}e^{-s(\sum_{k=1}^{n}\frac{1}{k}-\log n)}\prod_{k=1}^{n}\left(1+\frac{s}{k}\right)^{-1}e^{s/k},$$

which is an expression whose limit (if it exists) is the expression on the right-hand side of (2.3). This shows that representations (2.3) and (2.4) are equivalent, and from the discussion above, both of them hold at least for $\mathrm{Re}(s) > 0$.

We now check that the infinite product on the right-hand side of (2.3)—or rather its reciprocal, corresponding to the entire function $\Gamma(s)^{-1}$, which is slightly more convenient—satisfies the assumptions of Proposition 1.60 (with $\Omega = \mathbb{C}$) and therefore defines an entire function. Indeed, if K is a compact subset of $\mathbb{C}$, then, for $s \in K$, we have

$$\sum_{n=1}^{\infty}\left|\left(1+\frac{s}{n}\right)e^{-s/n}-1\right| = \sum_{n=1}^{\infty}\left|\left(1+\frac{s}{n}\right)\left(1-\frac{s}{n}+O\left(\frac{1}{n^2}\right)\right)-1\right|$$

$$= \sum_{n=1}^{\infty}\left|O\left(\frac{1}{n^2}\right)\right| < \infty.$$

(Here the big-O notation hides a constant that depends on K but not on n.)

Therefore the infinite product $\prod_{n=1}^{\infty}(1 + \frac{s}{n})e^{-s/n}$ indeed defines an entire function, and relations (2.3) and (2.4) must hold for all $s \in \mathbb{C}$ by the principle of analytic continuation.

The last property that remains to be proved from the list of properties in Theorem 2.2 is the reflection formula (2.6). To prove this, we use the functional equation to transform the factor $\Gamma(1 - s)$ as $(-s)\Gamma(-s)$ and then apply the infinite product formulas (2.3) and (1.72) for the gamma and sine functions, respectively, to get that

$$\frac{1}{\Gamma(s)\Gamma(1-s)} = \frac{1}{\Gamma(s)\cdot(-s)\Gamma(-s)}$$

$$= \frac{-1}{s}\cdot se^{\gamma s}\prod_{n=1}^{\infty}\left(1+\frac{s}{n}\right)e^{-s/n}\cdot(-s)e^{-\gamma s}\prod_{n=1}^{\infty}\left(1-\frac{s}{n}\right)e^{s/n}$$

$$= s\prod_{n=1}^{\infty}\left(1-\frac{s^2}{n^2}\right) = s\frac{\sin(\pi s)}{\pi s} = \frac{\sin(\pi s)}{\pi},$$

as claimed.

An alternative method for proving (2.6) avoids the use of the infinite product formulas. Assume that s is real and satisfies $0 < s < 1$ (proving the identity for such s implies it for all s by analytic continuation). Then we have that

$$\Gamma(s)\Gamma(1-s) = \int_0^{\infty}e^{-t}t^{-s}\Gamma(s)\,dt = \int_0^{\infty}e^{-t}t^{-s}\left(t\int_0^{\infty}e^{-vt}(vt)^{s-1}\,dv\right)dt$$

$$= \int_0^\infty \int_0^\infty e^{-t(1+v)} v^{s-1} \, dv \, dt = \int_0^\infty \left(\int_0^\infty e^{-t(1+v)} \, dt \right) v^{s-1} \, dv$$

$$= \int_0^\infty \frac{v^{s-1}}{1+v} \, dv = \int_{-\infty}^\infty \frac{e^{sx}}{1+e^x} \, dx \quad \text{(by setting } v = e^x\text{)}.$$

So the claim reduces to the definite integral evaluation

$$\int_{-\infty}^\infty \frac{e^{sx}}{1+e^x} \, dx = \frac{\pi}{\sin(\pi s)} \quad (0 < s < 1).$$

This definite integral appeared in Exercise 1.47 and can be evaluated in a straightforward manner using contour integration techniques.

Suggested exercises for Section 2.2. 2.1, 2.2, 2.3, 2.4, 2.5, 2.6, 2.7.

2.3 The Riemann zeta function: definition and basic properties

The **Riemann zeta function** (often referred to simply as the zeta function when there is no risk of confusion), like the Euler gamma function is considered one of the most important special functions in "higher" mathematics. However, the Riemann zeta function is a lot more mysterious than the gamma function and remains the subject of many famous open problems, including the most famous of them all, the **Riemann hypothesis**, widely regarded as one of the most important open problem in mathematics today.

The main reason for the importance of the zeta function is its connection with prime numbers and other concepts and quantities from number theory. Its study and in particular the attempts to prove the Riemann hypothesis have also stimulated an unusually large number of important developments in many areas of mathematics.

As with the gamma function, the Riemann zeta function is usually defined on only part of the complex plane, and its definition is then extended by analytic continuation, which can be done in many different ways. Again, this strikes me as in some sense "missing the point" of the Riemann zeta function as a natural mathematical object that exists independently of which of the many formulas for it you choose as your definition. I will present the function in the form of a theorem summarizing its most important formulas and properties.

Theorem 2.5 (Riemann zeta function). *There exists a unique function, denoted $\zeta(s)$, of a complex variable s, having the following properties:*
1. *$\zeta(s)$ is a meromorphic function on $\mathbb{C}$.*
2. ***Series formula:*** *for $\mathrm{Re}(s) > 1$, $\zeta(s)$ is given by the series*

$$\zeta(s) = \sum_{n=1}^\infty \frac{1}{n^s} = 1 + \frac{1}{2^s} + \frac{1}{3^s} + \cdots. \tag{2.8}$$

3. **Euler product formula:** *for* $\mathrm{Re}(s) > 1$, $\zeta(s)$ *also has an infinite product representation*

$$\zeta(s) = \prod_p \frac{1}{1 - p^{-s}}, \tag{2.9}$$

where the product ranges over the prime numbers $p = 2, 3, 5, 7, 11, \ldots$.
4. $\zeta(s)$ *has no zeros in the region* $\mathrm{Re}(s) > 1$.
5. **The "trivial" zeros:** *the zeros of* $\zeta(s)$ *in the region* $\mathrm{Re}(s) < 0$ *are precisely at* $s = -2, -4, -6, \ldots$.
6. $\zeta(s)$ *has a unique pole, located at* $s = 1$. *It is a simple pole with residue* 1.
7. **The "Basel problem" and its generalizations:** *the values of* $\zeta(s)$ *at even positive integers are given by Euler's formula*

$$\zeta(2n) = \frac{(-1)^{n-1}(2\pi)^{2n}}{2(2n)!} B_{2n} \quad (n = 1, 2, \ldots), \tag{2.10}$$

where $(B_m)_{m=0}^{\infty}$ *are the Bernoulli numbers, defined as the coefficients in the Taylor expansion*

$$\frac{z}{e^z - 1} = \sum_{m=0}^{\infty} \frac{B_m}{m!} z^m.$$

Some of the properties of these remarkable numbers were discussed in Exercise 1.15.
8. **Values at negative integers:** *we have*

$$\zeta(-n) = -\frac{B_{n+1}}{n + 1} \quad (n = 1, 2, 3, \ldots).$$

(Note that for negative even integers, this coincides with the property stated above about the trivial zeros at $s = -2, -4, -6, \ldots$, *since the Bernoulli numbers satisfy* $B_{2k+1} = 0$ *for integer* $k \geq 1$. *However, this formula adds information about the values of* $\zeta(s)$ *at negative odd integers.)*
9. **Functional equation:** *the zeta function satisfies*

$$\zeta^*(1 - s) = \zeta^*(s) \quad (s \in \mathbb{C}), \tag{2.11}$$

where we denote by $\zeta^*(s)$ *the* **symmetrized zeta function**

$$\zeta^*(s) = \pi^{-s/2} \Gamma\left(\frac{s}{2}\right) \zeta(s). \tag{2.12}$$

An equivalent form for the functional equation is

$$\zeta(s) = 2^s \pi^{s-1} \sin\left(\frac{\pi s}{2}\right) \Gamma(1 - s) \zeta(1 - s). \tag{2.13}$$

10. **Integral representation:** *an expression for* $\zeta(s)$ *valid for all* $s \in \mathbb{C}$ *is*

$$\pi^{-s/2}\Gamma\left(\frac{s}{2}\right)\zeta(s) = -\frac{1}{1-s} - \frac{1}{s} + \frac{1}{2}\int_1^\infty (t^{-\frac{s+1}{2}} + t^{\frac{s-2}{2}})(\theta(t) - 1)\,dt, \qquad (2.14)$$

where $\theta(t)$ *is the **Jacobi theta function**[2] defined as*

$$\theta(t) = \sum_{n=-\infty}^{\infty} e^{-\pi n^2 t} = 1 + 2\sum_{n=1}^{\infty} e^{-\pi n^2 t}. \qquad (2.15)$$

To begin the proof of Theorem 2.5, we take as the definition of $\zeta(s)$ the standard infinite series representation (2.8). Since $\sum_n |n^{-s}| = \sum_n n^{-\operatorname{Re}(s)}$, we see that the series converges absolutely precisely when $\operatorname{Re}(s) > 1$ and that the convergence is uniform on any half-plane of the form $\operatorname{Re}(s) > \alpha$ with $\alpha > 1$. In particular, it is uniform on compact subsets, so $\zeta(s)$ is holomorphic in this region.

We now prove the Euler product formula (2.9). Intuitively, the remarkable identity between the infinite series (2.8) and the product (2.9) is often described as an analytic restatement of the fact that any positive integer has a unique factorization into primes. Indeed, observe that each of the factors $\frac{1}{1-p^{-s}}$ in the product can be expanded as a geometric series in powers of p^{-s}. Setting aside issues of convergence for a moment, the product can therefore be written as

$$\prod_p \frac{1}{1-p^{-s}} = \prod_p (1 + p^{-s} + p^{-2s} + p^{-3s} + \cdots) = \sum_{\substack{n = p_1^{j_1} \cdots p_k^{j_k} \\ p_1, \ldots, p_k \text{ primes}}} \frac{1}{n^s}. \qquad (2.16)$$

This last summation is in fact a sum over all positive integers n (with each n being summed over precisely once) by the fundamental theorem of arithmetic. So the sum is equal to $\sum_{n=1}^{\infty} \frac{1}{n^s} = \zeta(s)$.

This calculation is appealing and memorable but lacking in rigor, since we have not said anything about the assumptions about s, nor justified our expansion of an infinite product of infinite series into a single infinite series. A fully rigorous (though more tedious) version of the same calculation proceeds as follows. Define the holomorphic function

$$Z(s) = \prod_p (1 - p^{-s})^{-1}$$

and note that this product converges absolutely if and only if the series $\sum_p |p^{-s}| = \sum_p p^{-\operatorname{Re}(s)}$ converges and in particular if $\operatorname{Re}(s) > 1$. It follows that $Z(s)$ is well-defined

---

2 The same name is also used to refer to several other closely related functions; see Section 5.13, where some of those functions are discussed.

and nonzero for $\text{Re}(s) > 1$. We now prove that in this region, $Z(s) = \zeta(s)$. This can be done by manipulating the partial products associated with the infinite product defining $Z(s)$ in a similar vein to (2.16): if we denote by $\zeta_N(s)$ the product $\prod_{p \leq N} \frac{1}{1-p^{-s}}$ (still a product over primes), then

$$\zeta_N(s) = \prod_{p \leq N} \frac{1}{1 - p^{-s}} = \prod_{p \leq N} (1 + p^{-s} + p^{-2s} + p^{-3s} + \cdots).$$

This is a product of a *finite* number of infinite series, each of them absolutely convergent in $\text{Re}(s) > 1$. By the standard fact from analysis that in such a product, the summands can be rearranged and summed in any order we desire, we see that the product can be expanded as

$$\sum_{\substack{n = p_1^{j_1} \cdots p_k^{j_k} \\ p_1, \ldots, p_k \text{ primes } \leq N}} \frac{1}{n^s}.$$

So we have represented $\zeta_N(s)$ as a series of a similar form to (2.8) but involving terms of the form n^{-s} only for those positive integers n whose prime factorization contains only primes $\leq N$. This set of integers in particular contains all the integers in $[1, N]$. It follows that

$$\left| \zeta(s) - \zeta_N(s) \right| \leq \sum_{n > N} \frac{1}{n^s}.$$

Taking the limit as $N \to \infty$ shows that $Z(s) = \lim_{N \to \infty} \zeta_N(s) = \zeta(s)$. This proves the validity of the Euler product formula. As a corollary, we also get that $\zeta(s)$ has no zeros in the region $\text{Re}(s) > 1$ (Property 4 in Theorem 2.5) since we already noted that $Z(s)$ has this property.

Next, we prove that $\zeta(s)$ can be analytically continued to a meromorphic function on $\mathbb{C}$ that has a pole at $s = 1$ and is holomorphic everywhere else. In the process of doing so, we will also obtain a proof of the functional equation (2.11). We will be aided by an important result from harmonic analysis, the **Poisson summation formula**.

Theorem 2.6 (Poisson summation formula). *Let $f : \mathbb{R} \to \mathbb{C}$ be differentiable infinitely many times, and assume that $\sup_{x \in \mathbb{R}} |x^n f^{(k)}(x)| < \infty$ for all $k, n \geq 0$.*[3] *Then*

$$\sum_{n=-\infty}^{\infty} f(n) = \sum_{k=-\infty}^{\infty} \hat{f}(k), \tag{2.17}$$

where

---

3 A function satisfying these assumptions is called a **Schwartz function**. See Section A.6.

$$\hat{f}(u) = \int\limits_{-\infty}^{\infty} f(x)e^{-2\pi i u x}\, dx \quad (u \in \mathbb{R})$$

is the Fourier transform of f.

Proof. Define the function $g : [0,1] \to \mathbb{C}$ by

$$g(x) = \sum_{n=-\infty}^{\infty} f(x+n). \tag{2.18}$$

By the assumptions on f the series defining $g(x)$ converges, and g is differentiable. Note that $g(0) = g(1)$, so that g can also be interpreted as a periodic function on $\mathbb{R}$ or, equivalently, as a function on the circle $\mathbb{R}/\mathbb{Z}$; consequently, it is sometimes referred to as the "periodicization" of f. Now, since g is periodic and differentiable, a standard result from harmonic analysis [67, Thm. 2.1, p. 81] states that $g(x)$ will have a pointwise convergent Fourier series of the form

$$g(x) = \sum_{k=-\infty}^{\infty} \hat{g}(k)e^{2\pi i k x}, \tag{2.19}$$

where $\hat{g}(k)$ are the Fourier coefficients of g given by

$$\hat{g}(k) = \int\limits_{0}^{1} g(x)e^{-2\pi i k x}\, dx.$$

In particular, the particular case $x = 0$ of (2.19) is the relation

$$g(0) = \sum_{k=-\infty}^{\infty} \hat{g}(k). \tag{2.20}$$

Moreover, the Fourier coefficient $\hat{g}(k)$ can be expressed in terms of the Fourier coefficients of the original function $f(x)$:

$$\hat{g}(k) = \int\limits_{0}^{1} g(x)e^{-2\pi i k x}\, dx = \int\limits_{0}^{1} \sum_{n=-\infty}^{\infty} f(x+n)e^{-2\pi i k x}\, dx$$

$$= \sum_{n=-\infty}^{\infty} \int\limits_{0}^{1} f(x+n)e^{-2\pi i k x}\, dx = \sum_{n=-\infty}^{\infty} \int\limits_{n}^{n+1} f(u)e^{-2\pi i k u}\, du$$

$$= \int\limits_{-\infty}^{\infty} f(u)e^{-2\pi i k u}\, du = \hat{f}(k). \tag{2.21}$$

Combining (2.20) and (2.21), we get that

$$g(0) = \sum_{k=-\infty}^{\infty} \widehat{f}(k),$$

the quantity on the right-hand side of (2.17). On the other hand, setting $x = 0$ in (2.18) gives

$$g(0) = \sum_{n=-\infty}^{\infty} f(n),$$

so (2.17) follows. □

Theorem 2.7 (Functional equation for the Jacobi theta function). *The Jacobi theta function $\theta(t)$ satisfies the functional equation*

$$\theta\left(\frac{1}{t}\right) = \sqrt{t}\,\theta(t) \quad (t > 0). \tag{2.22}$$

We remark that equations of the form (2.22) and its variants are studied in the theory of **modular forms**, which is the subject of Chapter 5. Indeed, when we learn about this more general theory, we will see that $\theta(t)$ can be seen as belonging to a more general class of Jacobi theta functions, which are special functions with many applications in number theory and other areas of mathematics. See Section 5.13.1 and also Chapter 6.

Proof of Theorem 2.7. Fix $t > 0$, and define the function $f : \mathbb{R} \to \mathbb{R}$ (depending on the parameter t) by

$$f(x) = e^{-\pi t x^2}. \tag{2.23}$$

The function f clearly satisfies the assumptions of (2.6), so (2.17) holds. Note that the Fourier transform of f is given by

$$\widehat{f}(u) = t^{-1/2} e^{-\pi u^2 / t}. \tag{2.24}$$

Indeed, for $t = 1$, it is the standard integral

$$\int_{-\infty}^{\infty} e^{-\pi x^2} e^{-2\pi i x u}\, dx = e^{-\pi u^2} \tag{2.25}$$

(that is, the well-known fact that the function $e^{-\pi x^2}$ is its own Fourier transform; see Exercise 1.47), and for general $t > 0$, this follows from (2.25) by a linear change of variables. Now substituting (2.23)–(2.24) into (2.17) immediately gives (2.22). □

An alternative method of proving (2.22) using purely complex-analytic arguments is discussed in Exercise 2.15.

Lemma 2.8. *The asymptotic behavior of* $\theta(t)$ *near* $t = 0$ *and* $t = +\infty$ *is given by*

$$\theta(t) = O\left(\frac{1}{\sqrt{t}}\right) \qquad (t \to 0+), \tag{2.26}$$

$$\theta(t) = 1 + O(e^{-\pi t}) \quad (t \to \infty). \tag{2.27}$$

Proof. The claim about the behavior of $\theta(t)$ as $t \to \infty$ is immediate from

$$\theta(t) - 1 = 2 \sum_{n=1}^{\infty} e^{-\pi n^2 t} \le 2 \sum_{n=1}^{\infty} e^{-\pi n t} = \frac{2e^{-\pi t}}{1 - e^{-\pi t}},$$

which is bounded by $3e^{-\pi t}$ if $t > 1$. This gives (2.27). Using (2.22) now gives that $\theta(t) = t^{-1/2}\theta(1/t) = t^{-1/2}(1 + O(e^{-\pi/t})) = O(t^{-1/2})$ as $t \to 0+$, which proves (2.26). $\qquad \square$

We are now ready to prove that $\zeta(s)$ can be analytically continued to a meromorphic function on $\mathbb{C}$. This will be done by deriving representation (2.14) for $\mathrm{Re}(s) > 1$ and showing that the expression on the right-hand side of (2.14) in fact defines a meromorphic function on $\mathbb{C}$. Start with the identity

$$\Gamma\left(\frac{s}{2}\right) = \int_0^{\infty} e^{-x} x^{s/2-1}\, dx$$

for $\mathrm{Re}(s) > 0$. A linear change of variables $x = \pi n^2 t$ brings this to the form

$$\pi^{-s/2}\Gamma\left(\frac{s}{2}\right) n^{-s} = \int_0^{\infty} e^{-\pi n^2 t} t^{s/2-1}\, dt. \tag{2.28}$$

Summing the left-hand side over $n = 1, 2, \ldots$ gives $\pi^{-s/2}\Gamma(\frac{s}{2})\zeta(s)$—the function we denoted $\zeta^*(s)$—except that in order for this sum to converge, we now make the more restrictive assumption that $\mathrm{Re}(s) > 1$. Similarly, performing the same summation on the right-hand side of (2.28), we have that

$$\sum_{n=1}^{\infty} \int_0^{\infty} e^{-\pi n^2 t} t^{s/2-1}\, dt. = \int_0^{\infty} \left(\sum_{n=1}^{\infty} e^{-\pi n^2 t}\right) t^{s/2-1}\, dt = \int_0^{\infty} \frac{\theta(t) - 1}{2} t^{s/2-1}\, dt.$$

Here we again assume that $\mathrm{Re}(s) > 1$; by Lemma 2.8 this ensures that the integral in the last expression is absolutely convergent and therefore also, by the dominated convergence theorem, that it is permissible to interchange the order of summation and integration as we did.

Summarizing the above discussion, we have obtained the representation

$$\zeta^*(s) = \frac{1}{2} \int_0^{\infty} (\theta(t) - 1) t^{s/2-1}\, dt \quad (\mathrm{Re}(s) > 1)$$

for the symmetrized zeta function $\zeta^*(s)$ defined in (2.12). It is convenient to rewrite this as

$$\zeta^*(s) = \int_0^\infty \varphi(t)t^{s/2-1}\,dt \quad (\mathrm{Re}(s) > 1),$$

where we denote $\varphi(t) = \frac{1}{2}(\theta(t) - 1)$. Next, we use the functional equation (2.22) for $\theta(t)$ to bring this integral to a new form, which is well-defined for all $s \in \mathbb{C}$ except $s = 0, 1$. More specifically, note that (2.22) can be expressed in the equivalent form

$$\varphi(t) = t^{-1/2}\varphi(1/t) + \frac{1}{2}t^{-1/2} - \frac{1}{2}.$$

We can therefore write, still assuming that $\mathrm{Re}(s) > 1$,

$$\zeta^*(s) = \int_0^1 \varphi(t)t^{s/2-1}\,dt + \int_1^\infty \varphi(t)t^{s/2-1}\,dt$$

$$= \int_0^1 \left(t^{-1/2}\varphi(1/t) + \frac{1}{2}t^{-1/2} - \frac{1}{2}\right)t^{s/2-1}\,dt + \int_1^\infty \varphi(t)t^{s/2-1}\,dt$$

$$= -\frac{1}{1-s} - \frac{1}{s} + \int_1^\infty \left(t^{(1-s)/2-1} + t^{s/2-1}\right)\varphi(t)\,dt. \tag{2.29}$$

This is representation (2.14). Now observe that since $\varphi(t) = O(e^{-\pi t})$ as $t \to \infty$, the integral $\int_1^\infty (t^{(1-s)/2-1} + t^{s/2-1})\varphi(t)\,dt$ satisfies the assumptions of Exercise 1.26 and therefore defines an entire function of s. Thus we have derived a formula for $\zeta^*(s)$ that defines a meromorphic function on all of $\mathbb{C}$, whose only poles are the simple poles at $s = 0, 1$ (due to the two terms $-1/s$ and $1/(s-1)$ in (2.29)). This concludes the proof that $\zeta(s)$ can be analytically continued to a meromorphic function on $\mathbb{C}$. The functional equation (2.11) also follows trivially: simply observe that the representation we derived for $\zeta^*(s)$ is manifestly symmetric with respect to replacing each occurrence of s by $1 - s$.

It is straightforward to verify that the two forms (2.11) and (2.13) of the functional equation are equivalent (Exercise 2.8).

The claims from Theorem 2.5 that remain to be proved are properties 5–8. Property 7 was proved in Chapter 1 as one of the consequences of the partial fraction expansion of the cotangent function (see Exercise 1.39). The remaining properties will now follow as a sequence of easy corollaries to the results we already proved.

Corollary 2.9. *The only pole of $\zeta(s)$ is a simple pole at $s = 1$ with residue 1.*

Proof. Our representation for $\zeta^*(s)$ expresses it as a sum of $-\frac{1}{s}, \frac{1}{s-1}$, and an entire function. Thus the poles of $\zeta^*(s)$ are simple poles at $s = 0, 1$ with residues -1 and 1, respectively. It follows that

$$\zeta(s) = \pi^{s/2}\Gamma(s/2)^{-1}\zeta^*(s)$$

has a pole at $s = 1$ with residue $\pi^{1/2}\Gamma(1/2)^{-1} = 1$ and a pole (that turns out to be a removable singularity) at $s = 0$ with residue $\pi^0\Gamma(0)^{-1} = 0$. (That is, the pole of $\zeta^*(s)$ at $s = 0$ is canceled out by the zero of $\Gamma(s/2)$.) □

Corollary 2.10. $\zeta(-n) = -B_{n+1}/(n+1)$ *for* $n = 1, 2, 3, \ldots$.

Proof. Let $n \geq 1$. Using version (2.13) of the functional equation, we have that

$$\zeta(-n) = 2^{-n}\pi^{-n-1}\sin(-\pi n/2)\Gamma(n+1)\zeta(n+1)$$
$$= 2^{-n}\pi^{-n-1}\sin(-\pi n/2)n!\zeta(n+1).$$

If $n = 2k$ is even, then $\sin(-\pi n/2) = 0$, so we get that $\zeta(-2k) = 0$ (that is, $n = 2k$ is one of the so-called "trivial zeros"). We also know from Exercise 1.15 that $B_{2k+1} = 0$ for $k = 1, 2, 3, \ldots$, so the formula $\zeta(-n) = B_{n+1}/(n+1)$ is satisfied in this case.

If on the other hand $n = 2k - 1$ is odd, then $\sin(-\pi(2k-1)/2) = (-1)^k$, and therefore using (2.10), we get that

$$\zeta(-n) = (-1)^k 2^{-2k+1}\pi^{-2k}(2k-1)!\zeta(2k)$$
$$= (-1)^k 2^{-2k+1}\pi^{-2k}(2k-1)!\frac{(-1)^{k-1}(2\pi)^{2k}}{2(2k)!}B_{2k}$$
$$= -\frac{B_{2k}}{2k} = -\frac{B_{n+1}}{n+1},$$

so again the formula is satisfied. □

Corollary 2.11. *The zeros of* $\zeta(s)$ *in the region* $\mathrm{Re}(s) < 0$ *are precisely the trivial zeros* $s = -2, -4, -6, \ldots$.

Proof. We have already established the existence of the trivial zeros. We leave to you to verify that the fact that there are no other zeros follows immediately from the functional equation. □

Suggested exercises for Section 2.3. 2.8, 2.9, 2.10, 2.11, 2.12, 2.13, 2.14, 2.15, 2.16, 2.17, 2.18.

2.4 A theorem on the zeros of the Riemann zeta function

Next, we prove a subtle and very important fact about the zeta function, which will play a crucial role in our proof of the prime number theorem.

Theorem 2.12. $\zeta(s)$ *has no zeros on the line* $\mathrm{Re}(s) = 1$.

Proof. For this proof, denote $s = \sigma + it$, where we assume that $\sigma > 1$ and t is real and nonzero. The proof is based on investigating simultaneously the behavior of $\zeta(\sigma + it)$,

$\zeta(\sigma + 2it)$, and $\zeta(\sigma)$ for fixed t as $\sigma \searrow 1$. Consider the following somewhat mysterious quantity:

$$X = \log\left|\zeta(\sigma)^3 \zeta(\sigma + it)^4 \zeta(\sigma + 2it)\right|.$$

Using the Euler product formula (2.9), we can evaluate X as

$$X = 3\log\left|\zeta(\sigma)\right| + 4\log\left|\zeta(\sigma + it)\right| + \log\left|\zeta(\sigma + 2it)\right|$$

$$= 3\log\left(\prod_{p \text{ prime}} \left|1 - p^{-\sigma}\right|^{-1}\right) + 4\log\left(\prod_{p \text{ prime}} \left|1 - p^{-\sigma - it}\right|^{-1}\right)$$

$$+ \log\left(\prod_{p \text{ prime}} \left|1 - p^{-\sigma - 2it}\right|^{-1}\right)$$

$$= \sum_{p \text{ prime}} \left(-3\log\left|1 - p^{-\sigma}\right| - 4\log\left|1 - p^{-\sigma - it}\right| - \log\left|1 - p^{-\sigma - 2it}\right|\right)$$

$$= \sum_{p \text{ prime}} \left(-3\operatorname{Re}\left[\operatorname{Log}(1 - p^{-\sigma})\right] - 4\operatorname{Re}\left[\operatorname{Log}(1 - p^{-\sigma - it})\right]\right)$$

$$- \operatorname{Re}\operatorname{Log}\left[1 - p^{-\sigma - 2it}\right]\right),$$

where in the last expression, $\operatorname{Log}(\cdot)$ denotes the principal branch of the logarithm function. Now note that for $z = a + ib$ with $a > 1$ and an arbitrary prime number p, we have $|p^{-z}| = p^{-a} < 1$, so by the Taylor expansion (1.93) of the $\operatorname{Log}(\cdot)$ function,

$$-\operatorname{Log}(1 - p^{-z}) = \sum_{m=1}^{\infty} \frac{p^{-mz}}{m},$$

and therefore

$$-\operatorname{Re}\left[\operatorname{Log}(1 - p^{-z})\right] = \sum_{m=1}^{\infty} \frac{p^{-ma}}{m} \operatorname{Re}\left[\cos(mb\log p) + i\sin(mb\log p)\right]$$

$$= \sum_{m=1}^{\infty} \frac{p^{-ma}}{m} \cos(mb\log p).$$

This means that if we define quantities β_n and c_n for $n \geq 1$ by

$$\beta_n = t\log n, \quad c_n = \begin{cases} 1/m & \text{if } n = p^m \text{ for some prime } p, \\ 0 & \text{otherwise,} \end{cases}$$

then we can rewrite X as

$$X = \sum_{n=1}^{\infty} c_n n^{-\sigma}(3 + 4\cos\beta_n + \cos(2\beta_n)).$$

We can now use the simple trigonometric identity

$$3 + 4\cos\beta + \cos(2\beta) = 2(1 + \cos\beta)^2$$

to rewrite X yet again as

$$X = 2\sum_{n=1}^{\infty} c_n n^{-\sigma}(1 + \cos\beta_n)^2.$$

We have proved a crucial fact that $X \geq 0$ or, equivalently, that

$$e^X = \left|\zeta(\sigma)^3 \zeta(\sigma + it)^4 \zeta(\sigma + 2it)\right| \geq 1. \tag{2.30}$$

We now claim that this innocent-looking inequality is incompatible with the existence of a zero of $\zeta(s)$ on the line $\mathrm{Re}(s) = 1$. Indeed, assume by contradiction that $\zeta(1 + it) = 0$ for some real $t \neq 0$. Then the three quantities $\zeta(\sigma)$, $\zeta(\sigma + it)$, and $\zeta(\sigma + 2it)$ have the following asymptotic behavior as $\sigma \searrow 1$:

$$\left|\zeta(\sigma)\right| = \frac{1}{\sigma - 1} + O(1) \quad \text{(since } \zeta(s) \text{ has a pole at } s = 1\text{),}$$
$$\left|\zeta(\sigma + it)\right| = O(\sigma - 1) \quad \text{(since } \zeta(s) \text{ has a zero at } s = 1 + it\text{),}$$
$$\left|\zeta(\sigma + 2it)\right| = O(1) \quad \text{(since } \zeta(s) \text{ is holomorphic at } s = 1 + 2it\text{).}$$

Combining these results, we have that

$$e^X = \left|\zeta(\sigma)^3 \zeta(\sigma + it)^4 \zeta(\sigma + 2it)\right| = O((\sigma - 1)^{-3}(\sigma - 1)^4) = O(\sigma - 1).$$

Thus $e^X \to 0$ as $\sigma \searrow 1$, in contradiction to (2.30). This finishes the proof. $\qquad\square$

2.5 Proof of the prime number theorem

The prime number theorem (Theorem 2.1) was proved in 1896 by Jacques Hadamard and independently by Charles Jean de la Vallée Poussin using the groundbreaking ideas from Riemann's famous 1859 paper, in which he introduced the use of the Riemann zeta function as a tool for counting prime numbers. The history of these developments is described in great detail (both historical and technical) in the book [25].

The original proofs of the prime number theorem were very complicated and relied on the "explicit formula of number theory" and some its variants (see the box on p. 109). Throughout the twentieth century, mathematicians worked hard to find simpler ways to derive the prime number theorem. This resulted in several important developments (such as the Wiener Tauberian theorem and the Hardy–Littlewood Tauberian theorem) that advanced not just the state of analytic number theory but also of complex analysis, harmonic analysis, and functional analysis. Despite all the efforts and the discovery of several paths to a proof that were simpler than the original approach, all known proofs remained quite difficult. A minor breakthrough occurred in 1980 when Donald Newman discovered a surprisingly simple way to derive the theorem using relatively elementary complex-analytic arguments. The proof presented here is adapted from of a version of Newman's proof due to Zagier [74]; see also [44, 49, 70].

Recall that the prime number theorem concerns the so-called prime-counting function $\pi(x)$ defined as the number of primes that are less than or equal to x. It is helpful to write this in the form of a sum over primes, namely

$$\pi(x) = \#\{p \text{ prime} : p \leq x\} = \sum_{p \leq x} 1$$

with the convention that the symbol p in summations always refers to primes. We also define the **Chebyshev function** $\psi(x)$ as a closely related *weighted* sum

$$\psi(x) = \sum_{p^k \leq x} \log p = \sum_{p \leq x} \log p \left\lfloor \frac{\log x}{\log p} \right\rfloor.$$

In this definition, the first sum is over prime *powers* p^k (with integer $k \geq 1$); the second sum is an alternative and trivially equivalent way of writing $\psi(x)$ as a sum over primes rather than over prime powers. Another customary and equivalent way to write the function $\psi(x)$ is as

$$\psi(x) = \sum_{n \leq x} \Lambda(n),$$

where the function $\Lambda(n)$, called the **von Mangoldt function**, is defined by

$$\Lambda(n) = \begin{cases} \log p & \text{if } n = p^k \text{ with } p \text{ prime}, k \geq 1, \\ 0 & \text{otherwise.} \end{cases}$$

Lemma 2.13. *The prime number theorem $\pi(x) \sim \frac{x}{\log x}$ is equivalent to the statement that $\psi(x) \sim x$.*

Proof. The functions $\psi(x)$ and $\pi(x)$ can be related to each other in an approximate sense through two simple inequalities. First, observe that

$$\psi(x) = \sum_{p \leq x} \log p \left\lfloor \frac{\log x}{\log p} \right\rfloor \leq \sum_{p \leq x} \log p \frac{\log x}{\log p} = \sum_{p \leq x} \log x = \log x \cdot \pi(x). \tag{2.31}$$

Second, in the opposite direction, we have that for any $0 < \epsilon < 1$ and $x \geq 2$,

$$\psi(x) \geq \sum_{p \leq x} \log p \geq \sum_{x^{1-\epsilon} < p \leq x} \log p \geq \sum_{x^{1-\epsilon} < p \leq x} \log(x^{1-\epsilon})$$

$$= (1 - \epsilon) \log x \left(\pi(x) - \pi(x^{1-\epsilon}) \right) \geq (1 - \epsilon) \log x \left(\pi(x) - x^{1-\epsilon} \right). \tag{2.32}$$

Now assume that $\psi(x) \sim x$ as $x \to \infty$. Then (2.31) implies that $\pi(x) \geq \frac{\psi(x)}{\log x}$, and therefore

$$\liminf_{x \to \infty} \frac{\pi(x)}{x/\log x} \geq 1. \tag{2.33}$$

On the other hand, (2.32) gives that $\pi(x) \leq \frac{1}{1-\epsilon} \cdot \frac{\psi(x)}{\log x} + x^{1-\epsilon}$, which then implies that

$$\limsup_{x \to \infty} \frac{\pi(x)}{x/\log x} \leq \frac{1}{1-\epsilon} + \limsup_{x \to \infty} \frac{\log x}{x^{\epsilon}} = \frac{1}{1-\epsilon}.$$

Since ϵ was an arbitrary number in $(0, 1)$, it follows that

$$\limsup_{x \to \infty} \frac{\pi(x)}{x/\log x} \leq 1. \tag{2.34}$$

Combining (2.33) and (2.34) gives that $\pi(x) \sim x/\log x$. This proves one of the two implications claimed in the theorem.

To prove the reverse implication, assume that $\pi(x) \sim \frac{x}{\log x}$, and note that, by (2.31),

$$\limsup_{x \to \infty} \frac{\psi(x)}{x} \leq \limsup_{x \to \infty} \frac{\pi(x)}{x/\log x} = 1. \tag{2.35}$$

On the other hand, (2.32) implies that

$$\liminf_{x \to \infty} \frac{\psi(x)}{x} \geq \liminf_{x \to \infty} \left(\frac{\pi(x)}{x/\log x} - \frac{\log x}{x^{\epsilon}} \right) = 1 - \epsilon.$$

Again, since $\epsilon \in (0, 1)$ was arbitrary, it follows that $\liminf_{x \to \infty} \frac{\psi(x)}{x} = 1$. When combined with (2.35), we have shown that $\lim_{x \to \infty} \frac{\psi(x)}{x} = 1$, as claimed. $\qquad \square$

A hint of the significance of the Chebyshev function and the equivalent form $\psi(x) \sim x$ of the prime number theorem is offered by the next lemma.

Lemma 2.14. *For* $\mathrm{Re}(s) > 1$ *we have*

$$-\frac{\zeta'(s)}{\zeta(s)} = \sum_{n=1}^{\infty} \Lambda(n) n^{-s}. \tag{2.36}$$

Proof. Using the Euler product formula and taking the logarithmic derivative (which is an operation that works as it should when applied to infinite products of holomorphic functions that are uniformly convergent on compact subsets), we have

$$-\frac{\zeta'(s)}{\zeta(s)} = \sum_{p} \frac{\frac{d}{ds}(1 - p^{-s})}{1 - p^{-s}} = \sum_{p} \frac{\log p \cdot p^{-s}}{1 - p^{-s}}$$

$$= \sum_p \log p \, (p^{-s} + p^{-2s} + p^{-3s} + \cdots) = \sum_{p \text{ prime}} \sum_{k=1}^{\infty} \log p \cdot p^{-ks}$$

$$= \sum_{n=1}^{\infty} \Lambda(n) n^{-s}. \qquad \qquad \square$$

At this point in the discussion, we can already outline a plausible-sounding heuristic explanation for why the prime number theorem *might* be true. Consider the two sequences $a_n = \Lambda(n)$ and $b_n = 1$. By Lemma 2.13 the prime number theorem is equivalent to the claim that

$$\frac{1}{x} \sum_{n \leq x} a_n \sim \frac{1}{x} \sum_{n \leq x} b_n \quad \text{as } x \to \infty, \tag{2.37}$$

that is, that the sequences a_n and b_n exhibit similar *average* asymptotic behavior. On the other hand, if we are willing to be a bit more flexible about interpreting what we mean by "average", that is, replacing the straightforward arithmetic averages by a certain class of *weighted* averages, then there is a statement of this type that is easily seen to be true, namely, the statement that

$$\frac{1}{\zeta(\sigma)} \sum_{n=1}^{\infty} \frac{a_n}{n^\sigma} \sim \frac{1}{\zeta(\sigma)} \sum_{n=1}^{\infty} \frac{b_n}{n^\sigma} \quad \text{as } \sigma \searrow 1. \tag{2.38}$$

Indeed, the right-hand side of this relation is equal to 1, and the left-hand side is, by (2.36), equal to $\frac{-\zeta'(\sigma)/\zeta(\sigma)}{\zeta(\sigma)}$, which converges to 1 as $\sigma \searrow 1$ due to the fact that both the numerator and the denominator in this fraction have a simple pole with residue 1 at $\sigma = 1$.

The above argument raises the question of whether this heuristic explanation can be turned into a proof. That is, is it generally true that an asymptotic equivalence of the form (2.38) can be used to deduce the more natural equivalence (2.37)? Or, if it is not true in unrestricted generality, what additional assumptions are needed to make such a deduction correct, and are these assumptions satisfied for our particular case of interest? The general area in which such questions belong is that of **Tauberian theorems** (a name honoring an 1897 result of the mathematician Alfred Tauber, who proved an important early result of this type). These questions turn out to be quite delicate, and although this approach does in fact offer a viable route toward a proof of the prime number theorem (see [47, p. 261]), following this route requires rather involved ideas from Fourier analysis. Here we take a slightly different path that, although also in line with the general philosophy of Tauberian theorems, starts by further reducing the problem into that of showing the convergence of a certain improper integral. The following lemma gives the details of this simple reduction.

Lemma 2.15. *Assume that the improper integral*

$$\int_1^{\infty} \left(\frac{\psi(x)}{x} - 1 \right) \frac{dx}{x} \tag{2.39}$$

converges. Then the prime number theorem follows.

Proof. Keeping in mind Lemma 2.13, we will prove the contrapositive claim that if $\frac{\psi(x)}{x}$ does not converge to 1 as $x \to \infty$, then the integral (2.39) cannot converge.

Assume that $\frac{\psi(x)}{x} \not\to 1$. In this scenario, either $L_+ = \limsup_{x\to\infty} \frac{\psi(x)}{x} > 1$, or $L_- = \liminf_{x\to\infty} \frac{\psi(x)}{x} < 1$. In the first case, observe that there are arbitrarily large values of x for which $\frac{\psi(x)}{x} > 1 + 2\epsilon$, where we denote $\epsilon = \frac{L_+ - 1}{4} > 0$. For a value of x with that property, using the fact that $\psi(x)$ is weakly monotone increasing, we see that

$$\int_x^{(1+\epsilon)x} \left(\frac{\psi(t)}{t} - 1 \right) \frac{dt}{t} \geq \int_x^{(1+\epsilon)x} \left(\frac{(1+2\epsilon)x}{(1+\epsilon)x} - 1 \right) \frac{dt}{(1+\epsilon)x} = \left(\frac{\epsilon}{1+\epsilon} \right)^2 =: C.$$

Thus we have shown that the function $I(T) = \int_1^T (\frac{\psi(x)}{x} - 1)\frac{dx}{x}$ has infinitely many intervals over which it changes value by at least the fixed positive constant C, which implies that the improper integral (2.39) cannot converge.

Similarly, in the second case in which $L_- < 1$, we again note that there are arbitrarily large values of x for which $\frac{\psi(x)}{x} < 1 - 2\epsilon$, where ϵ is defined as the constant $\epsilon = \frac{1 - L_-}{4}$ (which is positive and trivially bounded from above by 1/4). For such x, again from the monotonicity of $\psi(x)$ we get that

$$\int_{(1-\epsilon)x}^x \left(\frac{\psi(t)}{t} - 1 \right) \frac{dt}{t} \leq \int_{(1-\epsilon)x}^x \left(\frac{(1-2\epsilon)x}{(1-\epsilon)x} - 1 \right) \frac{dt}{(1-\epsilon)x} = -\left(\frac{\epsilon}{1-\epsilon} \right)^2.$$

This is again inconsistent with the possibility that the integral (2.39) converges. □

One additional ingredient of our proof is the following elementary bound on the Chebyshev function.

Lemma 2.16. *There is a constant $C > 0$ such that $\psi(x) \leq Cx$ for all $x \geq 1$.*

Proof. The idea of the proof is that the binomial coefficient $\binom{2n}{n}$ is not too large on the one hand but is divisible by many primes (at least all primes between $n+1$ and $2n$) on the other hand; hence it follows that there cannot be too many primes, and in particular the weighted prime-counting function $\psi(x)$ can be easily bounded from above using such an argument. More precisely, we have that

$$2^{2n} = (1+1)^{2n} = \sum_{k=0}^{2n} \binom{2n}{k} > \binom{2n}{n} \geq \prod_{n<p\leq 2n} p = \exp\left(\sum_{n<p\leq 2n} \log p \right)$$

$$= \exp\left(\psi(2n) - \psi(n) - \sum_{n<p^k\leq 2n,\, k\geq 2} \log p \right). \tag{2.40}$$

The sum in the last expression is easily bounded as

$$\sum_{n<p^k\leq 2n,\, k\geq 2} \log p \leq 10\sqrt{n}\log^2 n + 10 \quad (n \geq 1) \tag{2.41}$$

(Exercise 2.19). Thus taking the logarithm of the first and last expressions in (2.40), we get the bound

$$\psi(2n) - \psi(n) \le 2n \log 2 + 10 \sqrt{n} \log n + 10 \le \frac{1}{2} Cn$$

for all $n \ge 1$ with some constant $C > 0$. For n of the form $n = 2^m$, $m \ge 0$, this allows us to write

$$\psi(2^m) = (\psi(2^m) - \psi(2^{m-1}))$$
$$+ (\psi(2^{m-1}) - \psi(2^{m-2})) + \cdots + (\psi(2^1) - \psi(2^0))$$
$$\le \frac{1}{2} C(2^{m-1} + \cdots + 2^0) \le C 2^{m-1},$$

thereby establishing the inequality $\psi(x) \le \frac{1}{2} Cx$ for any x that is a power of 2. Finally, for a general integer $x \ge 1$, we can represent x as $x = 2^m + \ell$ for some $m \ge 0$ and $0 \le \ell < 2^m$. We then observe that

$$\psi(x) = \psi(2^m + \ell) \le \psi(2^{m+1}) \le C 2^m \le Cx,$$

which is the desired bound. □

We are ready to state a Tauberian theorem, which in some sense forms the heart of the proof of the prime number theorem.

Theorem 2.17 (Newman's Tauberian theorem). *Let $f : [1, \infty) \to \mathbb{R}$ be a bounded function that is integrable on compact intervals. Define a function $g(s)$ of a complex variable s by*

$$g(s) = \int_1^\infty f(x) x^{-s-1} \, dx. \tag{2.42}$$

Clearly, $g(s)$ is defined and holomorphic in the open half-plane $\operatorname{Re}(s) > 0$. Assume that $g(s)$ has an analytic continuation to an open region Ω containing the closed half-plane $\operatorname{Re}(s) \ge 0$. Then the improper integral

$$\int_1^\infty \frac{f(x)}{x} \, dx \tag{2.43}$$

converges, and its value is equal to $g(0)$, the value at $s = 0$ of the analytic continuation of g.

Before we proceed with the proof, it is worth pausing to appreciate the subtlety of this result. The conclusion of the theorem about the existence of the improper integral (2.43) can be expressed as the statement that

$$\lim_{T \to \infty} \int_1^T \frac{f(x)}{x} \, dx = \lim_{\epsilon \searrow 0} \int_1^\infty \frac{f(x)}{x} x^{-\epsilon} \, dx.$$

This sort of equivalence of limits seems to fall readily within the realm of *real* analysis. It is remarkable that the condition needed for this conclusion to hold is a *complex*-analytic condition involving the existence of an analytic continuation for the function $g(s)$ (and, moreover, to a region that contains parts that extend arbitrarily far from the real axis). If you were not already convinced of the importance and relevance of complex analysis to the rest of mathematics, I hope this will make you rethink your skepticism!

Proof of Theorem 2.17. Define a truncated version of the integral defining $g(s)$, namely

$$g_T(s) = \int_1^T f(x) x^{-s-1} \, dx$$

for $T > 1$. We claim that $g_T(s)$ is an entire function of s for any fixed T. This can be proved using Morera's theorem: let γ be a closed contour in $\mathbb{C}$. Then

$$\oint_\gamma g_T(s) \, ds = \oint_\gamma \int_1^T f(x) x^{-s-1} \, dx \, ds = \int_1^T \oint_\gamma f(x) x^{-s-1} \, ds \, dx = \int_I 0 \, dx = 0.$$

In the above calculation, interchanging the order of the two integrals is justified by Fubini's theorem, which (as we can easily check) is applicable in the current situation. Since the integral of $g_T(s)$ over an arbitrary closed contour γ vanishes, g_T is entire by Morera's theorem.

Now our goal is to show that $\lim_{T \to \infty} g_T(0) = g(0)$. This will be achieved through an application of Cauchy's integral formula. Fix some large number $R > 0$ and a small number $\delta > 0$ (which depends on R in a way that will be explained shortly), and consider the contour C consisting of the part of the circle $|s| = R$ that lies in the half-plane $\mathrm{Re}(s) \geq -\delta$ together with the straight line segment along the line $\mathrm{Re}(s) = -\delta$ connecting the top and bottom intersection points of this circle with the line (see Fig. 2.1(a)). Assume that δ is small enough so that $g(s)$ (which by the assumptions of the theorem extends analytically at least slightly to the left of $\mathrm{Re}(s) = 0$) is holomorphic in an open set containing C and the region enclosed by it. Then by Cauchy's integral formula the difference $g(0) - g_T(0)$ can be expressed as

$$g(0) - g_T(0) = \frac{1}{2\pi i} \oint_C (g(s) - g_T(s)) T^s \left(1 + \frac{s^2}{R^2} \right) \frac{ds}{s}. \tag{2.44}$$

Note that this equation would still hold true if the integrand on the right-hand side were the simpler expression $\frac{g(s) - g_T(s)}{s}$; however, Newman's inspired observation was that the inclusion of the additional factors $T^s(1 + \frac{s^2}{R^2})$ actually helps by producing an integral that

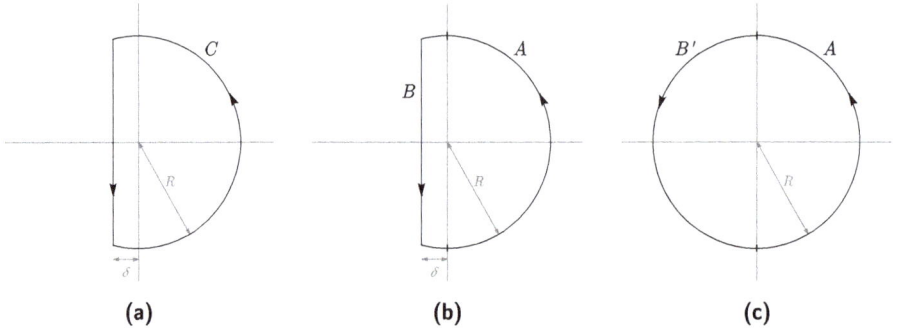

Figure 2.1: The contours C, A, B, and B'.

can be estimated effectively (while keeping the value of the integral the same). To see how this works, start by separating the contour C into two parts, a semicircular arc A that lies in the half-plane $\mathrm{Re}(s) > 0$ and the remaining part B in the half-plane $\mathrm{Re}(s) < 0$ (Fig. 2.1(b)). We can then write

$$g(0) - g_T(0) = I_1 + I_2, \tag{2.45}$$

where

$$I_1 = \frac{1}{2\pi i} \int_A (g(s) - g_T(s)) T^s \left(1 + \frac{s^2}{R^2}\right) \frac{ds}{s}, \tag{2.46}$$

$$I_2 = \frac{1}{2\pi i} \int_B (g(s) - g_T(s)) T^s \left(1 + \frac{s^2}{R^2}\right) \frac{ds}{s}. \tag{2.47}$$

We now bound I_1 and I_2 separately. Denote

$$M = \sup_{t \geq 1} |f(t)|$$

(and recall the assumption that this number is finite). For s with $\mathrm{Re}(s) > 0$, we are in the region where formula (2.42) is valid, so we can bound the expression $g(s) - g_T(s)$ as

$$
\begin{aligned}
|g(s) - g_T(s)| &= \left| \int_1^\infty f(x) x^{-s-1}\, dx - \int_1^T f(x) x^{-s-1}\, dx \right| \\
&= \left| \int_T^\infty f(x) x^{-s-1}\, dx \right| \leq M \int_T^\infty |x^{-s-1}|\, dx = \frac{MT^{-\mathrm{Re}(s)}}{\mathrm{Re}(s)}.
\end{aligned}
\tag{2.48}
$$

Note also that for s satisfying $|s| = R$, we have that

$$\left|T^s\left(1+\frac{s^2}{R^2}\right)\right| = T^{\mathrm{Re}(s)}\left|\frac{s}{R}\cdot\left(\frac{R}{s}+\frac{s}{R}\right)\right| = T^{\mathrm{Re}(s)}\left|(\overline{s}/R + s/R)\right|$$
$$= T^{\mathrm{Re}(s)}\frac{2\,|\,\mathrm{Re}(s)|}{R}. \tag{2.49}$$

The bounds (2.48)–(2.49) both apply on the subcontour A, so by combining them we get that

$$|I_1| \le \frac{1}{2\pi}(\pi R)\frac{2M}{R^2} = \frac{M}{R}. \tag{2.50}$$

Next, we bound I_2 by bounding the contributions from $g(s)$ and $g_T(s)$ separately, that is, further decomposing that integral as

$$I_2 = \frac{1}{2\pi i}\int_B g(s)T^s\left(1+\frac{s^2}{R^2}\right)\frac{ds}{s} - \frac{1}{2\pi i}\int_B g_T(s)T^s\left(1+\frac{s^2}{R^2}\right)\frac{ds}{s} =: J_1 - J_2. \tag{2.51}$$

In the case of J_2, since $g_T(s)$ is an entire function and the only singularity of the integrand is at $s = 0$, we can deform the integration contour B replacing it with the semicircular arc $B' = \{s : |s| = R, \mathrm{Re}(s) < 0\}$ (Fig. 2.1(c)). By Cauchy's theorem the value of the integral remains the same. On the new contour B' the bound (2.49) holds, and there we also have the estimate

$$|g_T(s)| = \left|\int_1^T f(x)x^{-s-1}\,dx\right| \le M\int_0^T |x^{-s-1}|\,dx = \frac{MT^{-\mathrm{Re}(s)}}{|\mathrm{Re}(s)|}.$$

Therefore, similarly to (2.50), we have the bound

$$|J_2| \le \frac{1}{2\pi}(\pi R)\frac{2M}{R^2} = \frac{M}{R}. \tag{2.52}$$

The remaining integral J_1 tends to 0 as $T \to \infty$ (with R fixed), since the dependence on T is only through the factor T^s, which converges to 0 uniformly on compact sets in $\mathrm{Re}(s) < 0$ as $T \to \infty$.

Combining this last observation with (2.45), (2.50), (2.51), and (2.52), we have therefore shown that

$$\limsup_{T\to\infty}|g(0) - g_T(0)| \le \frac{2M}{R}.$$

Since R was an arbitrary positive number, the lim sup must be 0, and the theorem is proved. $\qquad\square$

Consider now the following application of Theorem 2.17 to a specific function: take

$$f(x) = \frac{\psi(x)}{x} - 1 \quad (x \ge 1)$$

as our function $f(x)$. Note that $f(x)$ is bounded by Lemma 2.16. The associated function $g(s)$ is then

$$
\begin{aligned}
g(s) &= \int_1^\infty \left(\frac{\psi(x)}{x} - 1 \right) x^{-s-1}\, dx \\
&= \int_1^\infty \psi(x) x^{-s-2}\, dx - \frac{1}{s} = \int_1^\infty \left(\sum_{n \le x} \Lambda(n) \right) x^{-s-2}\, dx - \frac{1}{s} \\
&= \sum_{n=1}^\infty \Lambda(n) \left(\int_n^\infty x^{-s-2}\, dx \right) - \frac{1}{s} = \sum_{n=1}^\infty \Lambda(n) \left. \frac{x^{-s-1}}{-s-1} \right|_n^\infty - \frac{1}{s} \\
&= \frac{1}{s+1} \sum_{n=1}^\infty \Lambda(n) n^{-s-1} - \frac{1}{s} = -\frac{1}{s+1} \cdot \frac{\zeta'(s+1)}{\zeta(s+1)} - \frac{1}{s} \quad (\mathrm{Re}(s) > 0)
\end{aligned}
$$

by (2.36). Recall that $-\zeta'(s)/\zeta(s)$ has a simple pole at $s = 1$ with residue 1 (because $\zeta(s)$ has a simple pole at $s = 1$; it is useful to remember the more general fact that if a holomorphic function $h(z)$ has a zero of order k at $z = z_0$, then the logarithmic derivative $h'(z)/h(z)$ has a simple pole at $z = z_0$ with residue k). So $-\frac{1}{s+1} \cdot \frac{\zeta'(s+1)}{\zeta(s+1)}$ has a simple pole with residue 1 at $s = 0$, and therefore $-\frac{1}{s+1} \cdot \frac{\zeta'(s+1)}{\zeta(s+1)} - \frac{1}{s}$ has a *removable* singularity at $s = 0$. Thus the identity $g(s) = -\frac{1}{s+1} \cdot \frac{\zeta'(s+1)}{\zeta(s+1)} - \frac{1}{s}$ shows that $g(s)$ extends analytically to a holomorphic function in the region

$$
\{ s \in \mathbb{C} : \zeta(s+1) \ne 0 \}.
$$

By Theorem 2.12, $g(s)$ in particular extends holomorphically to an open set containing the half-plane $\mathrm{Re}(s) \ge 0$.

We have therefore shown that $f(x)$ satisfies the assumption of Newman's Tabuerian theorem. We conclude from the theorem that the improper integral

$$
\int_0^\infty \frac{f(x)}{x}\, dx = \int_1^\infty \left(\frac{\psi(x)}{x} - 1 \right) \frac{dx}{x}
$$

converges. By Lemma 2.15 the prime number theorem follows. □

Suggested exercises for Section 2.5. 2.19, 2.20, 2.21.

The explicit formulae of number theory and the Riemann hypothesis

The proof of the prime number theorem presented in this chapter made crucial use of the fact that $\zeta(s)$ has no zeros on the line $\mathrm{Re}(s) = 1$, but when following this approach, the connection between those two facts seems somewhat opaque and mysterious.

Another, more advanced, approach to the prime number theorem that draws a clearer conceptual line between the location of the zeros of $\zeta(s)$ and the validity of the asymptotic formula $\psi(x) \sim x$ is based on the so-called "explicit formulae of number theory." This is the name given to a family of identities, the simplest of which being

$$\psi(x) = x - \sum_{\rho} \frac{x^{\rho}}{\rho} - \log(2\pi) \quad (x > 1, x \text{ noninteger}). \tag{2.53}$$

In this formula the sum on the right-hand side ranges over all zeros ρ of the Riemann zeta function counted with their respective multiplicities. (In most textbooks the sum is separated into two sums, one ranging over the trivial zeros, which can be evaluated explicitly, and the other ranging over the zeros in the strip $0 < \mathrm{Re}(s) < 1$. Also, the sum is only conditionally convergent; refer to [47, p. 397] for the proper way to interpret it to get a convergent sum.) Note that this is an *exact identity*, not an asymptotic result. To convert it to an asymptotic result, the key observation is that each of the power terms x^{ρ} has magnitude $x^{\mathrm{Re}(\rho)}$. Thus, knowing that $\mathrm{Re}(\rho) < 1$ suggests that the term x^{ρ} is of a smaller order of magnitude than the "principal" term x and therefore plays a negligible role in the asymptotic behavior of $\psi(x)$. This leads directly to the asymptotic formula $\psi(x) \sim x$. (Note that this argument is incomplete, since there are infinitely many zeros, so we would be dropping infinitely many of these terms, which requires further justification.)

The same type of reasoning involving (2.53) also suggests that even if we had more precise bounds on the real parts of the zeros of $\zeta(s)$, we could prove quantitative versions of the prime number theorem with explicit error bounds. The strongest statement of this type that is believed to hold is the celebrated **Riemann hypothesis**.

Conjecture 2.18 (The Riemann hypothesis). *All the nontrivial zeros of $\zeta(s)$ are on the "critical line"* $\mathrm{Re}(s) = 1/2$.

For more details, see [25, 46, 47] and [W14].

Exercises for Chapter 2

2.1 Prove the following properties satisfied by the Euler gamma function:
  (a) Values at half-integers:

$$\Gamma\left(n + \frac{1}{2}\right) = \frac{(2n)!}{4^n n!} \sqrt{\pi} \quad (n = 0, 1, 2, \ldots).$$

  (b) The duplication formula:

$$\Gamma(s)\Gamma(s + 1/2) = 2^{1-2s} \sqrt{\pi} \Gamma(2s).$$

  (c) The multiplication theorem: for any $k \geq 1$,

$$\Gamma(s)\Gamma\left(s + \frac{1}{k}\right)\Gamma\left(s + \frac{2}{k}\right) \cdots \Gamma\left(s + \frac{k-1}{k}\right) = (2\pi)^{(k-1)/2} k^{1/2-ks} \Gamma(ks).$$

2.2 Prove the following representation for the gamma function:

$$\Gamma(s) = \sum_{n=0}^{\infty} \frac{(-1)^n}{n!(n + s)} + \int_1^{\infty} e^{-x} x^{s-1} \, dx \quad (s \in \mathbb{C}).$$

2.3 For $n \geq 1$, let V_n denote the volume of the unit ball in $\mathbb{R}^n$. By evaluating the n-dimensional integral

$$A_n = \iint \cdots \int_{\mathbb{R}^n} \exp\left(-\frac{1}{2} \sum_{j=1}^{n} x_j^2\right) dx_1 \, dx_2 \ldots dx_n$$

in two ways, prove the well-known formula

$$V_n = \frac{\pi^{n/2}}{\Gamma(\frac{n}{2} + 1)}.$$

Note. This problem requires applying a small amount of geometric intuition (or, alternatively, having some technical knowledge of spherical coordinates in $\mathbb{R}^n$). For the solution, see [W15].

2.4 The **beta function** is a function $B(s, t)$ of two complex variables, defined for $\mathrm{Re}(s), \mathrm{Re}(t) > 0$ by

$$B(s, t) = \int_0^1 x^{s-1}(1 - x)^{t-1} \, dx.$$

  (a) Show that the improper integral defining $B(s, t)$ converges absolutely if and only if $\mathrm{Re}(s), \mathrm{Re}(t) > 0$.

(b) Show that $B(s, t)$ can be expressed in terms of the gamma function as

$$B(s, t) = \frac{\Gamma(s)\Gamma(t)}{\Gamma(s + t)}.$$

Guidance. Start by writing $\Gamma(s)\Gamma(t)$ as a double integral on the positive quadrant $[0, \infty)^2$ of $\mathbb{R}^2$ (with integration variables, say, x and y); then make the change of variables $u = x + y$, $v = x/(x + y)$ and use the change-of-variables formula for two-dimensional integrals to show that the integral evaluates as $\Gamma(s + t)B(s, t)$.

2.5 The **digamma function** $\psi(s)$ is the logarithmic derivative

$$\psi(s) = \frac{\Gamma'(s)}{\Gamma(s)}$$

of the gamma function, also considered as a somewhat important special function in its own right.

(a) Show that $\psi(s)$ has the convergent series expansions

$$\psi(s) = -\gamma - \frac{1}{s} + \sum_{n=1}^{\infty} \frac{s}{n(n + s)}$$

$$= -\gamma + \sum_{n=0}^{\infty} \left(\frac{1}{n + 1} - \frac{1}{n + s} \right) \quad (s \neq 0, -1, -2, \ldots),$$

where γ is the Euler–Mascheroni constant.

(b) Equivalently, show that $\psi(s)$ can be expressed as

$$\psi(s) = -\lim_{n \to \infty} \left(\sum_{k=0}^{n} \frac{1}{k + s} - \log n \right).$$

(c) Show that $\psi(s)$ satisfies the functional equation

$$\psi(s + 1) = \psi(s) + \frac{1}{s} \quad (s \neq 0, -1, -2, \ldots).$$

(d) Show that

$$\psi(n + 1) = -\gamma + \sum_{k=1}^{n} \frac{1}{k} \quad (n = 0, 1, 2, \ldots).$$

That is, $\psi(x) + \gamma$ can be thought of as extending the definition of the **harmonic numbers** $H_n = \sum_{k=1}^{n} \frac{1}{k}$ to noninteger arguments.

(e) Show that $\psi(s)$ satisfies the reflection formula

$$\psi(1 - s) - \psi(s) = \pi \cot(\pi s).$$

(f) Here is a curious application of the digamma function. Consider the sequence of polynomials

$$P_n(x) = x(x-1)\ldots(x-n) \quad (n = 0,1,2,\ldots)$$

and their derivatives

$$Q_n(x) = P'_n(x).$$

By Rolle's theorem, $Q_n(x)$ has precisely one root in each interval $(k, k+1)$ for $0 \le k \le n-1$. Denote this root by $k + a_{n,k}$, so that the numbers $a_{n,k}$ (the fractional parts of the roots of $Q_n(x)$) are in $(0,1)$.

A curious phenomenon can now be observed by plotting the points $a_{n,k}$, $k = 0,\ldots,n-1$, numerically, say for $n = 50$. You will see that for large n, the plot appears to approximate a smooth limiting curve. The following precise statement can be proved.

Theorem 2.19 ([56]). *Let $t \in (0,1)$. Let $k = k(n)$ be a sequence such that $0 \le k(n) \le n-1$, $k(n) \to \infty$ as $n \to \infty$, $n - k(n) \to \infty$ as $n \to \infty$, and $k(n)/n \to t$ as $n \to \infty$. Then we have*

$$\lim_{n\to\infty} a_{n,k(n)} = \frac{1}{\pi} \operatorname{arccot}\left(\frac{1}{\pi}\log\left(\frac{1-t}{t}\right)\right).$$

In the above formula, $\operatorname{arccot}(\cdot)$ refers to the branch of the inverse cotangent function taking values between 0 and π.

Prove Theorem 2.19 using the facts you learned about the digamma function.

2.6 Given two integrable functions $f, g : \mathbb{R} \to \mathbb{C}$ of a real variable, their **convolution** is the function $h = f * g$ defined by the formula

$$h(x) = (f * g)(x) = \int\limits_{-\infty}^{\infty} f(t)g(x-t)\,dt \quad (x \in \mathbb{R}).$$

The convolution operation is extremely important in harmonic analysis, since it corresponds to a simple multiplication operation in the Fourier domain; in probability theory, where it corresponds to the addition of independent random variables; and in many other areas of mathematics, science, and engineering.

For $\alpha > 0$, define the **gamma density** with parameter α, denoted $\gamma_\alpha : \mathbb{R} \to \mathbb{R}$, as

$$\gamma_\alpha(x) = \frac{1}{\Gamma(\alpha)} e^{-x} x^{\alpha-1} \chi_{[0,\infty)}(x) \quad (x \in \mathbb{R})$$

(where χ_A denotes the characteristic function of a set $A \subset \mathbb{R}$). Note that $\gamma_\alpha(x)$ is a nonnegative function whose integral equals 1, so that it is a probability density function.

Show that for all $\alpha, \beta > 0$, we have

$$\gamma_\alpha * \gamma_\beta = \gamma_{\alpha+\beta},$$

that is, the family of density functions $(\gamma_\alpha)_{\alpha>0}$ is closed under the convolution operation. This fact is one of the reasons why the family of gamma densities plays an important role in probability theory and appears in many real-life applications.

2.7   Show that the initial terms in the Laurent expansion of $\Gamma(s)$ around $s = 0$ are of the form

$$\Gamma(s) = \frac{1}{s} - \gamma + \left(\frac{\gamma^2}{2} + \frac{\pi^2}{12}\right)s + O(s^2),$$

where γ is the Euler–Mascheroni constant.

2.8   Prove the equivalence of the two versions (2.11) and (2.13) of the functional equation for the Riemann zeta function.

2.9   Show that the initial terms in the Laurent expansion of $\zeta(s)$ around $s = 1$ are of the form

$$\zeta(s) = \frac{1}{s - 1} + \gamma + O(s - 1).$$

2.10 Define the function $\eta(s)$ of a complex variable s by

$$\eta(s) = \sum_{n=1}^{\infty} \frac{(-1)^{n-1}}{n^s} = 1 - \frac{1}{2^s} + \frac{1}{3^s} - \frac{1}{4^s} + \cdots.$$

This function, a close cousin of the Riemann zeta function, is known as the **Dirichlet eta function**.

(a)   Prove that the series defining $\eta(s)$ converges uniformly on any half-plane of the form $\mathrm{Re}(s) \geq \alpha$ withe $\alpha > 0$, and conclude that $\eta(s)$ is defined and holomorphic in the half-plane $\mathrm{Re}(s) > 0$.

(b)   Show that $\eta(s)$ is related to the Riemann zeta function by the formula

$$\eta(s) = (1 - 2^{1-s})\zeta(s) \quad (\mathrm{Re}(s) > 1).$$

(c)   Using this relation, deduce a new proof that the zeta function can be analytically continued to a meromorphic function on $\mathrm{Re}(s) > 0$ that has a simple pole at $s = 1$ with residue 1 and is holomorphic everywhere else in the region.

2.11 Now that you have learned about the Riemann zeta function and its properties, go back and look at identities (1.54)–(1.55). Can you make sense of what these formulas claim? How do they relate to $\zeta(s)$ and to the Dirichlet eta function $\eta(s)$ discussed in Exercise 2.10?

2.12 Show that the Taylor expansion of the digamma function $\psi(s) = \frac{\Gamma'(s)}{\Gamma(s)}$ (discussed in Exercise 2.5) around $s = 1$ is given by

$$\psi(s) = -\gamma + \sum_{n=1}^{\infty} (-1)^{n-1} \zeta(n+1)(s-1)^n \quad (|s-1| < 1),$$

where γ is the Euler–Mascheroni constant.

2.13 (a) Prove that for all $x \geq 1$,

$$\prod_{p \leq x} \frac{1}{1 - \frac{1}{p}} \geq \log x$$

(where the product is over all prime numbers $p \leq x$).

(b) Pass to the logarithm and deduce that for some constant $K > 0$, we have the bound

$$\sum_{p \leq x} \frac{1}{p} \geq \log \log x - K \quad (x \geq 1).$$

(It is also possible to show a matching upper bound of $\log \log x + K'$ for some constant $K' > 0$, that is, the *harmonic series of primes* $\sum_p \frac{1}{p}$ diverges as $\log \log x$, in contrast to the usual harmonic series, which diverges as $\log x$.)

2.14 **Riemann's contour integral representation for $\zeta(s)$.** Prove another expression for $\zeta(s)$ valid for all $s \in \mathbb{C}$:

$$\zeta(s) = \frac{\Gamma(1-s)}{2\pi i} \int_C \frac{(-x)^s}{e^x - 1} \frac{dx}{x}, \tag{2.54}$$

where C is a keyhole contour coming from $+\infty$ to 0 slightly above the positive x-axis, then circling the origin in a counterclockwise direction around a circle of small radius, and then going back to $+\infty$ slightly below the positive x-axis.

Note. Representation (2.54) is due to Riemann, who used it in his famous 1859 paper for his first proof of the analytic continuation and functional equation for his eponymous zeta function. In the same paper, he proceeded to give a second proof using the method described in Section 2.3. See [25, Ch. 1] for more details.

2.15 **An alternative proof of the functional equation of the Jacobi theta function.**

(a) Recall the definition of the Jacobi theta function $\theta(t)$ in (2.15). Use the residue theorem to evaluate the contour integral

$$\oint_{\gamma_N} \frac{e^{-\pi z^2 t}}{e^{2\pi i z} - 1} \, dz,$$

where γ_N is the rectangle with vertices $\pm(N+1/2) \pm i$ (with N a positive integer), then take the limit as $N \to \infty$ to derive the integral representation

$$\theta(t) = \int\limits_{-\infty-i}^{\infty-i} \frac{e^{-\pi z^2 t}}{e^{2\pi i z} - 1} \, dz - \int\limits_{-\infty+i}^{\infty+i} \frac{e^{-\pi z^2 t}}{e^{2\pi i z} - 1} \, dz \quad (t > 0) \qquad (2.55)$$

for $\theta(t)$.

(b) In representation (2.55), expand the factor $(e^{2\pi i z} - 1)^{-1}$ as a geometric series in $e^{-2\pi i z}$ (for the first integral) and as a geometric series in $e^{2\pi i z}$ (for the second integral). Evaluate the resulting infinite series, rigorously justifying all steps, to obtain an alternative proof of the functional equation (2.22).

2.16 Define the following arithmetic functions taking an integer argument n:

$$d(n) = \sum_{d|n} 1 \quad \text{(the number of divisors function)},$$

$$\sigma(n) = \sum_{d|n} d \quad \text{(the sum of divisors function)},$$

$$\phi(n) = \#\{1 \le k \le n-1 : \gcd(k,n) = 1\}$$
$$\text{(the Euler totient function)},$$

$$\Lambda(n) = \begin{cases} \log p & \text{if } n = p^k, \ p \text{ prime}, \\ 0 & \text{otherwise}, \end{cases}$$
$$\text{(the von Mangoldt } \Lambda\text{-function)},$$

$$\mu(n) = \begin{cases} (-1)^k & \text{if } n = p_1 p_2 \cdots p_k \text{ is a product of } k \text{ distinct primes}, \\ 0 & \text{otherwise}, \end{cases}$$
$$\text{(the Möbius } \mu\text{-function)},$$

$$\lambda(n) = (-1)^k \quad \text{if } n = p_1 p_2 \cdots p_k \text{ is a product of } k \text{ primes},$$
$$\text{(the Liouville } \lambda\text{-function)}.$$

We saw that the zeta function and its logarithmic derivative have the series representations

$$\zeta(s) = \sum_{n=1}^{\infty} n^{-s}, \quad -\frac{\zeta'(s)}{\zeta(s)} = \sum_{n=1}^{\infty} \Lambda(n) n^{-s}.$$

Both these series are of the general form

$$\sum_{n=1}^{\infty} \frac{c_n}{n^s}$$

for some sequence $(c_n)_{n=1}^{\infty}$. A series of this type is called a **Dirichlet series**. Prove the following additional identities (valid for $\text{Re}(s) > 1$) expressing various functions related to $\zeta(s)$ as Dirichlet series:

$$-\zeta'(s) = \sum_{n=1}^{\infty} \log n \cdot n^{-s},$$

$$\frac{1}{\zeta(s)} = \sum_{n=1}^{\infty} \mu(n) n^{-s},$$

$$\frac{\zeta(s)}{\zeta(2s)} = \sum_{n=1}^{\infty} |\mu(n)| n^{-s},$$

$$\frac{\zeta(2s)}{\zeta(s)} = \sum_{n=1}^{\infty} \lambda(n) n^{-s},$$

$$\zeta(s)^2 = \sum_{n=1}^{\infty} d(n) n^{-s},$$

$$\frac{\zeta(s-1)}{\zeta(s)} = \sum_{n=1}^{\infty} \phi(n) n^{-s},$$

$$\zeta(s)\zeta(s-1) = \sum_{n=1}^{\infty} \sigma(n) n^{-s}.$$

2.17 Evaluate the following infinite products:

(a) $\prod_{p \text{ prime}} \frac{p^2-1}{p^2} = \frac{3}{4} \cdot \frac{8}{9} \cdot \frac{24}{25} \cdot \frac{48}{49} \cdot \dots = ?$

(b) $\prod_{p \text{ prime}} \frac{p^2+1}{p^2} = \frac{5}{4} \cdot \frac{11}{9} \cdot \frac{26}{25} \cdot \frac{50}{49} \cdot \dots = ?$

(Compare with the products in Exercise 1.42.)

2.18 Show that the infinite product $K := \prod_{p \text{ prime}} \frac{p^2-1}{p^2}$ whose value you computed in Exercise 2.17 can be given the following geometric interpretation as "the fraction of lattice points in $\mathbb{Z}^2$ visible from the origin." That is, assume that you are standing at the origin point $(0, 0)$ of an infinite grove of trees, positioned at the lattice points $(m, n) \in \mathbb{Z}^2 \setminus \{(0,0)\}$. These are idealized trees that have zero thickness, so you will be able to see the tree at (m, n) from your vantage point if and only if there is no other tree obscuring the view from some position $(m/k, n/k)$, where k is a common divisor of m and n, that is, if and only if m and n are relatively prime. Define

$$K_N = \frac{\#\{(m, n) \in \mathbb{Z}^2 \setminus \{(0,0)\} : |m|, |n| \le N, \ m, n \text{ are relatively prime}\}}{\#\{(m, n) \in \mathbb{Z}^2 \setminus \{(0,0)\} : |m|, |n| \le N\}}$$

for $N \ge 1$. Prove that $K_n \to K$ as $N \to \infty$. This gives a precise asymptotic meaning to the above informal description of K as the fraction of lattice points visible from the origin.

2.19 Prove the bound (2.41).

2.20 Let p_n denote the nth prime number. Prove that the prime number theorem is equivalent to the statement that

$$p_n \sim n \log n \quad \text{as } n \to \infty.$$

2.21 Define a sequence of numbers $(\beta(n))_{n=1}^{\infty}$ by

$$\beta(n) = \mathrm{lcm}(1, 2, \ldots, n),$$

where for integers $a_1, \ldots, a_k$, $\mathrm{lcm}(a_1, \ldots, a_k)$ denotes the least common multiple of $a_1, \ldots, a_k$. This natural number-theoretic sequence of integers [W16] has the numbers $1, 2, 6, 12, 60, 60, 420, 840, 2\,520, 2\,520, 27\,720$ as its first few values.

(a) Prove that $\beta(n) = \exp(\psi(n))$, where $\psi(x) = \sum_{p^k \leq x} \log p$ denotes Chebyshev's weighted prime counting function.

(b) Conclude using the equivalent formulation of the prime number theorem in terms of Chebyshev's function that

$$\beta(n) = e^{(1+o(1))n} \quad \text{as } n \to \infty.$$

3 Conformal mapping

Second Hypothesis: That small regions of the Earth should be displayed as similar figures in the plane.

Leonhard Euler, "On the mapping of spherical surfaces onto the plane" (1777)

3.1 Motivation: classifying complex regions up to conformal equivalence

As we discussed in Chapter 1, the notion of a conformal mapping is a highly appealing geometric idea that can be explained to anyone without any requirement that they ever heard of complex analysis, let alone understand any of the mathematics underlying it. Anyone who can appreciate the art of M. C. Escher (see Fig. 1.2 on p. 8) will intuitively grasp that there is something special and beautiful about conformal maps.

Conformal maps are also an important tool in the toolkit of applied mathematicians. They have many applications for solving important partial differential equations that show up in physics, engineering, and in other areas as diverse as cartography [68] and medical imaging [37].

In this chapter, we will approach the area of conformal mapping from a purely complex-analytic direction. We will see that this side of the theory has a beauty all its own, which, while subtle and requiring patience and contemplation to appreciate, equals and perhaps surpasses the more obvious aspects appreciated by art lovers and equation solvers.

Let $\Omega \subset \mathbb{C}$ be a complex region. In complex analysis, we often wish to understand the classes of functions $\mathcal{H}(\Omega)$ and $\mathcal{M}(\Omega)$ of holomorphic and meromorphic functions on Ω, respectively. You might think that the structures of these classes of functions would depend in some highly sensitive way on the particular choice of the region Ω. As it turns out, this is largely untrue: although the structure of such a family does vary somewhat, there are large families of regions Ω for which the structure of $\mathcal{H}(\Omega)$ (respectively, $\mathcal{M}(\Omega)$) is the same across all members of a given family, so that it is in practice enough to understand what is happening in one representative region of each family. Moreover, the question of which family a particular region Ω belongs to can in many cases be answered using *topological* properties of Ω.

To make this idea precise, we define an equivalence relation on regions that captures the notion that for two regions Ω and Ω', $\mathcal{H}(\Omega)$ and $\mathcal{H}(\Omega')$ "have the same structure." This relation is called **biholomorphism** or **conformal equivalence**. We say that Ω and Ω' are **conformally equivalent** if there is a bijective holomorphic map $g : \Omega \to \Omega'$ whose inverse is also holomorphic. Such a map g is called a **biholomorphism**, **biholomorphic map**, or **conformal map**. Note that a conformal map must satisfy $g'(z) \neq 0$ for any $z \in \Omega$, by Corollary 1.58. It is trivial to check that the relation of conformal equivalence is, as its name suggests, an equivalence relation.[1]

If Ω and Ω' are conformally equivalent and related by a conformal map $g : \Omega \to \Omega'$, then each holomorphic function (respectively, meromorphic function) $f : \Omega \to \mathbb{C}$ can be used to define a holomorphic (respectively, meromorphic) function $\tilde{f} : \Omega' \to \mathbb{C}$ by

$$\tilde{f} = f \circ g^{-1}.$$

It is immediate to check that the correspondence $f \mapsto \tilde{f}$ defines a bijection between $\mathcal{H}(\Omega)$ and $\mathcal{H}(\Omega')$ (respectively, between $\mathcal{M}(\Omega)$ and $\mathcal{M}(\Omega')$). Thus the conformal map allows us to translate any question about holomorphic or meromorphic functions on Ω' to a question about holomorphic or meromorphic functions on Ω. The definition of conformal equivalence therefore captures precisely the notion of equivalence we were interested in.

In many areas of mathematics, when we find an interesting equivalence relation, this immediately leads to a standard set of interesting questions: how do we determine equivalence? Can we describe all equivalence classes, or at least some particularly simple or important ones? Do there exist some canonical representatives in each of those equivalence classes? How can we construct a map demonstrating equivalence, and to what extent is it unique? And so on. Asking such questions for this particular equivalence relation turns out to be very fruitful and is what the area of conformal mapping is about.

Examples. Here are some regions that seem worth thinking about from the point of view of conformal mapping, both theoretically and because they arise in applications (for example, in the study of Laplace's equation in mathematical physics, electrostatics, hydrodynamics, etc):
1. the complex plane $\mathbb{C}$
2. the punctured plane $\mathbb{C} \setminus \{0\}$
3. the unit disc $\mathbb{D} = \{z \in \mathbb{C} : |z| < 1\}$
4. the upper half-plane $\mathbb{H} = \{z \in \mathbb{C} : \text{Im}(z) > 0\}$
5. the Riemann sphere[2] $\widehat{\mathbb{C}} = \mathbb{C} \cup \{\infty\}$

---

1 In this chapter, we use the term "conformal map" with a slightly different meaning than the sense in which this term was used in Subsection 1.3.4. That subsection was concerned with understanding the property of being conformal as a *local* property; here we develop the conceptually much richer set of ideas related to understanding maps that are *globally* conformal—that is, conformal everywhere in the local sense but also bijective. Moreover, the conformal maps from Subsection 1.3.4 were not assumed to be orientation preserving. Here we focus on conformal maps that are holomorphic, which in particular means that they are orientation preserving (see (1.25)).

2 The Riemann sphere is not quite a complex region in the usual sense; technically, it is a Riemann surface, but we will still count it and trust that you understand how the various definitions apply in that situation; refer to Section 1.11. Actually, the same classification questions we are addressing in the context

6. the slit plane $\mathbb{C} \setminus (-\infty, 0]$
7. a strip $S(x_1, x_2) = \{z \in \mathbb{C} : 0 < \mathrm{Re}(z) < 1\}$
8. a rectangle $\{z \in \mathbb{C} : 0 < \mathrm{Re}(z) < 1, a < \mathrm{Im}(z) < b\}$
9. an annulus $A(r_1, r_2) = \{z \in \mathbb{C} : r_1 < |z| < r_2\}$
10. a quadrant $\{z : \mathrm{Re}(z) > 0, \mathrm{Im}(z) > 0\}$
11. an ellipse $\{z = x + iy : (\frac{x}{A})^2 + (\frac{y}{B})^2 < 1\}$
12. the plane with an interval removed, $\mathbb{C} \setminus [-1, 1]$
13. the upper half-plane with an interval removed, $\mathbb{H} \setminus [0, i]$
14. a "blob" (Fig. 3.1)

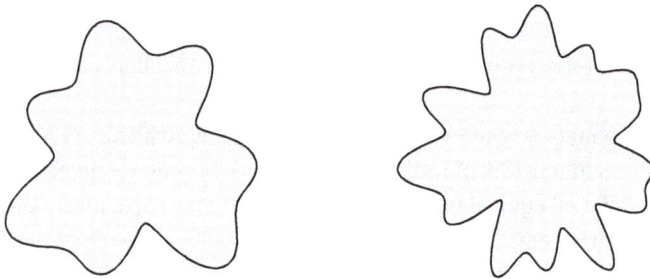

Figure 3.1: Two blob-shaped regions. Are they conformally equivalent?

Can you guess what is the correct grouping of these regions according to conformal equivalence? (Note: in example 9 of the annulus, we in fact have a *family* of regions, which may not all be conformally equivalent to each other.) By the end of this chapter, you will know the answers.

Since conformal maps are continuous, the relation of conformal equivalence is a stronger notion of equivalence than topological equivalence (a. k. a. homeomorphism). We record this obvious but important fact as a lemma.

Lemma 3.1. *If regions Ω and Ω' are conformally equivalent, then they are homeomorphic.*

Next, if regions Ω and Ω' are conformally equivalent, with the conformal map $g : \Omega \to \Omega'$ relating them, then is g unique? If not, can the extent to which it is not unique be made precise? The answer to these questions is described in terms of the **automorphism group** of a complex region. More precisely, if $\tilde{g} : \Omega \to \Omega'$ is another conformal map, then the map $h : \Omega \to \Omega$ defined by

$$h = g^{-1} \circ \tilde{g}$$

---

of conformal equivalence apply more generally in the theory of Riemann surfaces. We will encounter an interesting example of the classification of a class of Riemann surfaces up to conformal equivalence in Chapters 4 and 5; see Sections 4.15, 5.5, and 5.11.

is a conformal equivalence map between Ω and itself. We call such a map a (conformal) **automorphism** of Ω. Conversely, if $g : \Omega \to \Omega'$ is a conformal map and $h : \Omega \to \Omega$ is a conformal automorphism, then $\tilde{g} : \Omega \to \Omega'$ defined by

$$\tilde{g} = g \circ h$$

is also a conformal map from Ω to Ω', and clearly every conformal map $\tilde{g} : \Omega \to \Omega'$ can be represented in such a way for some automorphism $h : \Omega \to \Omega$ (just define h as above). Thus the family of automorphisms of Ω precisely measures the extent of the nonuniqueness of the conformal map $g : \Omega \to \Omega'$ for any Ω' that is conformally equivalent to Ω. This family has the algebraic structure of a group, with the group operation being composition of maps, and is thus referred to as the automorphism group of Ω. We denote this group by $\mathrm{Aut}(\Omega)$. We will seek to give explicit descriptions of automorphism groups whenever this is possible.

To conclude this general discussion, we note one additional useful fact about conformal maps.

Lemma 3.2. *In the definition of conformal equivalence, the condition that g^{-1} is holomorphic can be dropped, that is, if $g : \Omega \to \Omega'$ is holomorphic and bijective, then g^{-1} is automatically holomorphic.*

Proof. Since g satisfies $g'(z_0) \neq 0$ for any $z_0 \in \Omega$, the inverse function theorem (Theorem 1.56) implies that the inverse map g^{-1} exists locally in a neighborhood of $g(z_0)$ as a holomorphic function for any $z_0 \in \Omega$. Since g is a bijection, the inverse function exists globally (in the sense of set theory) as a function $g^{-1} : \Omega' \to \Omega$. The fact that g^{-1} is locally holomorphic implies that the global inverse function g^{-1} is holomorphic, which is the claim of the lemma. □

In the next few sections, we begin to classify some of the main conformal equivalence classes that every complex analyst should be familiar with. The most important classification result in this chapter is the Riemann mapping theorem, which is formulated in Section 3.4.

Suggested exercises for Section 3.1. 3.1.

3.2 First singleton conformal equivalence class: the complex plane

The first conformal equivalence class we discuss contains just a single element, the complex plane. This is explained by the following theorem.

Theorem 3.3. *Let $g : \mathbb{C} \to \Omega$ be a conformal map between $\mathbb{C}$ and a region Ω. Then $\Omega = \mathbb{C}$, $g(z)$ is a conformal automorphism, and $g(z)$ has the form*

$$g(z) = az + b$$

for some complex numbers a, b with $a \neq 0$.

Proof. Let $g : \mathbb{C} \to \Omega$ be a conformal equivalence map. We will prove that $g(z)$ is of the form $g(z) = az + b$ with $a \neq 0$ just based on the assumption that it is an entire function and that it is injective; the additional claims that $\Omega = \mathbb{C}$ and $g(z)$ is an automorphism will then follow.

Since $g(z)$ is an entire function, it is either a polynomial, or it is not. We treat each of those two cases separately (proving that $g(z)$ is of the desired form in the first case and proving that the second case cannot occur).

If $g(z)$ is a polynomial, it cannot be a constant since those certainly are not injective maps. We claim that it also cannot be a polynomial of degree $k \geq 2$, which if true would leave only the option of a linear function $g(z) = az + b$ with $a \neq 0$. The fact that polynomials of degree higher than 1 are not injective is easy to see: a polynomial of degree k has k roots counting with multiplicity, which means that either there are at least two distinct zeros (contradicting the assumption of injectivity), or there is a single zero of multiplicity k, which means that the polynomial is of the form $g(z) = c(z - a)^k$. This polynomial is clearly also not injective since in that case the equation $g(z) = 1$ has k distinct solutions.

It remains to consider the other possibility of an entire function that is not a polynomial. In that scenario, we claim that $g(z)$ has an essential singularity at $z = \infty$. For otherwise, by our classification of singularities (Section 1.12), $g(z)$ must have a pole of some order k at infinity. However, having such a pole implies that the rate of growth of $|g(z)|$ is restricted by the order of the pole; specifically, $g(z)$ satisfies a bound of the form $|g(z)| \leq A + B|z|^k$ for all z, where A and B are positive real constants. Now a well-known argument from basic complex analysis (Exercise 1.25) implies that $g(z)$ is actually a polynomial of degree at most k, which is a contradiction.

We are now in a good position to apply the Casorati–Weierstrass theorem (Theorem 1.46) about the behavior of functions near an essential singularity. Denote $w_0 = g(0)$. Since $g(z)$ is an open mapping by the open mapping theorem (Theorem 1.50), the image $g(\mathbb{D})$ of the unit disc under $g(z)$ contains an open neighborhood E of w_0. But by the Casorati–Weierstrass theorem the image $g(\mathbb{C} \setminus D_{\leq R}(0))$ of the complement of any closed disc around 0 (i. e., any neighborhood of ∞) is dense in $\mathbb{C}$ and therefore has a nonempty intersection with E. This intersection means that there exist points $z_1 \in \mathbb{D}$ and $z_2 \in \mathbb{C} \setminus D_R(0)$ for which

$$g(z_1) = g(z_2).$$

Now if $R > 1$, then $z_1 \neq z_2$. We have therefore shown that $g(z)$ is not injective, which contradicts our initial assumption. Thus the scenario of a conformal map on $\mathbb{C}$ that is not a polynomial is impossible, and the proof is complete. □

By Theorem 3.3 the group of conformal automorphisms of $\mathbb{C}$ is

$$\mathrm{Aut}(\mathbb{C}) = \{z \mapsto az + b : a, b \in \mathbb{C}, a \neq 0\}.$$

3.3 Second singleton conformal equivalence class: the Riemann sphere

There is a second conformal equivalence class that is a singleton, the Riemann sphere. The following result is the analogue of Theorem 3.3 for $\widehat{\mathbb{C}}$.

Theorem 3.4. *If $g : \widehat{\mathbb{C}} \to \Omega$ is a conformal map between $\widehat{\mathbb{C}}$ and a region Ω, then $\Omega = \widehat{\mathbb{C}}$, $g(z)$ is a conformal automorphism, and $g(z)$ has the form*

$$g(z) = \frac{az + b}{cz + d} \tag{3.1}$$

for some complex numbers a, b, c, d with $ad - bc \neq 0$.

Proof of Theorem 3.4. We start by proving that $\Omega = \widehat{\mathbb{C}}$. Assume that this is not the case, i. e., that there is at least one point $w \in \widehat{\mathbb{C}}$ that is not in the image $g(\widehat{\mathbb{C}})$. We can assume without loss of generality that $w = \infty$; otherwise, replace the map $g(z)$ with $\tilde{g}(z) = \frac{1}{g(z)-w}$. Once $\tilde{g}(z)$ is shown to be of the desired form (3.1), solving the equation $\tilde{g}(z) = \frac{1}{g(z)-w}$ for $g(z)$ shows that $g(z)$ is of that form as well.

Since $g(z)$ does not take the value ∞, it also cannot *approach* infinity, that is, there does not exist a sequence $(z_n)_{n=1}^{\infty}$ of points in $\widehat{\mathbb{C}}$ for which $g(z_n) \to \infty$. If such a sequence existed, we could use the fact that $\widehat{\mathbb{C}}$ is compact to extract a convergent subsequence $z_{n_k} \to Z \in \widehat{\mathbb{C}}$, whence it would follow, since $g(z)$ is a continuous function, that $g(Z) = \infty$, which cannot happen since ∞ is not in the image of $g(z)$.

The fact that $g(z)$ does not approach ∞ means simply that $g(z)$ is a bounded function and a holomorphic one at that (our a priori assumption that allows Ω to contain the point ∞ only means it is meromorphic). Thus it is a bounded entire function and hence constant by Liouville's theorem, a contradiction.

Having established that $\Omega = \widehat{\mathbb{C}}$, we now know that $g(z)$ is a genuine automorphism of $\widehat{\mathbb{C}}$. Denote $w = g(\infty)$. Once again, we can assume without loss of generality that $w = \infty$; otherwise, replace the map $g(z)$ with $\tilde{g}(z) = \frac{1}{g(z)-w}$ as before. Under this assumption, the restriction of $g(z)$ to $\mathbb{C}$ is a conformal automorphism of $\mathbb{C}$, so from the discussion in the previous section we know that $g(z)$ is of the form $az + b$ for some $a, b \in \mathbb{C}$, $a \neq 0$. □

By Theorem 3.4 the group of conformal automorphisms of $\widehat{\mathbb{C}}$ is

$$\text{Aut}(\widehat{\mathbb{C}}) = \left\{ z \mapsto \frac{az + b}{cz + d} \ : \ a, b, c, d \in \mathbb{C}, ad - bc \neq 0 \right\}. \tag{3.2}$$

The elements of this group are known as **Möbius transformations**. An important and easy-to-check property of such transformations is that they act as 2×2 linear transformations; more precisely, given two Möbius transformations

$$T_1(z) = \frac{a_1 z + b_1}{c_1 z + d_1} \quad \text{and} \quad T_2(z) = \frac{a_2 z + b_2}{c_2 z + d_2}, \tag{3.3}$$

their composition is given by

$$(T_1 \circ T_2)(z) = \frac{\alpha z + \beta}{\gamma z + \delta}, \tag{3.4}$$

where $\alpha, \beta, \gamma, \delta$ are the entries of the matrix

$$\begin{pmatrix} \alpha & \beta \\ \gamma & \delta \end{pmatrix} = \begin{pmatrix} a_1 & b_1 \\ c_1 & d_1 \end{pmatrix} \begin{pmatrix} a_2 & b_2 \\ c_2 & d_2 \end{pmatrix}. \tag{3.5}$$

For this reason, Möbius transformations are also known as **fractional linear transformations**.

The group (3.2) is also sometimes referred to as the **projective linear group** (of order 2 over the complex numbers) and denoted PSL(2, $\mathbb{C}$). The reason for this terminology is as follows. If we define the **special linear group** (of order 2 over the complex numbers) by

$$SL(2, \mathbb{C}) = \left\{ \begin{pmatrix} a & b \\ c & d \end{pmatrix} : a, b, c, d \in \mathbb{C},\ ad - bc = 1 \right\},$$

then we can easily check that the association mapping a matrix $\left(\begin{smallmatrix} a & b \\ c & d \end{smallmatrix} \right) \in SL(2, \mathbb{C})$ to the Möbius transformation $z \mapsto \frac{az+b}{cz+d}$ is a surjective group homomorphism, which has the subgroup $\{\pm \left(\begin{smallmatrix} 1 & 0 \\ 0 & 1 \end{smallmatrix} \right)\}$ as its kernel. Thus, by the first isomorphism theorem in group theory, the group Aut($\widehat{\mathbb{C}}$) can be identified with the quotient group

$$SL(2, \mathbb{C})/\{\pm \left(\begin{smallmatrix} 1 & 0 \\ 0 & 1 \end{smallmatrix} \right)\}.$$

The quotienting operation in this context is often referred to as *projectivization*, which leads to the name projective linear group both for the quotient group and the occasional use of the same name and notation for the group of Möbius transformations.

The group PSL(2, $\mathbb{C}$) is an important group in mathematics and even has interesting connections to physics; see the box overleaf.

Suggested exercises for Section 3.3. 3.2.

3.4 The Riemann mapping theorem

We have seen two conformal equivalence classes consisting of a single element each. Obviously, if all other equivalence classes were also singletons, the situation would be extremely boring, and the notion of conformal equivalence would not even deserve its own name. It is easy to see however that the true situation is, at least, more complicated than this simplistic scenario (see Exercise 3.3).

The group $\mathrm{PSL}(2, \mathbb{C})$ and the night sky of a relativistically moving observer

Suppose you get into a spaceship and speed away from Earth, reaching a velocity of αc, where c is the speed of light, and the fraction α is substantial (say, higher than 5 %). We know from science fiction movies that your view of the stars as you peer through the spaceship window will appear distorted. But how, exactly? This problem has a delightful connection to complex analysis and the automorphism group $\mathrm{PSL}(2, \mathbb{C})$ of the Riemann sphere. In fact, your view of the celestial sphere of stars gets transformed by a Möbius transformation acting on the celestial sphere precisely as if it were the Riemann sphere.

Mathematically, the connection is roughly as follows: it is well known from the theory of special relativity that an observer moving at relativistic velocity **v** relative to the Earth (which for the sake of discussion we assume is an inertial frame of reference) will have their time and space coordinates transformed from the Earth's time and space coordinate system according to a type of linear transformation known as a **proper, orthochronous Lorentz transformation**. The group of such transformations can be represented as the group of 4×4 real matrices

$$L_+^\uparrow = \left\{ T \in \mathrm{Mat}_{4\times4}(\mathbb{R}) \ : \ \det(T) = 1, \ T_{1,1} < 0, \ T^\top X T = X \right\},$$

where X is the 4×4 diagonal matrix with diagonal entries $-1, 1, 1, 1$. In fact, it can be shown that $L_+^\uparrow$ is isomorphic to $\mathrm{PSL}(2, \mathbb{C})$ and that the isomorphism $\rho : L_+^\uparrow \to \mathrm{PSL}(2, \mathbb{C})$ is such that for the moving observer with a given associated Lorentz transformation T, the distortion of the moving observer's celestial sphere relative to the celestial sphere of the static frame of reference is described precisely by the Möbius transformation $\rho(T)$, under the obvious identification between the celestial sphere and the Riemann sphere. See [53, Appendix B] and [55, Ch. 1] for the details of this surprising result.

On this optimistic note, it looks like there ought to be some interesting phenomena for us to explore. This brings us to one of the most fundamental results on conformal mapping, the **Riemann mapping theorem**, which identifies the first nontrivial conformal equivalence class and the one that undoubtedly plays the most central role in complex analysis.

Theorem 3.5 (Riemann mapping theorem: simple version). *Let* $\Omega, \Omega' \subset \mathbb{C}$ *be simply connected complex regions with* $\Omega, \Omega' \neq \mathbb{C}$. *Then* Ω *and* Ω' *are conformally equivalent.*

As an immediate corollary, we get an interesting result in topology, an illustration of the principle that the often symbiotic relationship between complex analysis and topology involves a flow of ideas in both directions.

Corollary 3.6. *Any two simply connected regions in the plane are homeomorphic.*

This well-known result can also be proved without the use of complex analysis. See [W17] for a related discussion.

To prove Theorem 3.5, we will need to develop some new theoretical ideas (which are also interesting in their own right and are of broader applicability). A more precise version of the theorem is stated in Section 3.7.

Tangentially to that effort, we also wish to understand the structure of the automorphism groups $\mathrm{Aut}(\Omega)$ for regions Ω belonging to the conformal equivalence class described by the theorem. By Exercise 3.1 all such groups are isomorphic in such a way that the isomorphism between any two can be described in terms of conformal equivalence maps $g : \Omega \to \Omega'$ relating different class members. Thus, to understand the automorphism groups, it is in fact sufficient to classify the automorphisms for just one representative member of the class. There are two fairly canonical choices for such a member, the unit disc $\mathbb{D}$ and the upper half-plane $\mathbb{H}$ (and those two are easy to relate to each other, though doing so is still interesting). We discuss these regions in the next two sections.

Suggested exercises for Section 3.4. 3.3.

3.5 The unit disc and its automorphisms

The next result, known as the Schwarz lemma, is a simple yet powerful result about holomorphic functions from the unit disc to itself that keep the origin fixed. It is an important tool on the path to characterizing the automorphisms of the unit disc.

If $g : \mathbb{D} \to \mathbb{D}$, then we say that $g(z)$ is a **rotation map**, or simply a **rotation**, if it is of the form $g(z) = e^{i\theta}z$ for some $\theta \in [0, 2\pi)$.

Lemma 3.7 (The Schwarz lemma). *Let $g : \mathbb{D} \to \mathbb{D}$ be a holomorphic function that satisfies $g(0) = 0$. Then:*
1. $|g(z)| \le |z|$ *for all $z \in \mathbb{D}$.*
2. *If $|g(z)| = |z|$ for some $z \ne 0$, then $g(z)$ is a rotation.*
3. $|g'(0)| \le 1$.
4. *If $|g'(0)| = 1$, then $g(z)$ is a rotation.*

Proof. Since $g(z)$ has a zero at $z = 0$, we know that it satisfies $|g(z)| \le C|z|$ for some $C > 0$ and all z in some neighborhood of 0. This is a weaker inequality than the one we are trying to prove, but in fact it is a helpful observation, as it can be restated as the claim that $h(z) = g(z)/z$ satisfies $|h(z)| \le C$ for all $z \in \mathbb{D} \setminus \{0\}$; that is, $h(z)$ is *bounded* in a punctured neighborhood of 0 and of course holomorphic there. By Riemann's removable singularity theorem (Theorem 1.38), $h(z)$ therefore has a removable singularity at 0 and can be extended to a holomorphic function on all of $\mathbb{D}$ (which we still denote $h(z)$, as per the usual convention when talking about analytic continuation). Now let $z \in \mathbb{D} \setminus \{0\}$, and let r be a real number with $|z| < r < 1$. By the maximum modulus principle (Theorem 1.51) the maximum modulus of $h(z)$ in the closed disc of radius r around 0 is attained at the boundary of that disc. Therefore we have that

$$\left| \frac{g(z)}{z} \right| = |h(z)| \le \max_{|w| \le r}|h(w)| \le \max_{0 \le t < 2\pi} \left| h(re^{it}) \right| = \max_{0 \le t < 2\pi} \frac{|g(re^{it})|}{r} \le \frac{1}{r}.$$

(In the last step, we used the fact that $g(z)$ maps $\mathbb{D}$ into itself, so $|g(w)| \le 1$ for all $w \in \mathbb{D}$.) Since this is true for all $|z| < r < 1$, we then have that

$$\left|\frac{g(z)}{z}\right| \le \inf_{|z|<r<1}\frac{1}{r} = 1,$$

that is, $|g(z)| \le |z|$, which was the first claim of the lemma. Now claim 3 also follows by taking an additional limit of these inequalities as $z \to 0$, since $|g'(0)| = |\lim_{z\to 0}\frac{g(z)-g(0)}{z}| = \lim_{z\to 0}|\frac{g(z)}{z}|$.

Now, for the claim 2, note that an equality for some $z \in \mathbb{D}$ in the bound $|h(z)| \le 1$ means that $|h(z)|$ attains its maximal value in the interior of the disc. By the condition for equality in the maximum modulus principle, $h(z)$ must be a constant, which is of unit magnitude (since we know that $|h(z)| = 1$ for some z). That is, we have shown that $h(z) \equiv e^{i\theta}$ for some θ or, equivalently, that $g(z)$ is a rotation, giving claim 2.

Similarly, for the fourth claim, if $1 = |g'(0)| = \lim_{z\to 0}|\frac{g(z)}{z}| = \lim_{z\to 0}|h(z)| = |h(0)|$, then again we see that $|h(z)|$ attains its maximum value in the interior of the disc (in this case at $z = 0$) and infer using the same argument as above that $g(z)$ is a rotation. □

Corollary 3.8 (Automorphisms of the unit disc that fix 0). *The automorphisms $g : \mathbb{D} \to \mathbb{D}$ of the unit disc that fix 0 (that is, satisfy $g(0) = 0$) are precisely the rotations.*

Proof. Obviously, a rotation is a conformal automorphism of $\mathbb{D}$ that fixes 0. Conversely, let $g : \mathbb{D} \to \mathbb{D}$ be an automorphism that fixes 0. Then both $g(z)$ and its inverse function $g^{-1}(z)$ satisfy the assumptions of the Schwarz lemma. It follows that $|g(z)| \le z$ and $|g^{-1}(w)| \le w$ for all $z, w \in \mathbb{D}$; or, setting $w = g(z)$ for an arbitrary $z \in \mathbb{D}$ in the second inequality,

$$\left|g(z)\right| \le z \text{ and } |z| \le \left|g(z)\right| \quad \Longrightarrow \quad \left|g(z)\right| = |z|$$

for all $z \in \mathbb{D}$. By part 2 of the Schwarz lemma, $g(z)$ is a rotation. □

We can now exhibit a more general two-parameter family of automorphisms of $\mathbb{D}$, which are obtained by composing rotations with an additional family of automorphisms that *do not* fix 0. As a first step, for $w \in \mathbb{D}$, we define the Möbius transformation

$$\varphi_w(z) = \frac{w - z}{1 - \overline{w}z}. \tag{3.6}$$

Lemma 3.9. *The transformation φ_w is an automorphism of $\mathbb{D}$. Moreover, it has the following properties:* (a) $\varphi_w(0) = w$; (b) $\varphi_w(w) = 0$; (c) $\varphi_w^{-1} = \varphi_w$.

Proof. Properties (a)–(c) are trivial to check through a direct calculation, which I leave as an exercise. For the claim that φ_w is an automorphism, note that if $|z| = 1$, then

$$\left|\varphi_w(z)\right| = \frac{|w - z|}{|1 - \overline{w}z|} = \frac{|w - z|}{|1 - \overline{w}z| \cdot |\overline{z}|} = \frac{|w - z|}{|\overline{z} - \overline{w}z\overline{z}|} = \frac{|w - z|}{|\overline{z} - \overline{w}|} = 1.$$

Thus φ_w maps the unit circle into itself. It is also injective (as a meromorphic function on $\mathbb{C}$) since it is a Möbius transformation. Therefore either φ maps the unit disc $\mathbb{D}$ into itself and maps the complement $\widetilde{\mathbb{D}} = \{|z| > 1\}$ of the closed unit disc into itself, or φ_w

maps $\mathbb{D}$ into $\widetilde{\mathbb{D}}$ and maps $\widetilde{\mathbb{D}}$ into $\mathbb{D}$. However, we know that $\varphi_w(0) = w$ and $w \in \mathbb{D}$, so that rules out the latter possibility. Finally, since we have established that $\varphi_w(\mathbb{D}) \subset \mathbb{D}$, and we know that $\varphi_w^{-1} = \varphi_w$, the mapping of $\mathbb{D}$ into itself by φ is bijective, and φ_w is a conformal equivalence. □

The composition of an arbitrary member of the family of rotations (specified by a real-valued parameter $\theta \in [0, 2\pi)$) and an arbitrary member of the family φ_w, specified by the point $w \in \mathbb{D}$, is a map of the form

$$z \mapsto e^{i\theta} \frac{w - z}{1 - \overline{w}z}.$$

It turns out that all automorphisms of the unit disc are of this form. This is the well-known characterization of the automorphism group $\mathrm{Aut}(\mathbb{D})$, given in the following theorem.

Theorem 3.10 (Automorphisms of the unit disc). *A function $g : \mathbb{D} \to \mathbb{D}$ is an automorphism of $\mathbb{D}$ if and only if it is of the form*

$$g(z) = e^{i\theta} \frac{w - z}{1 - \overline{w}z} \tag{3.7}$$

for some $\theta \in [0, 2\pi)$ and $w \in \mathbb{D}$. The pair (θ, w) in this representation is unique.

Proof. The "if" part was already explained above. To prove the "only if" claim, let $g : \mathbb{D} \to \mathbb{D}$ be an automorphism. Denote $w = g^{-1}(0) \in \mathbb{D}$, and let $h = g \circ \varphi_w$. As the composition of two automorphisms of $\mathbb{D}$, $h(z)$ is itself an automorphism of $\mathbb{D}$. It also leaves $z = 0$ fixed. By Corollary 3.8 it is a rotation and can be expressed as $h(z) = e^{i\theta}z$ for some $\theta \in [0, 2\pi)$. Therefore $g(z) = (h \circ \varphi_w)(z)$ is of the desired form (3.7).

For the uniqueness claim, note that (3.7) implies that $w = g^{-1}(0)$, which determines w uniquely for a given automorphism g. Now if $w \neq 0$, then we have $g(0) = e^{i\theta}w$, which can be written as $e^{i\theta} = g(0)/w$, and thus θ is also determined uniquely from the map g. In the second case where $w = 0$, we are back to the scenario of an automorphism that fixes 0, which we have seen must be a rotation $g(z) = e^{i\theta}z$, with θ again clearly being uniquely determined. □

An alternative, but less frequently used, characterization of the automorphisms of the unit disc is given in the next result. The proof is left as an exercise (Exercise 3.4).

Theorem 3.11 (Automorphisms of the unit disc: alternative representation). *A function $g : \mathbb{D} \to \mathbb{D}$ is an automorphism of $\mathbb{D}$ if and only if it is of the form*

$$g(z) = \frac{\mu z + v}{\overline{v}z + \overline{\mu}} \tag{3.8}$$

for some $\mu, v \in \mathbb{C}$ satisfying $|\mu|^2 - |v|^2 = 1$. The pair (μ, v) is unique.

The explicit description of the automorphisms of $\mathbb{D}$ in terms of the representations (3.7)–(3.8), involving formulas that one rarely encounters outside of complex analysis, masks the fact that the group of such automorphisms bears a close relationship with a standard matrix group you may be familiar with from linear algebra, the theory of Lie groups, topology, and other areas. As we will see in the next section, the connection becomes apparent when we switch from the unit disc to its "conformal sibling," the upper half-plane.

Suggested exercises for Section 3.5. 3.4.

3.6 The upper half-plane and its automorphisms

Lemma 3.12. *The unit disc $\mathbb{D}$ and the upper half-plane $\mathbb{H}$ are conformally equivalent. The pair of maps $\Phi : \mathbb{H} \to \mathbb{D}$ and $\Psi : \mathbb{D} \to \mathbb{H}$ given by*

$$\Phi(z) = \frac{z - i}{z + i} \quad and \quad \Psi(z) = -i\frac{z + 1}{z - 1} \tag{3.9}$$

give an explicit pair of mutually inverse conformal maps mapping each of the regions onto the other.

Proof. Note that if $z = x + iy$, then $|\Phi(z)|^2 = \frac{|z-i|^2}{|z+i|^2} = \frac{x^2+(y-1)^2}{x^2+(y+1)^2}$, which is < 1 if and only if $\text{Im}(z) = y > 0$ (the geometric meaning of this statement is simply that $\Phi(z)$ is the ratio of the distances of z to i and $-i$, and the upper half-plane is precisely the locus of points that are closer to i than to $-i$). Thus Φ maps $\mathbb{H}$ into $\mathbb{D}$ and the complement of $\mathbb{H}$ into the complement of $\mathbb{D}$. Since we know that Φ is a conformal map when regarded as a map from $\widehat{\mathbb{C}}$ to itself, this is enough to imply that it maps $\mathbb{H}$ surjectively and conformally onto $\mathbb{D}$. Finally, it is trivial to verify by direct calculation that the inverse map to $\Phi(z)$ is given by the formula defining $\Psi(z)$. □

Theorem 3.13 (Conformal automorphisms of the upper half-plane). *A function $g : \mathbb{H} \to \mathbb{H}$ is a conformal automorphism if and only if it is of the form*

$$g(z) = \frac{az + b}{cz + d} \tag{3.10}$$

for real numbers a, b, c, d satisfying $ad - bc = 1$. The numbers a, b, c, d in this representation are unique up to a single choice of sign, in the sense that if a, b, c, d and a', b', c', d' are coefficients in two distinct representations, then $(a', b', c', d') = \pm(a, b, c, d)$.

Proof. "If": assume that $g(z)$ has the stated form (3.10) with a, b, c, d real and $ad - bc = 1$. As we already know from Theorem 3.4, $g(z)$ is a conformal automorphism of $\mathbb{C}$. Moreover, since $a, b, c, d \in \mathbb{R}$, we have

$$\mathrm{Im}\left(\frac{az+b}{cz+d}\right) = \mathrm{Im}\left(\frac{(az+b)(c\bar{z}+d)}{|cz+d|^2}\right)$$

$$= \frac{1}{|cz+d|^2}\,\mathrm{Im}\left(ac|z|^2 + bd + adz + bc\bar{z}\right) = \frac{ad-bc}{|cz+d|^2}\,\mathrm{Im}(z). \qquad (3.11)$$

This immediately implies that $\mathrm{Im}(g(z)) > 0$ if and only if $\mathrm{Im}(z) > 0$, that is, g is an automorphism of $\mathbb{H}$.

"Only if": assume that $g \in \mathrm{Aut}(\mathbb{H})$. Then $f = \Phi \circ g \circ \Psi$ is an automorphism of the unit disc, where Φ and Ψ are given in (3.9). By Theorem 3.11, f can be expressed as

$$f(z) = \frac{\mu z + v}{\bar{v}z + \bar{\mu}}$$

for some $\mu, v \in \mathbb{C}$ with $|\mu|^2 - |v|^2 = 1$. To calculate what this means for $g = \Psi \circ f \circ \Phi$, we switch to the notation of matrix multiplication, which, as we know from (3.3)–(3.5), is a way to represent the action of Möbius transformations. The matrices associated with the action of Φ, Ψ, and f are

$$\Phi = \begin{pmatrix} 1 & -i \\ 1 & i \end{pmatrix}, \quad \Psi = \begin{pmatrix} -i & -i \\ 1 & -1 \end{pmatrix}, \quad f = \begin{pmatrix} \mu & v \\ \bar{v} & \bar{\mu} \end{pmatrix}.$$

Therefore the map $\Psi \circ f \circ \Phi$ is represented by the matrix product

$$\Psi f \Phi = \begin{pmatrix} -i & -i \\ 1 & -1 \end{pmatrix}\begin{pmatrix} \mu & v \\ \bar{v} & \bar{\mu} \end{pmatrix}\begin{pmatrix} 1 & -i \\ 1 & i \end{pmatrix}.$$

More explicitly, if we denote $\mu = x + iy$ and $v = u + iv$ to represent μ, v in terms of their real and imaginary parts, then this matrix product is

$$\Psi f \Phi = \begin{pmatrix} -i & -i \\ 1 & -1 \end{pmatrix}\begin{pmatrix} x + iy & u + iv \\ u - iv & x - iy \end{pmatrix}\begin{pmatrix} 1 & -i \\ 1 & i \end{pmatrix}$$

$$= 2i\begin{pmatrix} -x - u & -y + v \\ y + v & -x + u \end{pmatrix} =: 2i\begin{pmatrix} a & b \\ c & d \end{pmatrix}.$$

The numbers a, b, c, d thus defined are real, and moreover it is easy to check that $ad-bc = 1$ (hint: determinants). Note that the scalar factor $2i$ multiplying the matrix is irrelevant when we go back to considering g as a Möbius transformation instead of a matrix, that is, we see that $g(z)$ is indeed of the form $\frac{az+b}{cz+d}$ with a, b, c, d as claimed in the theorem. $\square$

The automorphism group

$$\mathrm{Aut}(\mathbb{H}) = \left\{ z \mapsto \frac{az+b}{cz+d} \;:\; a, b, c, d \in \mathbb{R},\; ad-bc = 1 \right\}$$

is known as the **projective special linear group** (of order 2 over the real numbers) and sometimes denoted PSL(2, $\mathbb{R}$). By the natural association between 2×2 matrices and

Möbius transformations discussed in Section 3.3, it can be identified with the quotient group

$$SL(2, \mathbb{R})/\{\pm I\},$$

where $SL(2, \mathbb{R})$ is the special linear group of order 2 over $\mathbb{R}$ (the group of invertible 2×2 real matrices with determinant 1), and $\{\pm I\}$ is its subgroup with two elements containing the identity matrix and its negation.

3.7 The Riemann mapping theorem: a more precise formulation

We formulated in Section 3.4 a version of the Riemann mapping theorem that identifies an interesting conformal equivalence class of complex regions. Conceptually, this is what I regard as the main content of the theorem. Note that this formulation is carefully "neutral" in the sense of not singling out any member of the equivalence class as being more important or worthy of attention than others. However, in practice, we already discussed the fact that the unit disc and upper half-plane are each in their own way somewhat canonical members of the class. By contrast, other member regions such as, say, the unit square, seldom play a particularly important role in the theory, although from a purely geometric point of view, they may be just as natural, and they may appear in specific applications.

Furthermore, as we inch our way toward a proof of the theorem, it does in fact become convenient to fix a specific member of the class—the unit disc—as the target region for the conformal maps we will construct. Another small conceptual advance is to add more information about the conformal map mapping a given region Ω to $\mathbb{D}$ so as to ensure uniqueness. This leads us to the following more detailed version of the theorem.

Theorem 3.14 (Riemann mapping theorem: detailed version). *Let $\Omega \subset \mathbb{C}$ be a simply connected complex region with $\Omega \neq \mathbb{C}$, and let $z_0 \in \Omega$. Then there exists a unique biholomorphism $F : \Omega \to \mathbb{D}$ with the property that*
1. *$F(z_0) = 0$*
2. *$F'(z_0)$ is a positive real number.*

Proof of uniqueness. Let F_1 and F_2 be two biholomorphisms with the properties described in the theorem. Then the conformal map $\Phi = F_2 \circ F_1^{-1}$ is an automorphism of $\mathbb{D}$ that fixes 0, so by Corollary 3.8 it is a rotation, that is, of the form $\Phi(z) = az$ for some a with $|a| = 1$. On the other hand, the constant a can be expressed as

$$a = \Phi'(0) = F_2'(F_1^{-1}(0))(F_1^{-1})'(0) = \frac{F_2'(z_0)}{F_1'(z_0)},$$

which shows that it is a positive real number. It follows that $a = 1$ and $\Phi(z) \equiv z$, that is, $F_1 \equiv F_2$. $\qquad \square$

The history of the Riemann mapping theorem

The Riemann mapping theorem was formulated by the great Bernhard Riemann in 1851 as part of his PhD thesis. Riemann stated the result for regions with a piecewise smooth boundary and gave a proof that contained useful ideas but was later realized to be flawed. Later nineteenth-century mathematicians worked hard to fill in the gaps in Riemann's argument, with varying levels of success. The first proof considered to be fully correct by modern standards was given by Osgood in 1900. Osgood's proof, like others before it, relied on the "potential-theoretic" approach (related to Dirichlet's principle and the study of Laplace's equation) advocated by Riemann rather than on ideas of a more conceptually complex-analytic nature. This approach, while interesting, has since fallen out of fashion as an approach to proving the Riemann mapping theorem because of various technical shortcomings it has.

The proof of the theorem we present in Sections 3.8–3.9 is described in Walsh's historical survey [72] as the "standard modern proof." You will find it described in most complex analysis textbooks, as it appears to be the simplest proof known today. For additional details on the interesting history of Riemann's famous theorem and the ideas developed out of it, see the historical reviews [33, 72].

The more difficult part of Theorem 3.14 is the existence claim. As we will see, the key insight needed for the proof is that the problem of mapping Ω conformally to $\mathbb{D}$ can be formulated as a maximization problem for a certain functional. Specifically, in the class $\mathcal{F}$ consisting of all the *injective* maps from Ω into $\mathbb{D}$ that map z_0 to 0 and for which $F'(z_0)$ is a positive real number, we will see that the one map that is also surjective (and thus establishes the required conformal equivalence of Ω to $\mathbb{D}$) is the one for which the number $F'(z_0)$ is maximal. This will be shown in a somewhat constructive way by arguing that if $F(z)$ is not surjective, then we can exploit the point that is "missing" from the image to produce a new conformal map $G : \Omega \to \mathbb{D}$ with a larger value of $G'(z_0)$. Although the basic idea of how this is done is fairly simple (see Lemma 3.21), there are a few technical issues that need to be addressed to turn it into a complete proof, namely showing that the class $\mathcal{F}$ is nonempty, that the functional $F \mapsto F'(z_0)$ attains a maximum, and so on. The details are given in the next two sections.

3.8 Proof of the Riemann mapping theorem, part I: technical background

In this section, we prove a few auxiliary results needed for the proof of the Riemann mapping theorem. Two of the results, Montel's and Hurwitz's theorems, are theorems in complex analysis. The third, the Arzelà–Ascoli theorem, is a theorem in real analysis.

Let $\mathcal{F}$ be a family of complex-valued continuous functions on a complex region Ω. We say that $\mathcal{F}$ is **locally uniformly bounded** if for any compact set $K \subset \Omega$, we have

$$\sup_{f \in \mathcal{F},\, z \in K} |f(z)| < \infty. \tag{3.12}$$

We say that $\mathcal{F}$ is **locally uniformly equicontinuous** if for any compact $K \subset \Omega$ and any $\varepsilon > 0$, there exists $\delta > 0$ such that if $z_1, z_2 \in K$ and $|z_1 - z_2| < \delta$, then $\sup_{f \in \mathcal{F}} |f(z_1) - f(z_2)| < \varepsilon.$ (3.13)

The following is a version of the well-known Arzelà–Ascoli theorem, a staple of real and functional analysis, slightly adapted to our setting.

Theorem 3.15 (Arzelà–Ascoli theorem). *Let $\mathcal{F}$ be a family of continuous complex-valued functions on Ω. Assume that the family is locally uniformly equicontinuous and locally uniformly bounded. Then any sequence $(f_n)_{n=1}^{\infty}$ of functions in $\mathcal{F}$ has a subsequence $(f_{n_k})_{k=1}^{\infty}$ that converges uniformly on compacts in Ω to some continuous function f.*

Proof. Let $Q = (z_m)_{m=1}^{\infty}$ be a dense countable set of points in Ω (ordered as a sequence according to some arbitrary enumeration). The sequence $(f_n(z_1))_{n=1}^{\infty}$ is a sequence of complex numbers taking values in a compact set $\{|z| \leq M_1\}$, where we denote $M_1 = \sup_{f \in \mathcal{F}} |f(z_1)| < \infty$ (guaranteed to be finite by (3.12)). By compactness this sequence therefore has a convergent sequence, which we denote by $(f_n^{(1)}(z_1))_{n=1}^{\infty}$ (instead of the more traditional subsequence notation $f_{n_k}(z_1)$). That is, $f_n^{(1)}$ is the notation for the nth function in the extracted *subsequence* of the original sequence of functions $(f_n(z))_n$.

Now we extract a further subsequence of this subsequence, noting that the sequence $(f_n^{(1)}(z_2))_{n=1}^{\infty}$ is a sequence of complex numbers taking values in a compact set $\{|z| \leq M_2\}$, where

$$M_2 = \sup_{f \in \mathcal{F},\ z \in \{z_1, z_2\}} |f(z)|.$$

(Again, the local uniform boundedness assumption guarantees that $M_2 < \infty$.) So again by compactness, this sequence has a convergent sequence, which we denote by $(f_n^{(2)}(z_1))_{n=1}^{\infty}$.

Continuing in this way, we proceed to successively extract nested subsequences $(f_n^{(3)})_{n=1}^{\infty}$, $(f_n^{(4)})_{n=1}^{\infty}$, ... of the original sequence of functions, where each subsequence is extracted as a further subsequence of the previous one. These subsequences have the property that for each $j \geq 1$, the jth sequence $(f_n^{(j)})_{n=1}^{\infty}$ is a subsequence of the original sequence $(f_n)_n$ for which $f_n^{(j)}(z_m)$ converges to a limit as $n \to \infty$ for $m = 1, 2, \ldots, j$.

Now consider the "diagonal" sequence in this nested sequence of subsequences: we let $g_n = f_n^{(n)}$. Then $(g_n)_{n=1}^{\infty}$ is a subsequence of $(f_n)_n$ with the property that $g_n(z_m)$ converges to a limit as $n \to \infty$ for *all* $m \geq 1$.

We claim that the sequence of functions $(g_n(z))_{n=1}^{\infty}$ converges uniformly on compacts in Ω. Let $K \subset \Omega$ be compact, and let $\varepsilon > 0$. Let $\delta > 0$ be a number, guaranteed to exist by the assumption of local uniform equicontinuity, with the property that

if $z_1, z_2 \in K$ and $|z_1 - z_2| < \delta$, then $\sup_{f \in \mathcal{F}} |f(z_1) - f(z_2)| < \dfrac{\varepsilon}{3}.$

(Compare with (3.13): we merely replaced ε there with $\varepsilon/3$, with the usual goal in mind that some other bound later will end up smaller than ε.) The containment

$K \subset \cup_{\xi \in K} D_{\delta/2}(\xi)$ gives an open covering of K, which by compactness has a finite sub-covering $(D_{\delta/2}(\xi_j))_{j=1}^q$. Select a point z_{v_j} of the countable dense set Q from each of the subcovering discs $D_{\delta/2}(\xi_j)$. For any $1 \leq j \leq q$, $(g_k(z_{v_j}))_{k=1}^\infty$ is a convergent sequence or, equivalently, is a Cauchy sequence; therefore there exists an index $N_j \geq 1$ such that

$$\left| g_\ell(z_{v_j}) - g_k(z_{v_j}) \right| < \frac{\varepsilon}{3}$$

whenever $k, \ell \geq N_j$. Set $N = \max(N_1, N_2, \ldots, N_q)$. Then for any $w \in K$, we have that $w \in D_{\delta/2}(\xi_j) \subset D_\delta(z_{v_j})$ for some $1 \leq j \leq q$. It follows that, for $k, \ell \geq N$,

$$\left| g_\ell(w) - g_k(w) \right| \leq \left| g_\ell(w) - g_\ell(z_{v_j}) \right| + \left| g_\ell(z_{v_j}) - g_k(z_{v_j}) \right|$$
$$+ \left| g_k(z_{v_j}) - g_k(w) \right| < \frac{\varepsilon}{3} + \frac{\varepsilon}{3} + \frac{\varepsilon}{3} = \varepsilon.$$

This establishes that $(g_k(z))_{k=1}^\infty$ is a Cauchy sequence uniformly on K and hence (by a standard fact from real analysis) converges uniformly on K. The compact K was arbitrary, so we proved the existence of a subsequence that converges uniformly on compacts; the fact that the limiting function must be continuous is standard, and the proof of the theorem is complete. □

Returning to the realm of complex analysis, we now introduce the concept of a **normal family** of functions. Let Ω be a complex region as before. A family $\mathcal{F}$ of holomorphic functions on Ω is called **normal**, or a **normal family**, if every sequence $(f_n)_{n=1}^\infty$ in the family has a subsequence $(f_{n_k})_{k=1}^\infty$ such that f_{n_k} converges uniformly on compacts to a holomorphic function g.

Theorem 3.16 (Montel's theorem). *Let $\mathcal{F}$ be a family of holomorphic functions on a region Ω that is locally uniformly bounded. Then $\mathcal{F}$ is a normal family.*

Proof. We claim that the added assumption of holomorphicity of the members of $\mathcal{F}$, together with local uniform boundedness, implies that the family is uniformly locally equicontinuous. Once we show this, the Arzelà–Ascoli theorem will imply that every sequence $(F_n)_{n=1}^\infty$ of elements in the family has a subsequence F_{n_k} that converges uniformly on compacts to a limiting function F. Then it would follow that F is holomorphic by standard properties of uniform convergence on compacts (Theorem 1.39 on p. 45), and we would be done.

We start by showing a weaker version of the required property that does not include uniformity over compact subsets. Fix a point $a \in \Omega$ and a radius $\rho > 0$ such that $D_{2\rho}(a) \subset \Omega$. Later we will need to emphasize the dependence of ρ on a, so we will then denote it by $\rho(a)$. If $z_1, z_2 \in D_\rho(a)$, then by Cauchy's integral formula we have, uniformly over all $f \in \mathcal{F}$,

$$\left| f(z_1) - f(z_2) \right| = \left| \frac{1}{2\pi i} \oint_{|w-a|=2\rho} f(w) \left(\frac{1}{w - z_1} - \frac{1}{w - z_2} \right) dw \right|$$

$$= \left| \frac{z_1 - z_2}{2\pi i} \oint_{|w-a|=2\rho} \frac{f(w)}{(w - z_1)(w - z_2)} \, dw \right|$$

$$\leq \frac{1}{2\pi} |z_1 - z_2| \cdot \sup_{|w-a|=2\rho} |f(w)| \cdot 2\pi(2\rho) \frac{1}{\rho^2} \leq \frac{2M}{\rho} |z_1 - z_2|, \tag{3.14}$$

where we denote $M = \sup_{f \in \mathcal{F}, |w-a|=2\rho} |f(w)|$, a finite number by the local uniform boundedness assumption.

Now fix a number $\varepsilon > 0$. If we define the number

$$\eta = \min\left(\rho, \frac{\rho \varepsilon}{4M}\right) > 0,$$

then by (3.14) we have the property that

$$\text{if } z_1, z_2 \in D_\eta(a), \quad \text{then } \sup_{f \in \mathcal{F}} |f(z_1) - f(z_2)| < \varepsilon. \tag{3.15}$$

This is the nonuniform local equicontinuity property alluded to above. Note that the parameter η depends on the point a, so we will now redenote it by $\eta(a)$ to emphasize this dependence. (η also depends on ε, but the value of ε will remain fixed throughout the discussion.)

Finally, we can derive the uniform-over-compacts version of local equicontinuity. Let $K \subset \Omega$ be a compact set, and let $\varepsilon > 0$ be the same as above. Consider the covering of K by open sets given by

$$K \subset \bigcup_{a \in K} D_{\eta(a)/2}(a).$$

By compactness there exists a finite subcovering

$$K \subset \bigcup_{j=1}^{n} D_{\eta(a_j)/2}(a_j)$$

for some points $a_1, \ldots, a_n \in K$. Denote $\delta = \frac{1}{2} \min(\eta(a_1), \ldots, \eta(a_n))$. Then we claim that for all $z_1, z_2 \in K$ such that $|z_1 - z_2| < \delta$,

$$\sup_{f \in \mathcal{F}} |f(z_1) - f(z_2)| < \varepsilon. \tag{3.16}$$

Indeed, z_1 must belong to $D_{\eta(a_j)/2}(a_j)$ for some $1 \leq j \leq n$ by the defining property of the subcovering. This also implies that

$$|z_2 - a_j| \leq |z_2 - z_1| + |z_1 - a_j| < \delta + \frac{\eta(a_j)}{2} \leq \frac{\eta(a_j)}{2} + \frac{\eta(a_j)}{2} = \eta(a_j),$$

so altogether we see that both z_1, z_2 are in $D_{\eta(a_j)}(a_j)$. Relation (3.16) therefore follows from (3.15). To summarize, we proved that for any compact set $K \subset \Omega$ and $\varepsilon > 0$, (3.13) is satisfied without choice of δ as defined above; this proves that the family $\mathcal{F}$ is locally uniformly equicontinuous and concludes the proof of the theorem. $\qquad\qquad\square$

Theorem 3.17 (Hurwitz's theorem). *Let $\Omega \subset \mathbb{C}$ be a region, and let $(f_n(z))_{n=1}^\infty$ and $g(z)$ be holomorphic functions on Ω such that $f_n(z) \to g(z)$ uniformly on compacts in Ω as $n \to \infty$, where $g(z)$ is not the zero function. If $z_0 \in \Omega$ is a zero of $g(z)$ of order $k \geq 0$, and $D_r(z_0) \subset \Omega$ is a disc centered at z_0 such that the punctured closed disc $D_{\leq r}(z_0) \setminus \{z_0\}$ contains no zeros of $g(z)$, then for any large enough n, $f_n(z)$ has precisely k zeros in $D_r(z_0)$ counting multiplicities.*

Proof. Recall that by the argument principle the order k of the zero of $g(z)$ at z_0 can be expressed as the contour integral

$$k = \frac{1}{2\pi i} \oint_{|z-z_0|=r} \frac{g'(z)}{g(z)} \, dz. \tag{3.17}$$

Denote by κ_n the number of zeros of $f_n(z)$ in $D_r(z_0)$ counting multiplicities. We wish to express κ_n similarly as a contour integral over the same circle. This can be done but requires first checking that $f_n(z)$ does not have any zeros on the circle, which is indeed true for large n. Let $M = \inf_{|z-z_0|=r} |g(z)|$ and note that $M > 0$ by the assumption that $g(z)$ has no zeros in the punctured disc $D_{\leq r}(z_0) \setminus \{z_0\}$ and, in particular, on the circle. By the uniform convergence of $f_n(z)$ to $g(z)$ on the circle there exists an index $N \geq 1$ such that for all $n \geq N$, $\inf_{|z-z_0|=r} |f_n(z)| \geq M/2$, so that, in particular, $f_n(z)$ also does not have any zeros on the circle $|z - z_0| = r$ as we wanted to show. Thus we have the expression

$$\kappa_n = \frac{1}{2\pi i} \oint_{|z-z_0|=r} \frac{f_n'(z)}{f_n(z)} \, dz \tag{3.18}$$

for all $n \geq N$.

Note also that on the circle $|z-z_0|$ we have not only the uniform convergence $f_n(z) \to g(z)$, but also that of the derivatives $f_n'(z) \to g'(z)$ (recall Theorem 1.39). Combining those facts, we deduce also that

$$\frac{f_n'(z)}{f_n(z)} \xrightarrow[n\to\infty]{} \frac{g'(z)}{g(z)}$$

uniformly on the circle $|z - z_0| = r$. Finally, this, together with (3.17) and (3.18), implies that

$$\kappa_n = \frac{1}{2\pi i} \oint_{|z-z_0|=r} \frac{f_n'(z)}{f_n(z)} \, dz \xrightarrow[n\to\infty]{} \frac{1}{2\pi i} \oint_{|z-z_0|=r} \frac{g'(z)}{g(z)} \, dz = k$$

Since k and κ_n are all integers, it follows that $\kappa_n = k$ for all sufficiently large n. $\qquad\square$

Corollary 3.18. *Let $\Omega \subset \mathbb{C}$ be a region, and as in Hurwitz's theorem, let $(f_n(z))_{n=1}^{\infty}$ and $g(z)$ be holomorphic functions on Ω such that $f_n(z) \to g(z)$ uniformly on compacts in Ω. If the functions $f_n(z)$ are all injective, then $g(z)$ is either injective or a constant.*

Proof. Assume by contradiction that $g(z)$ is not injective and also not a constant function. Then there exist distinct points $a, b, \in \Omega$ for which $g(a) = g(b)$. We have the convergence $f_n(a) \to g(a)$, and so, if we define functions $\psi(z)$ and $\varphi_n(z), n = 1, 2, \ldots$, by

$$\psi(z) = g(z) - g(a), \quad \varphi_n(z) = f_n(z) - f_n(a),$$

then $\varphi_n(z) \to \psi(z)$ uniformly on compacts in Ω. Moreover, $\psi(z)$ is not the zero function. Therefore we are in a position to apply Hurwitz's theorem. Specifically, note that $\psi(b) = 0$, and denote the order of the zero at b by $k \geq 1$. Let $r > 0$ be such that the punctured closed disc $D_{\leq r}(b) \setminus \{b\}$ does not contain any other zeros of $\psi(z)$ (so, in particular, it does not contain the point $z = a$). Applying Hurwitz's theorem, we conclude that for all sufficiently large n, $\varphi_n(z)$ has at least one zero in the disc $D_r(b)$. However, this is impossible, since $\varphi_n(z)$ already has one zero at $z = a$ and was assumed to be an injective function. We have reached a contradiction, and the proof is complete. $\qquad\square$

Suggested exercises for Section 3.8. 3.5, 3.6.

3.9 Proof of the Riemann mapping theorem, part II: the main construction

From now on, let Ω be a simply connected complex region with $\Omega \neq \mathbb{C}$ and $z_0 \in \Omega$, as in the statement of Theorem 3.14.

Lemma 3.19. *There exists an injective holomorphic function $G : \Omega \to \mathbb{D}$.*

Proof. We know that Ω is not the entire complex plane, so take some point $\alpha \in \mathbb{C} \setminus \Omega$. The function $z \mapsto z - \alpha$ has no zeros on Ω, so, since Ω is simply connected, by Theorem 1.53 there exists a branch of the logarithm function of $z - \alpha$ on it, that is, a holomorphic function $h(z)$ such that $e^{h(z)} = z - \alpha$ for all $z \in \Omega$.

Fix an arbitrary point $\beta \in \Omega$, and define a function $G : \Omega \to \mathbb{C}$ by

$$G(z) = \frac{1}{h(z) - h(\beta) - 2\pi i}. \tag{3.19}$$

We claim that $G(z)$ is holomorphic, injective, and bounded on Ω; this would imply that its scaled version $F(z) = cG(z)$ is injective and maps into $\mathbb{D}$ if c is a small enough positive constant, which would prove the result.

To establish these properties of $G(z)$, note first that $h(z)$ is injective, since $h(z) = h(w)$ implies $z - \alpha = e^{h(z)} = e^{h(w)} = w - \alpha$, so $z = w$. Clearly, $G(z) = G(w)$ also implies $h(z) = h(w)$, so similarly implies $z = w$, which shows that $G(z)$ is injective.

Now the claim that $G(z)$ is bounded is equivalent to the claim that

$$\inf_{z \in \Omega} |h(z) - (h(\beta) + 2\pi i)| > 0.$$

Assume by contradiction that this is not true. Then there is a sequence $(z_n)_{n=1}^{\infty}$ of points in Ω such that $h(z_n) \xrightarrow[n \to \infty]{} h(\beta) + 2\pi i$. Exponentiating, we get that

$$z_n - \alpha = e^{h(z_n)} \xrightarrow[n \to \infty]{} e^{h(\beta) + 2\pi i} = e^{h(\beta)} = \beta - \alpha.$$

In other words, z_n converges to β as $n \to \infty$. However, then we would have that $h(z_n)$ converges to $h(\beta)$ and not to $h(\beta) + 2\pi i$. This gives a contradiction and finishes the proof. □

Now define the family of functions

$$\mathcal{F} = \{F : \Omega \to \mathbb{D} : F(z) \text{ is holomorphic and injective}, F(z_0) = 0\}.$$

The family $\mathcal{F}$ is not empty: if $G(z)$ is an injective holomorphic function $G : \Omega \to \mathbb{D}$ guaranteed to exist by Lemma 3.19, then clearly $F(z) = c(G(z) - G(z_0))$ is an element of $\mathcal{F}$ if c is a small enough positive number. Define the number $\lambda \in [0, \infty]$ by

$$\lambda = \sup_{F \in \mathcal{F}} |F'(z_0)|.$$

Lemma 3.20. $0 < \lambda < \infty.$

Proof. Let $F \in \mathcal{F}$. To bound $|F'(z_0)|$ from above, observe that, by the Cauchy integral formula, if $r > 0$ is a number for which the closed disc $D_{\leq r}(z_0)$ is contained in Ω, then

$$|F'(z_0)| = \left| \frac{1}{2\pi i} \oint_{|w - z_0| = r} \frac{F(w)}{(w - z_0)^2} \, dw \right| \leq \frac{1}{2\pi} (2\pi r) \frac{1}{r^2} \sup_{w \in \Omega} |F(w)| \leq \frac{1}{r},$$

since F maps into the unit disc. Since this is true for all $F \in \mathcal{F}$, we get that $\lambda \leq \frac{1}{r}$. On the other hand, we claim that $|F'(z_0)| > 0$, which would show that $\lambda > 0$. Indeed, if $F'(z_0) = 0$, then $F(z)$ has a zero of order at least 2 in z_0. By Corollary 1.58, $F(z)$ is not locally injective in any neighborhood of z_0, in contradiction to the fact that F is injective. Thus $|F'(z_0)|$ must be positive. □

We now come to the most important lemma of this section, which contains the key idea behind our proof of the Riemann mapping theorem.

Lemma 3.21. *Given $F \in \mathcal{F}$, if $F(\Omega) \subsetneq \mathbb{D}$ (that is, the image of Ω under F does not cover all of $\mathbb{D}$), then there exists $G \in \mathcal{F}$ for which $|G'(z_0)| > |F'(z_0)|$.*

Proof. Take some $w \in \mathbb{D} \setminus F(\Omega)$, known to exist by the assumption. Since w is not in the image of Ω under F, the point 0 is not in the image of the composed map $\varphi_w \circ F : \Omega \to \mathbb{D}$, where (recall from (3.6) and Lemma 3.9) $\varphi_w(z) = \frac{w-z}{1-\bar{w}z}$ is the standard automorphism of $\mathbb{D}$ mapping 0 and w to each other. Since $\varphi_w \circ F$ does not take the value 0 and is defined on a simply connected region, by the construction of nth root functions described in Section 1.15 there exists a holomorphic branch of its square root, that is, a holomorphic function $S : \Omega \to \mathbb{D}$ satisfying

$$S(z)^2 = (\varphi_w \circ F)(z). \tag{3.20}$$

Now define $G : \Omega \to \mathbb{D}$ by the composition

$$G(z) = (\varphi_{S(z_0)} \circ S)(z). \tag{3.21}$$

We claim that $G(z)$ has the properties claimed by the lemma. First,

$$G(z_0) = (\varphi_{S(z_0)} \circ S)(z_0) = \varphi_{S(z_0)}(S(z_0)) = 0.$$

Second, note that $S(z)$ is injective since its square is injective as a composition of two injective maps. Therefore $G(z)$ is also injective. Both of those facts together show that $G \in \mathcal{F}$.

Third and crucially, we wish to show that $|G'(z_0)| > |F'(z_0)|$. To this end, note that by (3.20) and (3.21), $F(z)$ can be represented in terms of $G(z)$ as

$$F(z) = \varphi_w\big((\varphi_{S(z_0)} \circ G)(z)^2\big). \tag{3.22}$$

(This is a key relation that deserves to be digested properly. Take a minute or two to unwrap all the horrible notation and convince yourself that this relation is correct, and see if you can find some deeper meaning here.) Alternatively, if we define the function $W : \mathbb{D} \to \mathbb{D}$ by

$$W(z) = \varphi_w\big(\varphi_{S(z_0)}(z)^2\big),$$

then (3.22) can be rewritten as

$$F(z) = (W \circ G)(z). \tag{3.23}$$

Note that

$$W(0) = \varphi_w\big(\varphi_{S(z_0)}(0)^2\big) = \varphi_w\big(S(z_0)^2\big) = \varphi_w(\varphi_w(F(0))) = F(0) = 0.$$

Thus $W(z)$ satisfies the assumptions of Schwarz's lemma, and we conclude that $|W'(0)| \le 1$, and in fact the strict inequality $|W'(0)| < 1$ holds, since $W(z)$ is clearly not a rotation. This is what we want, since by (3.23)

$$|F'(z_0)| = |W'(G(z_0))G'(z_0)| = |W'(0)| \cdot |G'(z_0)|,$$

which gives the desired conclusion that $|G'(z_0)| > |F'(z_0)|$. □

Lemma 3.22. *The family $\mathcal{F}$ is a normal family.*

Proof. The functions in $\mathcal{F}$ all map into the unit disc, so they are uniformly bounded, and a fortiriori locally uniformly bounded. By Montel's theorem, $\mathcal{F}$ is normal. □

Lemma 3.23. *There exists an element $F \in \mathcal{F}$ for which $|F'(z_0)| = \lambda$, that is, the functional $G \mapsto |G'(z_0)|$ attains a maximum in the family $\mathcal{F}$.*

Proof. Let $(F_n)_{n=1}^{\infty}$ be a sequence of elements of $\mathcal{F}$ such that we have the convergence $|F_n'(z_0)| \to \lambda$. By Lemma 3.22 there is a subsequence $(F_{n_k})_{k=1}^{\infty}$ that converges uniformly on compacts in Ω to some limiting function $F : \Omega \to \mathbb{C}$, which moreover satisfies $F(z_0) = 0$, since $F_n(z_0) = 0$ for all n. Since uniform convergence on compacts implies convergence of the derivatives, we have that $|F'(z_0)| = \lambda$. Since the F_n are all injective, by Hurwitz's theorem, F either is a constant function or is injective, but we know from Lemma 3.20 that $|F'(z_0)| = \lambda > 0$, and hence F is not a constant and is therefore injective.

Let $z \in \Omega$. We know that $|F(z)| \leq 1$, since it is the limit of functions whose modulus is bounded by 1. However, F is holomorphic, and hence by the open mapping theorem, $F(\Omega)$ is an open set contained in the closed disc $\{z : |z| \leq 1\}$ and therefore is contained in the open disc $\mathbb{D}$. Thus we have shown that F is an element of $\mathcal{F}$, and the proof is complete. □

Proof of existence in Theorem 3.14. Take the element $F \in \mathcal{F}$, guaranteed to exist by Lemma 3.23, for which $|F'(z_0)| = \lambda$. By composing F with a rotation if necessary, we may assume that $F'(z_0)$ is real and positive. By Lemma 3.21, $F(z)$ must be surjective, which, together with the positivity of $F'(z_0)$ and the properties implied by belonging to $\mathcal{F}$, gives that $F(z)$ is the biholomorphism whose existence was claimed. □

Summarizing, we proved the uniqueness claim from Theorem 3.14 in Section 3.7, and the existence claim was proved above. This finishes the proof of the Riemann mapping theorem.

3.10 Annuli and doubly connected regions

The topic of conformal mapping does not end with the consideration of simply connected regions, where the problem of classifying complex regions up to conformal equivalence is now essentially settled (at least in principle) by the Riemann mapping theorem. To conclude this chapter, we give a brief taste of some of the interesting phenomena that arise when we try to classify conformal equivalence classes of regions that are *not* simply connected, starting with the next simplest case of regions that are

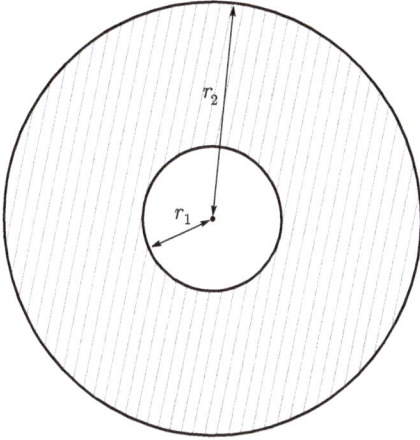

An annulus $A(r_1, r_2)$.

doubly connected. A region Ω is called doubly connected if the complement $\mathbb{C} \setminus \Omega$ has two connected components.[3]

One important class of doubly connected regions are the **annuli.** For $0 < r_1 < r_2$, we denote

$$A(r_1, r_2) = \{z : r_1 < |z| < r_2\},$$

an open annulus centered at 0 with internal radius r_1 and external radius r_2 (Fig. 3.2). It turns out that unlike the situation for simply connected regions, these annuli are not all in a single conformal equivalence class, despite being homeomorphic. The precise classification is given in the next result, sometimes known as Schottky's theorem.

Theorem 3.24 (Conformal classification of annuli). *Let $0 < r_1 < r_2$ and $0 < \rho_1 < \rho_2$. The annuli $A(r_1, r_2)$ and $A(\rho_1, \rho_2)$ are conformally equivalent if and only if*

$$\frac{r_1}{r_2} = \frac{\rho_1}{\rho_2}.$$

Proof. "If": assume that $\frac{r_1}{r_2} = \frac{\rho_1}{\rho_2}$. Then the map $z \mapsto \frac{\rho_1}{r_1} z = \frac{\rho_2}{r_2} z$ is a conformal equivalence between $A(r_1, r_2)$ and $A(\rho_1, \rho_2)$.

"Only if": this is the nontrivial direction. Assume that $A(r_1, r_2)$ and $A(\rho_1, \rho_2)$ are conformally equivalent. We start with a normalization that fixes the two inner radii at 1 to simplify things a bit: denote $\mu = r_2/r_1$ and $\nu = \rho_2/\rho_1$. Then $A(1, \mu)$ is conformally equivalent to $A(r_1, r_2)$ (by the scaling transformation mentioned in the "if" part), and

---

3 More generally, Ω is called k-**connected** if $\mathbb{C} \setminus \Omega$ has k connected components and **finitely connected** if it is k-connected for some $k \geq 1$.

similarly $A(1, v)$ is conformally equivalent to $A(\rho_1, \rho_2)$. Therefore $A(1, \mu)$ and $A(1, v)$ are conformally equivalent to each other. Let $f : A(1, \mu) \to A(1, v)$ be a conformal map. We can assume without loss of generality that f maps the inner boundary circle $|z| = 1$ to itself and maps the outer boundary circle $|z| = \mu$ of $A(1, v)$ to its counterpart $|z| = v$ in $A(1, v)$; otherwise, f maps the inner circle of $A(1, \mu)$ to the outer circle of $A(1, v)$ and vice versa, and in that case, we can get a conformal map that maps the inner circle to itself by replacing f by $f(\mu/z)$ (the composition of f with the inversion $z \mapsto \mu/z$, which is a conformal automorphism of $A(1, \mu)$).

For each $1 < r < \mu$, let γ_r denote the circular contour $\{|z| = r\}$, and let $\Gamma_r = f \circ \gamma_r$ denote its image under the map f. The curve Γ_r is a simple closed curve and hence encloses a well-defined region (see Theorem 1.26 and the discussion following it in Section 1.8), which we denote by Ω_r. The area enclosed by γ_r is, of course, πr^2. The area of Ω_r is a continuous increasing function of r, which we denote $a(r)$. Two important observations about $a(r)$ are that

$$\lambda_- := \lim_{r \searrow 1} a(r) = \pi \quad \text{and} \quad \lambda_+ := \lim_{r \nearrow \mu} a(r) = \pi v^2,$$

since λ_- and λ_+ are simply the areas enclosed by the inner and outer boundary circles of $A(1, v)$, respectively.

Now we claim that

$$a(r) \geq \pi r^2 \quad \text{for all } 1 < r < \mu. \tag{3.24}$$

This would imply, by taking the limit as $r \nearrow \mu$, that $\pi v^2 = \lambda_+ \geq \pi \mu^2$, so we would get that $v \geq \mu$. Reversing the roles of the two annuli would imply the reverse inequality $v \leq \mu$, and we would get that $\mu = v$, which is the claim we wanted, and the proof would be done.

To prove (3.24), we note that $a(r)$ can be evaluated as a contour integral using a complex-analytic version of Green's theorem from calculus. Specifically, appealing to the result of Exercise 3.7, we see that

$$a(r) = \frac{1}{2i} \oint_{\Gamma_r} \overline{z}\, dz = \frac{1}{2i} \int_0^{2\pi} \overline{f(re^{it})} \frac{d}{dt}(f(re^{it}))\, dt = \frac{r}{2} \int_0^{2\pi} \overline{f(re^{it})} f'(re^{it}) e^{it}\, dt. \tag{3.25}$$

Now let

$$f(z) = \sum_{n=-\infty}^{\infty} c_n z^n \tag{3.26}$$

be the Laurent expansion of f, which converges uniformly on compacts in the annulus $1 < |z| < \mu$ where f is holomorphic (see Theorem 1.65). Substituting (3.26) into (3.25), we get that

$$a(r) = \frac{1}{2} \int_0^{2\pi} \left(\sum_n \overline{c_n} r^n e^{-int} \right) \left(\sum_m mc_m r^m e^{i(m-1)t} \right) re^{it} \, dt$$

$$= \frac{1}{2} \sum_{n,m} mc_m \overline{c_n} r^{n+m} \int_0^{2\pi} e^{i(m-n)t} \, dt = \pi \sum_{n=-\infty}^{\infty} n|c_n|^2 r^{2n}.$$

Taking the limit as $r \searrow 1$ gives that

$$\sum_{n=-\infty}^{\infty} n|c_n|^2 = 1.$$

Now it follows that

$$a(r) - \pi r^2 = \pi \sum_{n=-\infty}^{\infty} n|c_n|^2 r^{2n} - \pi \sum_{n=-\infty}^{\infty} n|c_n|^2 = \pi \sum_{n=-\infty}^{\infty} n|c_n|^2 (r^{2n} - 1).$$

Since each summand in this last expression is nonnegative, we have that $a(r) - \pi r^2 \geq 0$, as claimed. □

Having classified the annuli up to conformal equivalence, we state without proof an additional result that explains why the family of annuli plays a role in the theory of conformal mapping of doubly connected regions that parallels the role of the unit disc in the case of simply connected regions. For the proof, see [2, 6].

Theorem 3.25 (Conformal classification of doubly connected regions). *The annuli $A(1, \rho)$, $\rho > 1$, form a complete set of conformal equivalence representatives for doubly connected complex regions. That is, if $\Omega \subset \mathbb{C}$ is a doubly connected region, then Ω is conformally equivalent to $A(1, \Lambda)$ for precisely one value of $\Lambda > 1$.*

The number $m_\Omega = \frac{1}{2\pi} \log(\Lambda)$, where Λ is the outer radius of the annulus to which Ω maps, is called the **conformal modulus of** Ω. Theorem 3.24 guarantees that if such a number exists, then it is unique, and the much stronger Theorem 3.25 guarantees that it exists. Thus m_Ω is an important example of what is known as a **conformal invariant**. Much more can be said about m_Ω, including a more direct way to define it that is intrinsic to Ω and does not rely on the idea of conformally mapping Ω to an annulus; consult the references mentioned above for details.

The final component in the discussion of conformal equivalence classes of doubly connected regions is the identification of the conformal automorphisms of such a region.

Theorem 3.26 (Conformal automorphisms of an annulus). *The conformal automorphism group of the annulus $A(r_1, r_2)$ is*

$$\text{Aut}(A(r_1, r_2)) = \{z \mapsto e^{i\theta} z \ : \ 0 \leq \theta < 2\pi\} \cup \left\{z \mapsto e^{i\theta} \frac{r_1 r_2}{z} \ : \ 0 \leq \theta < 2\pi\right\}.$$

That is, the automorphisms consist of the rotations $z \mapsto e^{i\theta}z$, together with the compositions of the inversion map $z \mapsto \frac{r_1 r_2}{z}$ with a rotation.

Proof. Exercise 3.9. □

Suggested exercises for Section 3.10. 3.7, 3.8, 3.9.

Exercises for Chapter 3

3.1   If Ω and Ω' are conformally equivalent with a conformal map $g : \Omega \to \Omega'$, then describe an explicit group isomorphism between $\mathrm{Aut}(\Omega)$ and $\mathrm{Aut}(\Omega')$.

3.2   Let $z_1, z_2, z_3, w_1, w_2, w_3$ be elements of $\widehat{\mathbb{C}}$. Prove that there is a unique Möbius transformation mapping z_j to w_j for $j = 1, 2, 3$.

3.3   Prove that besides the singleton conformal equivalence classes $\{\mathbb{C}\}$ and $\{\widehat{\mathbb{C}}\}$ described above, any other conformal equivalence class $\mathcal{K}$ is infinite and in fact contains an infinity of regions any two of which are not images of each other under an affine transformation $z \mapsto az + b$.

3.4   Prove Theorem 3.11.

3.5   Show that the assumption of holomorphicity in Montel's theorem (Theorem 3.16) cannot be removed; that is, the result properly belongs in complex analysis and does not have a real analysis analogue (at least not an obvious one).

3.6   Show that the real analysis analogue of Hurwitz's theorem is not true.

3.7   The complex-analytic version of Green's formula from multivariate calculus states that if γ is a simple closed contour in the plane, then the area A enclosed inside γ is given by

$$A = \frac{1}{2i} \oint_\gamma \bar{z}\, dz.$$

   Show that this follows from the usual Green's theorem in real-variable calculus.

3.8   Prove that the statement of Theorem 3.24 is also correct under the relaxed assumption $0 \le r_1 < r_2$ and $0 \le \rho_1 < \rho_2$, which addresses also the case of "degenerate" annuli with an inner radius of 0 (that is, punctured discs).

3.9   Prove Theorem 3.26.

4 Elliptic functions

The theory of elliptic functions is the fairyland of mathematics. The mathematician who once gazes upon this enchanting and wondrous domain crowded with the most beautiful relations and concepts is forever captivated.

Richard Bellman, "A Brief Introduction to Theta Functions" (1961)

4.1 Motivation: elliptic curves

Elliptic curves are fascinating objects studied in complex analysis, algebraic geometry, number theory, cryptography, and other areas of mathematics. An elliptic curve $\mathcal{E}$ is the set of solutions to an algebraic equation of the form

$$\mathcal{E}: \quad y^2 = ax^3 + bx^2 + cx + d \tag{4.1}$$

relating a cubic in x to a quadratic function of y, where the coefficients (and solutions) are assumed to be elements of some field $\mathbb{F}$, such as the rationals, reals, complex numbers, or a finite field. It is often helpful to assume further that the curve is **nondegenerate**, that is, that the cubic polynomial on the right-hand side of (4.1) has no multiple roots (see Section 4.11 for a related discussion).

To study elliptic curves, it is helpful to first bring equation (4.1) to a simpler canonical form, usually written as

$$\mathcal{E}: \quad y^2 = 4x^3 - g_2 x - g_3 \tag{4.2}$$

through a standard change of variables; I skip the details of such a reduction. From here on, we will take (4.2) as the definition of an elliptic curve.

A beautiful and surprising fact about elliptic curves that holds the key to many of their amazing properties is that they form an abelian group in a natural way. The group operation, denoted as a kind of "addition" operation $P \oplus Q$ for two points $P = (x_1, y_1)$ and $Q = (x_2, y_2)$ on the curve, can be defined algebraically using a messy and strange formula that you would never think to guess directly. However, the formula has a simple geometric interpretation, which is very easy to explain: the idea is that to compute $P \oplus Q$, you find the intersection point $R = (x_3, y_3)$ of the line passing through P and Q with the curve (other than the points P and Q themselves) and then reflect R in the y-coordinate to define $P \oplus Q = (x_3, -y_3)$; see Fig. 4.1. The fact that this construction is well-defined is tied to the subtle fact that a generic straight line intersects the curve at precisely three points. (I use the word "generic" because there are also technicalities involving degenerate cases where the line is tangent to the curve, which means that we have to be careful in interpreting this definition for a "doubling" operation $P + P$, or where one of the three intersection points is not actually there, in which case we add

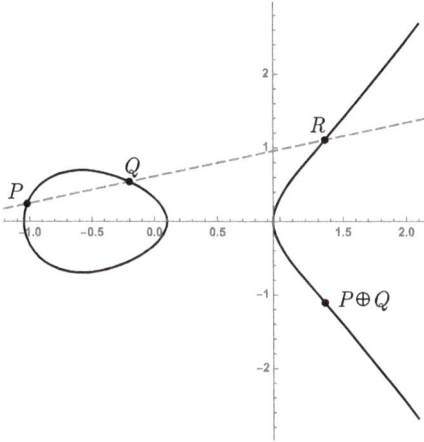

Figure 4.1: An elliptic curve and the group addition law, visualized here for the curve $y^2 = x^3 - x + \frac{1}{10}$ over the real numbers.

an additional "point at infinity" to serve in its place. I ignore such technical issues in the current informal discussion.)

Taking the above geometric construction, we can work out by an explicit calculation that the algebraic expression for the coordinates of the result $P \oplus Q = (x_3, -y_3)$ of the group addition of P and Q described above in geometric terms—again, in the generic situation—are given by the supremely unintuitive formulas

$$x_3 = \frac{1}{4}\left(\frac{y_1 - y_2}{x_1 - x_2}\right)^2 - x_1 - x_2, \tag{4.3}$$

$$-y_3 = -\frac{1}{4}\left(\frac{y_1 - y_2}{x_1 - x_2}\right)^3$$
$$+ \frac{(x_1^3 y_1 - x_2^3 y_2) - 2(x_1^3 y_2 - x_2^3 y_1) + 3x_1 x_2(x_1 y_2 - x_2 y_1)}{(x_1 - x_2)^3}. \tag{4.4}$$

It is far from clear why these formulas should define an associative operation, let alone a group law (at least the fact that the operation is commutative is easy to see). Even for the geometric construction, associativity requires some effort to explain (see [62, Ch. 1]).

All of this raises many intriguing questions about elliptic curves in the specific context of curves defined over the complex numbers:

1. Where does the group structure of elliptic curves "really" come from? That is, is there a conceptual way of thinking about them that makes it easy to see that such a group addition law should exist and that makes it possible to avoid the need for a cumbersome calculation to verify that (4.3)–(4.4) define a valid group operation?
2. What does an elliptic curve look like topologically?

3. Can we classify all elliptic curves up to conformal equivalence as Riemann surfaces? That is, how do we determine when two elliptic curves are conformally equivalent, and how do we parameterize the conformal equivalence classes of elliptic curves?
4. What additional roles exist for elliptic curves within complex analysis? What other topics or problems do they relate to?

It turns out that all these questions and more can be answered by studying a certain family of meromorphic functions in the complex plane, called **elliptic functions** or **doubly periodic functions**. In fact, all members of the family can be obtained from a single function, the so-called **Weierstrass $\wp$-function**, denoted $\wp(z)$, along with its derivative $\wp'(z)$; and the map $z \mapsto (\wp(z), \wp'(z))$ gives a convenient parameterization of the elliptic curve $\mathcal{E}$, which does much to explain what the elliptic curve and its group law "really" look like.

The situation is analogous to what happens in the case of a much simpler group arising from an algebraic equation, the circle group

$$S^1 = \{(x, y) \in \mathbb{R}^2 : x^2 + y^2 = 1\}.$$

There too we have an abelian group "addition" law $\boxplus$ given by

$$(x_1, y_1) \boxplus (x_2, y_2) = (x_1 x_2 - y_1 y_2, x_1 y_2 + x_2 y_1).$$

Although this formula can be easily verified to satisfy the properties of a commutative group operation through a purely formal calculation, to the uninitiated encountering it for the first time, the reason why such a group law exists may appear mysterious. Fortunately, there exists a "circular function" $C : \mathbb{R} \to \mathbb{R}$ that has the following properties:
1. The map $\varphi(t) = (C(t), C'(t))$ maps a real number to an element of S^1.
2. $\varphi(t + s) = \varphi(t) \boxplus \varphi(s)$ (that is, φ is a group homomorphism from $(\mathbb{R}, +)$ to $(S^1, \boxplus)$).
3. $\varphi(t + 2\pi) = \varphi(t)$, that is, φ is periodic with period 2π; equivalently, its kernel as a group homomorphism is the additive subgroup $2\pi\mathbb{Z}$ of $\mathbb{R}$.

These properties taken together imply that φ induces (by the first isomorphism theorem) a group isomorphism between the quotient group $\mathbb{R}/(2\pi\mathbb{Z})$ with "ordinary" addition of real numbers (which in the quotient group becomes "addition modulo 2π") on the one hand, and S^1 with the "exotic" addition law $\boxplus$ on the other hand. That is, the circular function $C(t)$ and the map φ derived from it "linearize" the group operation and make it apparent that the circle group is topologically a real interval with its two ends glued together (that is, a circle), with the group operation being addition modulo 2π. Of course, you may have realized by now that the "circular function" is nothing more than the familiar cosine function $C(t) = \cos t$. So in this point of view the cosine function and its derivative can be thought of as gadgets that help us understand the algebraic and topological structure of the circle group by parameterizing it in terms of a group that is easier to understand. As we will see, the situation with elliptic curves and the use of the elliptic functions $\wp(z)$ and $\wp'(z)$ to parameterize them is quite similar. Also, as happens with the case of the trigonometric functions, the functions we construct out of this group-theoretic motivation will end up being useful for many other things.

We now proceed to make precise these somewhat vague notions in a way that gives substance to the analogy described above. This will lead us to many new and beautiful ideas that will take us far beyond the familiar realm of trigonometric functions.

4.2 Doubly periodic functions

The cosine and sine functions in the example discussed above are periodic functions of a single real variable. We now double the dimensions and look for a meromorphic function of a *complex* variable that is "periodic" in two different directions in the plane. Such a function is called a **doubly periodic function** or an **elliptic function**. Formally, we say that $\omega \in \mathbb{C}$ is a **period** of a meromorphic function $f : \mathbb{C} \rightarrow \mathbb{C}$ if $f(z + \omega) = f(z)$ for all $z \in \mathbb{C}$. The set of periods of $f(z)$ is denoted Λ_f and is easily seen to be an additive subgroup of $\mathbb{C}$. We say that a meromorphic function f is **doubly periodic** if Λ_f contains two nonzero elements ω_1, ω_2 that are linearly independent when considered as elements of a vector space over the real numbers (this is equivalent to saying that the complex number ω_2/ω_1 is nonreal). Trivially, if f, g are doubly periodic with the same linearly independent periods ω_1, ω_2, then so are $f + g$, fg, $\frac{1}{f}$, and the derivative f'.

Note that the constant functions have every complex number as a period. This illustrates the fact that the pair ω_1, ω_2 of complex numbers attesting to the doubly periodic nature of a function f is not unique. To understand the less trivial scenario of a function f that is doubly periodic but not constant, observe that in that case Λ_f must be a topologically discrete additive subgroup of $\mathbb{C}$, for otherwise f can be seen to be constant by the uniqueness theorem for holomorphic functions (Corollary 1.36 on p. 42), since it takes the same value on a set of points with an accumulation point. It then follows (see Exercise 4.1) that Λ_f must be of the form $\omega_1\mathbb{Z} + \omega_2\mathbb{Z}$ with nonzero numbers ω_1, ω_2 that are linearly independent over $\mathbb{R}$; that is, Λ_f is a discrete rank-2 subgroup. A subgroup of $\mathbb{C}$ of this form is called a **lattice**. The subgroup Λ_f of periods of a nonconstant doubly periodic function f is called its **period lattice**.

If f is a nonconstant doubly periodic function with $\Lambda_f = \omega_1\mathbb{Z} + \omega_2\mathbb{Z}$, then we say that ω_1, ω_2 form a **fundamental period pair** for f. Not all pairs of periods are fundamental: for example, if ω_1, ω_2 is a fundamental period pair, then $2\omega_1, 2\omega_2$ is a pair of periods, which, while it attests to f being doubly periodic according to the above definition, is not fundamental since $2\omega_1\mathbb{Z} + 2\omega_2\mathbb{Z}$ is a proper sublattice of Λ_f. On the other hand, a nonconstant doubly periodic function has infinitely many fundamental period pairs, since it is easy to see that the representation $\omega_1\mathbb{Z} + \omega_2\mathbb{Z}$ of a lattice is far from unique; for example, $\omega_1\mathbb{Z} + \omega_2\mathbb{Z} = (\omega_1 + k\omega_2)\mathbb{Z} + \omega_2\mathbb{Z}$ for any $k \in \mathbb{Z}$. A more precise characterization of when two pairs (ω_1, ω_2) and (ω_1', ω_2') generate the same lattice is given in the following lemma.

Lemma 4.1. *Let $L = \omega_1\mathbb{Z} + \omega_2\mathbb{Z}$ and $L' = \omega_1'\mathbb{Z} + \omega_2'\mathbb{Z}$ be lattices. Then $L = L'$ if and only if ω_1' and ω_2' can be represented as*

$$\omega_1' = a\omega_1 + b\omega_2, \tag{4.5}$$
$$\omega_2' = c\omega_1 + d\omega_2, \tag{4.6}$$

where $\left(\begin{smallmatrix} a & b \\ c & d \end{smallmatrix}\right)$ is a 2×2 invertible matrix with integer entries, that is, $a, b, c, d \in \mathbb{Z}$, and $ad - bc = \pm 1$.

Proof. Proof of the "if" claim: assume that ω_1' and ω_2' have the form (4.5)–(4.6) with $a, b, c, d \in \mathbb{Z}$, $ad - bc = \pm 1$. Then $\omega_1', \omega_2' \in \omega_1\mathbb{Z} + \omega_2\mathbb{Z}$. This clearly implies that $L' \subseteq L$. For the reverse containment, invert relations (4.5)–(4.6) to see that

$$\omega_1 = \frac{d}{ad-bc}\omega_1' - \frac{b}{ad-bc}\omega_2',$$
$$\omega_2 = -\frac{c}{ad-bc}\omega_1' + \frac{a}{ad-bc}\omega_2',$$

which, because of the assumption that $ad - bc = \pm 1$, is a representation of the form (4.5)–(4.6) with coefficients satisfying the same conditions, but with the roles of the pairs (ω_1, ω_2) and (ω_1', ω_2') reversed. Therefore $L \subseteq L'$, and altogether we have shown that $L = L'$.

Proof of "only if": assume that $L = L'$, that is, $\omega_1\mathbb{Z} + \omega_2\mathbb{Z} = \omega_1'\mathbb{Z} + \omega_2'\mathbb{Z}$. In particular, $\omega_1, \omega_2 \in \omega_1'\mathbb{Z} + \omega_2'\mathbb{Z}$, and $\omega_1', \omega_2' \in \omega_1\mathbb{Z} + \omega_2\mathbb{Z}$. It follows that there exist integers $a, b, c, d, \alpha, \beta, \gamma, \delta$ such that

$$\omega_1' = a\omega_1 + b\omega_2, \quad \omega_1 = \alpha\omega_1' + \beta\omega_2',$$
$$\omega_2' = c\omega_1 + d\omega_2, \quad \omega_2 = \gamma\omega_1' + \delta\omega_2'.$$

Thus we have representation (4.5)–(4.6) with integer coefficients a, b, c, d. Moreover, since the matrices $\left(\begin{smallmatrix} a & b \\ c & d \end{smallmatrix}\right)$ and $\left(\begin{smallmatrix} \alpha & \beta \\ \gamma & \delta \end{smallmatrix}\right)$ are inverse to each other and have integer entries, their determinants are also mutually reciprocal integers, so we must have that $ad - bc = \pm 1$. $\qquad\square$

A doubly periodic function f with a fundamental period pair ω_1, ω_2 is determined uniquely by its values on the parallelogram

$$P_{z_0}(\omega_1, \omega_2) = \{z_0 + t\omega_1 + s\omega_2 \,:\, 0 \le t, s < 1\},$$

where $z_0 \in \mathbb{C}$ is an arbitrary point. This is geometrically obvious, since if we denote by $L = \omega_1\mathbb{Z} + \omega_2\mathbb{Z}$ the period lattice, then $\mathbb{C}$ is tiled perfectly by nonoverlapping L-translates of $P_{z_0}(\omega_1, \omega_2)$ (that is, shifted copies of the form $\omega + P_{z_0}(\omega_1, \omega_2)$ with $\omega \in L$), and the value of $f(z)$ for z in some L-translate $\omega + P_{z_0}(\omega_1, \omega_2)$ reduces by periodicity to the shifted value $f(z - \omega)$, which is in $P_{z_0}(\omega_1, \omega_2)$. We refer to $P_{z_0}(\omega_1, \omega_2)$ as a **fundamental parallelogram**

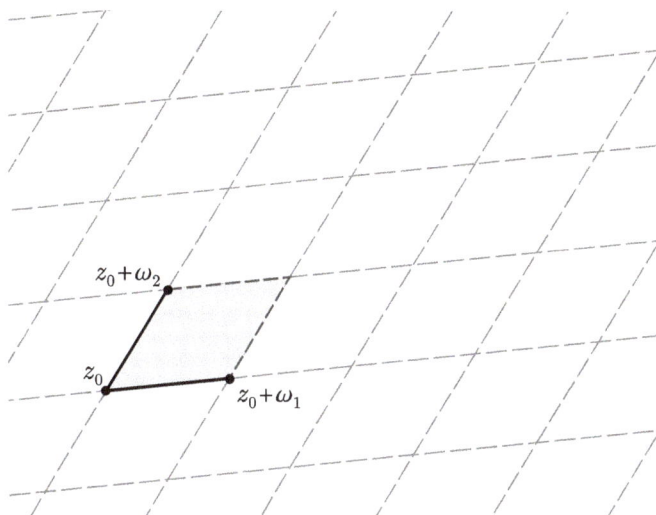

Figure 4.2: A fundamental parallelogram $P_{z_0}(\omega_1, \omega_2)$ and its L-translates.

for f; see Fig. 4.2. Note that the fundamental parallelogram depends on the choice of a fundamental period pair, so the choice of a parallelogram contains some arbitrariness in the same way that the choice of a fundamental period pair is arbitrary. Moreover, the additional (also arbitrary) parameter z_0 allows us to specify the "origin" of the parallelogram; it is convenient to have that extra degree of freedom to avoid slight technical complications in some of the results below.

Suggested exercises for Section 4.2. 4.1.

4.3 Poles and zeros; the order of a doubly periodic function

An obvious goal that we have is to construct some nontrivial doubly periodic functions, assuming that they exist.[1] To motivate our construction and help convince you that it is in a sense the simplest one that has any chance of working, it would be helpful to understand what sorts of constraints exist on doubly periodic functions. The next few results show that there are in fact rather rigid constraints that such functions must satisfy.

---

1 A tip for the reader: when you are reading a mathematical text and read a definition of a new and exotic class of mathematical objects, it is a good habit to always ask yourself right away: does such an object even exist? For, although in the case of a textbook the answer will usually be "yes," when you are reading research papers on topics at the forefront of human knowledge, the answer will occasionally be far from clear even to the writer of the text and may well turn out to be "no." Even for textbook-level mathematics, asking this question and spending a few minutes trying to answer it by yourself will often provide you with insight far beyond what a purely passive reading of the text can offer.

Proposition 4.2. *There are no entire doubly periodic functions other than the constant functions.*

Proof. If f is entire and doubly periodic, then in particular f is bounded on the parallelogram $\{t\omega_1 + s\omega_2 : t, s \in [0,1]\}$, which is a compact set. By periodicity, $f(z)$ is also bounded on all of $\mathbb{C}$ and is therefore constant by Liouville's theorem. $\square$

We see from Proposition 4.2 that a nonconstant doubly periodic function f must have poles; by applying the same result to $1/f$ we see that f must also have zeros. Note that since the sets of zeros and poles of a holomorphic function are discrete, f can have at most finitely many zeros and poles in any fundamental parallelogram. To avoid certain technical issues, it is helpful to choose the "origin point" z_0 for the fundamental parallelogram $P_{z_0}(\omega_1, \omega_2)$ in such a way that f does not have poles or zeros on the boundary of the parallelogram. We call a fundamental parallelogram with such a property **generic** (for the doubly periodic function f). It is easy to see that a generic fundamental parallelogram exists.

Proposition 4.3. *Let f be a doubly periodic function with fundamental period pair ω_1, ω_2. Let $P_{z_0}(\omega_1, \omega_2)$ be a generic fundamental parallelogram for f. Then*

$$\oint_{\partial P_{z_0}(\omega_1,\omega_2)} f(z)\, dz = 0, \tag{4.7}$$

where we consider the boundary $\partial P_{z_0}(\omega_1, \omega_2)$ as an integration contour oriented in the usual way in the positive mathematical direction.

Proof. Decompose the contour $\Gamma = \partial P_{z_0}(\omega_1, \omega_2)$ as the concatenation

$$\Gamma = \gamma_1 + \gamma_2 + \gamma_3 + \gamma_4$$

of four contours $\gamma_1, \gamma_2, \gamma_3, \gamma_4$ corresponding to the edges of the parallelogram, where γ_1 is the directed line segment from z_0 to $z_0 + \omega_1$, γ_2 is the directed line segment from $z_0 + \omega_1$ to $z_0 + \omega_1 + \omega_2$; γ_3 is the directed line segment from $z_0 + \omega_1 + \omega_2$ to $z_0 + \omega_2$, and γ_4 is the directed line segment from $z_0 + \omega_2$ to z_0. By the doubly periodic property of f we have

$$\int_{\gamma_1} f(z)\, dz = -\int_{\gamma_3} f(w)\, dw,$$

since the change of variables $w = z + \omega_2$ maps the integral on the left to the one on the right (including the minus sign). Thus, in the contour integral on Γ, the contributions from the integral over the two segments γ_1 and γ_3 cancel each other out. Similarly, by the change of variables $w = z + \omega_1$ we get a cancelation of the second and fourth segments:

$$\int_{\gamma_2} f(z)\, dz = -\int_{\gamma_4} f(w)\, dw,$$

so that in total we have

$$\oint_{\Gamma} f(z)\,dz = \int_{\gamma_1} f(z)\,dz + \int_{\gamma_2} f(z)\,dz + \int_{\gamma_3} f(z)\,dz + \int_{\gamma_4} f(z)\,dz = 0,$$

as claimed. □

Corollary 4.4. *Under the assumptions of Proposition 4.3, the sum of the residues of f over the poles of f in the fundamental parallelogram $P_{z_0}(\omega_1, \omega_2)$ is zero.*

Proof. By the residue theorem the integral on the left-hand side of (4.7) is equal to $2\pi i$ times the sum of the residues. □

Corollary 4.5. *A nonconstant doubly periodic function with a generic fundamental parallelogram $P_{z_0}(\omega_1, \omega_2)$ must have at least two poles, counting multiplicities, inside the parallelogram.*

Proposition 4.6. *Let $g : \mathbb{C} \to \mathbb{C}$ be a doubly periodic function with fundamental period pair ω_1, ω_2 and a generic fundamental parallelogram $P = P_{z_0}(\omega_1, \omega_2)$. The sum of the orders of the zeros of $g(z)$ inside P is equal to the sum of the orders of the poles of $g(z)$ in the parallelogram, counting with multiplicities.*

Proof. Apply Proposition 4.3 to $f(z) = \frac{g'(z)}{g(z)}$, and note that by the argument principle (Theorem 1.48) the resulting integral is $2\pi i$ times the number of zeros minus the number of poles of f in the interior of P. □

The last result enables us to define an important integer parameter associated with a doubly periodic function, called its order. This is made precise in the next result, which follows immediately from Proposition 4.6.

Corollary 4.7. *Let f be a nonconstant doubly periodic function. There exists a unique integer $m \geq 2$, called the **order** of f, with the following properties:*
1. *f has exactly m poles, counting with multiplicities, in any generic fundamental parallelogram $P_{z_0}(\omega_1, \omega_2)$.*
2. *For any $\alpha \in \mathbb{C}$, $f(z)$ assumes the value α exactly m times (that is, the function $z \mapsto f(z) - \alpha$ has m zeros), counting with multiplicities, in any fundamental parallelogram $P_{z_0}(\omega_1, \omega_2)$ that is generic for the doubly periodic function $f(z) - \alpha$.*

Proposition 4.8. *Let $g : \mathbb{C} \to \mathbb{C}$ be a nonconstant doubly periodic function with fundamental period pair ω_1, ω_2. Let $P = P_{z_0}(\omega_1, \omega_2)$ be a generic fundamental parallelogram for g. Denote by $z_1, \ldots, z_n$ the zeros of $g(z)$ in P, counting multiplicities, and let $w_1, \ldots, w_m$ be the poles of $g(z)$ in P, counting multiplicities. Then the number*

$$\sum_{j=1}^{n} z_j - \sum_{k=1}^{m} w_k \tag{4.8}$$

is a period of f.

Proof. Similarly to the proof of Proposition 4.3, we consider the contour integral

$$\oint_{\partial P} \frac{zg'(z)}{g(z)} \, dz,$$

which by the residue theorem is evaluated as $2\pi i(\sum_{j=1}^{n} z_j - \sum_{k=1}^{m} w_k)$. We use the same decomposition of the contour ∂P into four subcontours γ_j, $1 \le j \le 4$, as in the proof of Proposition 4.3. Note that by the periodicity of g the images of each of the subcontours γ_1 and γ_2 under $g(z)$ (denoted $g \circ \gamma_1$ and $g \circ \gamma_2$, respectively) are closed curves. Therefore we can use the same changes of variable as in the proof of Proposition 4.3 to write

$$\int_{\gamma_1} \frac{zg'(z)}{g(z)} \, dz + \int_{\gamma_3} \frac{wg'(w)}{g(w)} \, dw$$

$$= \int_{\gamma_1} \frac{zg'(z)}{g(z)} \, dz - \int_{\gamma_1} \frac{(z+\omega_2)g'(z+\omega_2)}{g(z+\omega_2)} \, dz$$

$$= -\omega_2 \int_{\gamma_1} \frac{g'(z)}{g(z)} \, dz = -\omega_2 \oint_{g \circ \gamma_1} \frac{d\xi}{\xi} = -\omega_2 \cdot 2\pi i m$$

for some integer m equal to the winding number (see Section 1.13) of the closed curve $g \circ \gamma_1$ around 0. (Note that $g \circ \gamma_1$ does not cross 0 because we chose P to be a generic parallelogram for g.) By similar reasoning,

$$\oint_{\gamma_2} \frac{zg'(z)}{g(z)} \, dz + \oint_{\gamma_4} \frac{wg'(w)}{g(w)} \, dw = \omega_1 \cdot 2\pi i n$$

with $n \in \mathbb{Z}$. Combining these results gives that the quantity in (4.8) is of the form $-m\omega_2 + n\omega_1$ for integer m, n and hence is a period. $\qquad\square$

4.4 Construction of the Weierstrass $\wp$-function

We are now ready to construct our first doubly periodic function, the **Weierstrass $\wp$-function** mentioned at the beginning of the chapter, which occupies a central place in the theory of elliptic functions. The construction is motivated by the following general principle that we see in many areas of mathematics: to construct an object with certain symmetry, it is often helpful to start with a nonsymmetric object and then symmetrize it by summing over its orbit under the action of the desired symmetry group. Our construction follows this template, although in practice we will need to deviate from it in a small way. In our situation the symmetry group is the group of translations $z \mapsto z + \omega$ where ω is a period, so this will involve an infinite summation over the elements of the period lattice L, which leads to slightly delicate issues of convergence. The next lemma clarifies what kind of summations are well-behaved enough to be useful.

The symbol $\wp$

The mathematical symbol $\wp$ (pronounced similarly to the name of the letter "p," or sometimes as "Weierstrass p" depending on the context) used for the Weierstrass elliptic function has an apparently unique status in mathematical notation as a symbol that is reserved for denoting one mathematical object and that object alone. Even the distinguished constants π, e, and i do not enjoy such an exclusivity! The symbol $\wp$ has its own code point in the Unicode string encoding system (U+2118) and its own escape string in the HTML standard (℘). It seems rather generous of the developers of these computing standards to go to such lengths to please the fairly small group of mathematicians who use elliptic functions in their work.

You may wonder how this quirky state of affairs came to be. It appears to have been little more than a historical accident. Both the function $\wp(z)$ and the notation for it were introduced by Weierstrass, who for this purpose used a stylized handwritten lowercase p bearing some resemblance to the Sütterlin alphabet used in handwritten German during that period in large parts of Prussia. Later authors ended up adopting not only Weierstrass's choice of the letter but also his particular stylization of it, and thus a new symbol was born. For more details, refer to the online discussion [W18].

Figure 4.3: Weierstrass' legacy in mathematical typography.

Lemma 4.9. *Let $L \subset \mathbb{C}$ be a lattice, and let $\beta > 0$. The infinite sum*

$$\sum_{\substack{\omega \in L \\ \omega \neq 0}} \frac{1}{|\omega|^\beta} \qquad (4.9)$$

converges if and only if $\beta > 2$.

Proof. Exercise 4.2. □

Theorem 4.10 (The Weierstrass $\wp$-function). *Fix a lattice $L \subset \mathbb{C}$. There exists a unique meromorphic function, called the Weierstrass $\wp$-function and denoted $\wp(z)$, with the following properties:*

1. $\wp(z)$ *is a doubly periodic function of order 2 with period lattice L.*
2. $\wp(z)$ *has a pole of order 2 at every period $\omega \in L$, with Laurent expansion around the pole beginning with*

$$\wp(z) = \frac{1}{(z-\omega)^2} + O(z-\omega) \quad (z \to \omega), \qquad (4.10)$$

   and no other poles.
3. $\wp(z)$ *is an even function.*

Moreover, the uniqueness already holds for a function satisfying the first two properties without assuming the even symmetry of $\wp(z)$.

Proof. Proof of uniqueness: if $\wp^{(1)}(z)$ and $\wp^{(2)}(z)$ are two meromorphic functions satisfying properties 1–2, then the function $f(z) = \wp^{(1)}(z) - \wp^{(2)}(z)$ is doubly periodic and has no poles. By Lemma 4.5 it must be a constant. However, its Laurent expansion around $z = 0$ has the constant term 0 by (4.10), so in fact $f(z) \equiv 0$ and $\wp^{(1)}(z) \equiv \wp^{(2)}(z)$.

Proof of existence: we define $\wp(z)$ as

$$\wp(z) = \frac{1}{z^2} + \sum_{\substack{\omega \in L \\ \omega \neq 0}} \left(\frac{1}{(z - \omega)^2} - \frac{1}{\omega^2} \right). \tag{4.11}$$

This is a doubly infinite sum that can be written more explicitly in terms of a fundamental pair of periods ω_1, ω_2 as

$$\wp(z) = \frac{1}{z^2} + \sum_{\substack{(m,n) \in \mathbb{Z}^2 \\ (m,n) \neq (0,0)}} \left(\frac{1}{(z - m\omega_1 - n\omega_2)^2} - \frac{1}{(m\omega_1 + n\omega_2)^2} \right).$$

We claim that for any compact $K \subset \mathbb{C}$, the series obtained from (4.11) by removing (if necessary) finitely many terms that have poles in K converges absolutely uniformly on K. This would show that (4.11) defines a meromorphic function on $\mathbb{C}$ with poles only at the points of L where individual summands of the series have poles. To prove the claim, fix a compact $K \subset \mathbb{C}$. For $z \in K$ and $\omega \in L \setminus K$, making the further assumption that $|\omega| > 2|z|$ (which applies to all but finitely many terms in the series), we have

$$\left| \frac{1}{(z - \omega)^2} - \frac{1}{\omega^2} \right| = \left| \frac{\omega^2 - (z - \omega)^2}{\omega^2 (z - \omega)^2} \right| = \left| \frac{2z\omega - z^2}{\omega^2 (z - \omega)^2} \right|$$

$$\leq \frac{2|z|}{|\omega|(|\omega| - |z|)^2} + \frac{|z|^2}{|\omega|^2(|\omega| - |z|)^2} \leq \frac{C}{|\omega|^3},$$

where $C > 0$ is a constant that depends only on K. The absolute convergence of the series now follows from Lemma 4.9.

Next, observe that $\wp(z)$ is trivially even, since $\omega \in L$ if and only if $\omega' = -\omega \in L$, so

$$\wp(-z) = \frac{1}{(-z)^2} + \sum_{\substack{\omega \in L \\ \omega \neq 0}} \left(\frac{1}{(-z - \omega)^2} - \frac{1}{\omega^2} \right)$$

$$= \frac{1}{z^2} + \sum_{\substack{\omega' \in L \\ \omega \neq 0}} \left(\frac{1}{(-z + \omega')^2} - \frac{1}{(-\omega')^2} \right)$$

$$= \frac{1}{z^2} + \sum_{\substack{\omega' \in L \\ \omega \neq 0}} \left(\frac{1}{(z - \omega')^2} - \frac{1}{\omega'^2} \right) = \wp(z).$$

Next, to prove that $\wp(z)$ is doubly periodic, differentiate (4.11) termwise to get

$$\wp'(z) = -\frac{2}{z^3} - 2 \sum_{\substack{\omega \in L \\ \omega \neq 0}} \frac{1}{(z - \omega')^3} = -2 \sum_{\omega \in L} \frac{1}{(z - \omega)^3}. \tag{4.12}$$

This infinite series is manifestly doubly periodic, as it is a true symmetrization with respect to the orbit of the L-action as discussed at the beginning of this section. (In fact, the expression $\sum_{\omega \in L}(z - \omega)^{-3}$ is probably the simplest possible formula we can write that defines a nontrivial doubly periodic function, except that the resulting function is of order 3 and thus not the "simplest" in the sense of having the smallest order possible.) Now let $\omega \in L$, and denote $g_\omega(z) := \wp(z + \omega) - \wp(z)$. Since

$$g_\omega'(z) = \wp'(z + \omega) - \wp'(z) \equiv 0,$$

that is, the derivative of g_ω is identically 0, we get that $g_\omega(z)$ is a constant. Taking $z = -\omega/2$ gives $g_\omega(z) = \wp(\omega/2) - \wp(-\omega/2) = 0$ since $\wp(z)$ is even. Thus $g_\omega(z) \equiv 0$ and $\wp(z+\omega) = \wp(z)$ for all z, which shows that $\wp(z)$ is doubly periodic.

Finally, note that $\wp(z)$ has a pole of order 2 at $z = 0$ with principal part $\frac{1}{z^2}$. After subtracting that principal part, we are left with

$$\wp(z) - \frac{1}{z^2} = \sum_{\substack{\omega \in L \\ \omega \neq 0}} \left(\frac{1}{(z - \omega)^2} - \frac{1}{\omega^2} \right),$$

which is holomorphic in the neighborhood of 0, with the constant term in its Taylor expansion obtained by setting $z = 0$ in this expression, which gives

$$\sum_{\substack{\omega \in L \\ \omega \neq 0}} \left(\frac{1}{(0 - \omega)^2} - \frac{1}{\omega^2} \right) = 0.$$

This proves the Laurent expansion (4.10) for the case $z = 0$, and the expansion around a general period $\omega \in L$ follows by periodicity. □

Note that the construction of the function $\wp(z)$ depends on the choice of lattice L. For the time being, we regard the lattice as fixed, but later on, we will start caring more about this dependence, and it will be helpful to have a notation that emphasizes it. To that end, two common ways to denote the function $\wp(z)$ associated with a specific lattice L are as $\wp_L(z)$ or as $\wp(z; L)$. At some point in the discussion, we will also replace L with a complex variable τ, called the **modular variable**, which parameterizes the space of lattices in a convenient way (see Section 4.14). In that context the notation $\wp(z; \tau)$ is used to denote the Weierstrass $\wp$-function including its dependence on both complex variables z and τ.

Suggested exercises for Section 4.4. 4.2.

4.5 Eisenstein series and the Laurent expansion of $\wp(z)$

Let $L \subset \mathbb{C}$ be a lattice. Define the quantities $G_n, n \geq 3$, associated with L by

$$G_n = \sum_{\omega \in L \setminus 0} \frac{1}{\omega^n}. \tag{4.13}$$

The G_n are known as the **Eisenstein series**. As with the remark above about $\wp(z)$, the value of G_n depends on the lattice L, and when we wants to emphasize that, the notation $G_n(L)$ can be used, or $G_n(\tau)$ once we switch to the point of view involving the modular variable τ. Note that $G_{2k-1} = 0$ for $k \geq 2$ because of each term associated with $\omega \in L$ canceling out the term associated with $-\omega$. Thus the interesting Eisenstein series are the even-indexed ones $G_4, G_6, G_8, \dots$. As the next result shows, these series are closely related to the Weierstrass $\wp$-function.

Theorem 4.11. *The Laurent expansion of $\wp(z)$ around $z = 0$ is given by*

$$\wp(z) = \frac{1}{z^2} + \sum_{n=1}^{\infty} (2n+1) G_{2n+2} z^{2n} = \frac{1}{z^2} + 3G_4 z^2 + 5G_6 z^4 + 7G_8 z^6 + \cdots. \tag{4.14}$$

Proof. Keeping in mind the standard Taylor expansion

$$\frac{1}{(1-x)^2} = 1 + 2x + 3x^2 + 4x^3 + \cdots,$$

we write

$$\wp(z) = \frac{1}{z^2} + \sum_{\substack{\omega \in L \\ \omega \neq 0}} \left(\frac{1}{(z-\omega)^2} - \frac{1}{\omega^2} \right) = \frac{1}{z^2} + \sum_{\substack{\omega \in L \\ \omega \neq 0}} \left(\frac{1}{\omega^2(1-(\frac{z}{\omega}))^2} - \frac{1}{\omega^2} \right)$$

$$= \frac{1}{z^2} + \sum_{\substack{\omega \in L \\ \omega \neq 0}} \frac{1}{\omega^2} \left(2\left(\frac{z}{\omega}\right) + 3\left(\frac{z}{\omega}\right)^2 + 4\left(\frac{z}{\omega}\right)^3 + 5\left(\frac{z}{\omega}\right)^4 + \cdots \right)$$

$$= \frac{1}{z^2} + 2\left(\sum_{\omega \in L \setminus 0} \frac{1}{\omega^3} \right) z + 3\left(\sum_{\omega \in L \setminus 0} \frac{1}{\omega^4} \right) z^2 + 4\left(\sum_{\omega \in L \setminus 0} \frac{1}{\omega^5} \right) z^3 + \cdots$$

$$= \frac{1}{z^2} + 2G_3 z + 3G_4 z^2 + 4G_5 z^3 + 5G_6 z^4 + \cdots$$

$$= \frac{1}{z^2} + 3G_4 z^2 + 5G_6 z^4 + 7G_8 z^6 + \cdots,$$

as claimed. Note that this calculation technically involved a rearrangement of terms in a double summation (the summation over $\omega \in L$ and the summation over the powers of z/ω in each of the hypergeometric series $1/(1 - (z/\omega))^2$ being expanded), which needs to be justified. This is easy to do and addressed in Exercise 4.3. $\square$

Suggested exercises for Section 4.5. 4.3.

4.6 The differential equation satisfied by $\wp(z)$

The first two Eisenstein series G_4 and G_6 play a special role in the theory of the Weierstrass $\wp$-function and of elliptic curves. It is traditional to define rescaled versions of them, labeled g_2 and g_3, by

$$g_2 = 60G_4, \quad g_3 = 140G_6. \tag{4.15}$$

The quantities g_2 and g_3 are known as the **elliptic invariants**. The role they play is hinted at by the following result (compare to (4.2)).

Theorem 4.12. *The function $\wp(z)$ satisfies the nonlinear differential equation*

$$\wp'(z)^2 = 4\wp(z)^3 - g_2\wp(z) - g_3. \tag{4.16}$$

Proof. The idea is to consider the behavior of each term in (4.16) near $z = 0$. Using (4.14), we have

$$\wp(z) = \frac{1}{z^2} + 3G_4 z^2 + 5G_6 z^4 + O(z^6),$$

$$\wp'(z) = -\frac{2}{z^3} + 6G_4 z + 20G_6 z^3 + O(z^5),$$

$$\wp'(z)^2 = \frac{4}{z^6} - \frac{24G_4}{z^2} - 80G_6 + O(z^2),$$

$$\wp(z)^3 = \frac{1}{z^6} + \frac{9G_4}{z^2} + 15G_6 + O(z^2).$$

We see that by taking an appropriate combination of $\wp'(z)^2, \wp(z)$, and $\wp(z)^3$ we can cancel the pole at $z = 0$ (and hence all the poles throughout the complex plane, since all of the functions involved are doubly periodic with poles only at periods). Specifically, we have the Taylor expansion

$$\wp'(z)^2 - 4\wp(z)^3 + 60G_4\wp(z) = -140G_6 + O(z) \tag{4.17}$$

around $z = 0$. This is a doubly periodic function without poles and therefore a constant by Proposition 4.2. The value of the constant must be equal to the constant coefficient on the right-hand side of (4.17), namely $-140G_6 = -g_3$. Thus the relation $\wp'(z)^2 - 4\wp(z)^3 + g_2\wp(z) = -g_3$ holds as an identity of meromorphic functions, proving (4.16). □

Corollary 4.13. *The function $\wp(z)$ also satisfies the second-order differential equation*

$$\wp''(z) = 6\wp(z)^2 - \frac{1}{2}g_2. \tag{4.18}$$

Proof. This follows immediately from (4.16) by differentiating both sides and dividing by $2\wp'(z)$. □

4.7 A recurrence relation for the Eisenstein series

Starting from the differential equations (4.16) or (4.18) and comparing Taylor coefficients on both sides, we get interesting identities relating the different Eisenstein series. For example, the coefficient of z^2 on the left-hand side of (4.16) is

$$-2 \cdot 2 \cdot 42 G_8 + 36 G_4^2 = -168 G_8 + 36 G_4^2,$$

whereas the coefficient of z^2 of the expression on the right-hand side of that equation is

$$4 \cdot 3 \cdot 7 G_8 + 4 \cdot 3 \cdot 3 \cdot 3 G_4^2 - 60 \cdot 3 G_4^2 = 84 G_8 - 72 G_4^2.$$

Equating the two and simplifying give the identity

$$G_8 = \frac{3}{7} G_4^2. \tag{4.19}$$

Similarly, inspecting the coefficients of z^4 and z^6 on both sides of (4.16) gives two additional identities of this type, namely

$$G_{10} = \frac{5}{11} G_4 G_6, \tag{4.20}$$

$$G_{12} = \frac{1}{143} \left(42 G_4 G_8 + 25 G_6^2 \right). \tag{4.21}$$

The above idea can be exploited systematically by extracting the coefficient for any power z^{2n}. In the general case, this results in a recurrence relation for the Eisenstein series.

Proposition 4.14. *The Eisenstein series can be computed recursively starting with the two initial values G_4, G_6. Specifically, for any $k \geq 4$, we have the recurrence relation*

$$G_{2k} = \frac{3}{(k-3)(2k-1)(2k+1)} \sum_{j=2}^{k-2} (2j-1)(2k-2j-1) G_{2j} G_{2(k-j)}. \tag{4.22}$$

Proof. Expand both sides of (4.18) as a Laurent series in z using (4.14). For the left-hand side, we have

$$\wp''(z) = \frac{6}{z^4} + 6 G_4 + \sum_{n=1}^{\infty} (2n+1)(2n+2)(2n+3) G_{2n+4} z^{2n}.$$

For the right-hand side,

$$6\wp(z)^2 - \frac{1}{2}g_2 = 6\left(\frac{1}{z^2} + \sum_{k=1}^{\infty}(2k+1)G_{2k+2}z^{2k}\right)^2 - 30G_4$$

$$= 6\sum_{n=1}^{\infty}\left[2(2n+3)G_{2n+4}\right.$$

$$\left. + \sum_{j=1}^{n-1}(2j+1)\big(2(n-j)+1\big)G_{2j+2}G_{2(n-j)+2}\right]z^{2n} + \frac{6}{z^4} + 36G_4 - 30G_4.$$

Equating the coefficients of z^{2n} in these expressions gives (4.22). $\square$

An alternative method of proving (4.22) that does not rely on doubly periodic functions is explored in Exercise 4.6; see also Exercise 4.7 for further applications of this method.

Corollary 4.15. *All the Eisenstein series G_{2k}, $k \geq 2$, can be expressed as polynomials in G_4 and G_6 with rational coefficients (that do not depend on the lattice L they are associated with).*

Suggested exercises for Section 4.7. 4.4, 4.5, 4.6, 4.7.

4.8 Half-periods; factorization of the associated cubic

Let ω_1, ω_2 be a fundamental period pair for our fixed lattice L. Denote by v_1, v_2, v_3 the numbers

$$v_1 = \frac{1}{2}\omega_1, \quad v_2 = \frac{1}{2}\omega_2, \quad v_3 = \frac{1}{2}(\omega_1 + \omega_2), \tag{4.23}$$

which we refer to as the **half-periods** associated with the fundamental period pair ω_1, ω_2.

Lemma 4.16. *The function $\wp'(z)$ is a doubly periodic function of order 3. Its zeros in any fundamental parallelogram $P_{z_0}(\omega_1, \omega_2)$ that is generic for $\wp'(z)$ are the unique three points in the parallelogram that are congruent modulo the lattice L to the half-periods v_1, v_2, v_3, respectively, and they are all simple zeros.*[2]

Proof. We know that $\wp'(z)$ is of order 3 since its poles are the periods, and each one is of order 3 (the principal part is $-2/(z - \omega)^3$; see (4.12)). Thus there are precisely three zeros counting multiplicities in a generic fundamental parallelogram, and if we identify three distinct zeros in such a parallelogram, then they are necessarily all simple. Now recall that $\wp(z)$ is an even function, so $\wp'(z)$ is odd. We also know that the values $\wp'(v_j)$ of $\wp'(z)$ at the half-periods are finite numbers (that is, each v_j is not a pole of $\wp'(z)$), since

---

2 We say that two complex numbers a and b are **congruent modulo** L if $a - b \in L$.

$\wp'(z)$ only has poles at periods. Combining these observations we see that for any of the half-periods v_j,

$$\wp'(v_j) = -\wp'(-v_j) = -\wp'(-v_j + 2v_j) = -\wp'(v_j). \tag{4.24}$$

Thus $\wp'(v_j) = 0$ for $j = 1, 2, 3$, and in any generic fundamental parallelogram $P_{z_0}(\omega_1, \omega_2)$, the three zeros of $\wp'(z)$ will be those three points that are congruent to v_1, v_2, v_3 modulo L. □

The values of $\wp(z)$ at the half-periods are also important. We denote them by e_1, e_2, e_3, that is,

$$e_1 = \wp\left(\frac{1}{2}\omega_1\right), \quad e_2 = \wp\left(\frac{1}{2}\omega_2\right), \quad e_3 = \wp\left(\frac{1}{2}(\omega_1 + \omega_2)\right). \tag{4.25}$$

Lemma 4.17. *The numbers e_1, e_2, e_3 are distinct and are the three roots of the cubic polynomial $4x^3 - g_2 x - g_3$ (where g_2 and g_3 are the elliptic invariants defined in (4.15)), that is, we have the factorization*

$$4x^3 - g_2 x - g_3 = 4(x - e_1)(x - e_2)(x - e_3).$$

Proof. If we denote $h(x) = 4x^3 - g_2 x - g_3$, then, by (4.16),

$$h(e_j) = h(\wp(v_j)) = 4\wp(v_j)^3 - g_2\wp(v_j) - g_3 = \left(\wp'(v_j)\right)^2 = 0.$$

Thus e_1, e_2, e_3 are zeros of $h(x)$. It remains to show that they are distinct. Assume by contradiction that $e_j = e_k$ for some $1 \le j < k \le 3$. This would mean that the function $\wp(z) - e_j$ has a zero of order at least 2 at $z = v_j$ (since $\wp'(v_j) = 0$ by Lemma 4.16) and also a zero of order at least 2 at $z = v_k$, counting multiplicities. So in total $\wp(z) - e_j$ would have at least 4 zeros in the fundamental parallelogram $P_0(\omega_1, \omega_2)$. This contradicts the fact that $\wp(z)$ is of order 2, and the proof is finished. □

The definitions of e_1, e_2, and e_3 makes it seem like they are dependent on the choice of a fundamental pair ω_1, ω_2. In fact, when regarded together, they depend only on the lattice itself, as the next result shows.

Corollary 4.18. *The numbers e_1, e_2, e_3, considered as an unordered triple of numbers, are independent of the choice of fundamental period pair ω_1, ω_2. That is, if ω_1', ω_2' is another fundamental period pair for L and e_1', e_2', e_3' are the numbers associated with it analogously to e_1, e_2, e_3, then*

$$\{e_1', e_2', e_3'\} = \{e_1, e_2, e_3\}.$$

Proof. The e_j are the roots of the cubic polynomial $4x^3 - g_2 x - g_3$, whose coefficients do not depend on the choice of fundamental pair. □

4.9 $\wp(z)$ and $\wp'(z)$ generate all doubly periodic functions

We say that a function f is L-**periodic** or **periodic with respect to** L if any $\omega \in L$ is a period of f. Our general discussion of doubly periodic functions earlier in the chapter motivated and complemented our explicit construction of the Weierstrass $\wp$-function, but it seems desirable to give an explicit way to generate *all* doubly periodic functions with respect to a fixed lattice L. The next two theorems give an elegant solution to this classification problem, which highlights the central role played by the Weierstrass $\wp$-function in the theory of doubly periodic functions.

Theorem 4.19. *Let $L \subset \mathbb{C}$ be a lattice. The set of even meromorphic functions that are periodic with respect to L coincides with the set of functions of the form*

$$f(z) = R(\wp(z)), \tag{4.26}$$

where $R(w)$ is a rational function.

Proof. If $f(z)$ is of the form (4.26), then clearly $f(z)$ is even, meromorphic, and L-periodic. Conversely, let $f(z)$ be even, meromorphic, and L-periodic. Assume that $f(z)$ is nonconstant, since otherwise there is nothing to prove. Fix a fundamental parallelogram $P = P_{z_0}(\omega_1, \omega_2)$ that is generic for f; as an extra precaution, choose this P in such a way that it does not contain any points of L on its boundary (it is easy to see that this is possible). Now define the even doubly periodic function

$$g(z) = \frac{\prod_{j=1}^{n}(\wp(z) - \wp(a_j))}{\prod_{k=1}^{m}(\wp(z) - \wp(b_k))}, \tag{4.27}$$

where $a_1, \ldots, a_n, b_1, \ldots, b_m$ are some points in P that will be specified shortly. The plan for the proof is as follows: we will find values for these points for which $g(z)$ defined by (4.27) has the same zeros and the same poles in $\mathbb{C} \setminus L$ as $f(z)$ (counting with multiplicities). We will then show that this property implies that $f(z) \equiv cg(z)$. Thus $f(z)$ would be of the form (4.26), and the claim would be proved.

 To show that points $a_1, \ldots, a_n, b_1, \ldots, b_m$ with the desired properties exist, consider the zeros first. The key property we need is the following claim: if the list of zeros of $f(z)$ in P that are not elements of L, counting with multiplicities, consists of points $c_1, \ldots, c_\nu$, then $\nu = 2n$ is an even number, and we can order the points in pairs c_{2j-1}, c_{2j} so that for each $1 \le j \le n$, c_{2j-1} is congruent to $-c_{2j}$ modulo L (that is, $c_{2j-1} + c_{2j} \in L$). To prove this, let α be any of zero of $f(z)$ that is not in L, and let μ denote its order. We consider two cases: first, if α is not a half-period, that is, α is not congruent to $-\alpha$ modulo L, then since $f(z)$ is even, $-\alpha$ is also a zero of $f(z)$ (and of the same order as α), so the list of zeros that are not in L has a number $\beta \in P$ that is congruent to $-\alpha$ modulo L, is distinct from α, and appears in the list of zeros the same number μ of times as α does. Thus we can pair up the μ appearances of α with the μ appearances of β as required.

Next, consider the case where a is a half-period. In that case, we claim that the multiplicity μ of a as a zero of $f(z)$ is an even number, so the required pairing would simply be $\mu/2$ pairs of a, a. The justification for this claim is that a being a zero of $f(z)$ of order μ means that

$$f(a) = f'(a) = f''(a) = \cdots = f^{(\mu-1)}(a) = 0, \quad f^{(\mu)}(a) \neq 0.$$

However, we also know that f is even, and therefore any derivative $f^{(2j)}(z)$ of f of even order is also even, and any derivative $f^{(2j-1)}(z)$ of f of odd order is an odd function. Then by a calculation similar to (4.24), taking into account that $2a \in L$, we get that

$$f^{(2j-1)}(a) = -f^{(2j-1)}(-a) = -f^{(2j-1)}(-a + 2a) = -f^{(2j-1)}(a),$$

whence $f^{(2j-1)}(a) = 0$ for $j \geq 1$. Since $f^{(\mu)}(a) \neq 0$, μ must be even.

Having shown that the zeros $c_1, \ldots, c_\nu$ can be matched in the way claimed above, we now define the numbers $a_1, \ldots, a_n$ by $a_j = c_{2j}$, $1 \leq j \leq n$, that is, we include in the list $a_1, \ldots, a_n$ a single representative from each pair c_{2j-1}, c_{2j}. The numbers $b_1, \ldots, b_m$ are now defined by repeating the same construction as with the zeros but for the function $1/f$ instead of f.

We defined the numbers $a_1, \ldots, a_n$ and $b_1, \ldots, b_m$. They were all chosen as elements of $P \setminus L$, so that $\wp(a_j)$ and $\wp(b_k)$ are all finite complex numbers; thus the right-hand side of (4.27) is a well-defined expression.

We now claim that $g(z)$ has the same zeros and poles as $f(z)$ in $P \setminus L$, counting multiplicities. Let $a \in P \setminus L$ be a zero of $f(z)$ of order μ. Denote by β the unique point in P for which β is congruent to $-a$ modulo L. Again, we consider the cases where a is a half-period or not a half-period separately. If a is not a half-period, then by our construction the list of numbers $a_1, \ldots, a_n$ includes μ numbers γ that are equal to either a or β. Each of them corresponds to a factor in the numerator of $g(z)$ of the form $\wp(z) - \wp(\gamma) = \wp(z) - \wp(a) = \wp(z) - \wp(\beta)$, which is a function that has simple zeros at a and at β and no other zeros or poles in $P \setminus L$. None of the other factors in the products that make up the numerator and denominator of $g(z)$ have a zero or pole at a. Thus the order of the zero of $g(z)$ at a is μ.

In the case where a is a half-period, we have $a = \beta$. The function $h_a(z) = \wp(z) - \wp(a)$ has a *double* zero at a (the point $z = a$ is a zero of h_a of order at least 2, since both h_a and its derivative vanish there, but h_a is a doubly periodic function of order 2, so the order of the zero is exactly 2) and no other zeros or poles in $P \setminus L$. This function was included $\mu/2$ times in the product in the numerator of $g(z)$, and again, none of the other factors in the products in the numerator and denominator of $g(z)$ has a zero or pole at a. So in this case, we also have shown that the order of the zero of $g(z)$ at a is μ.

We showed that the zeros of $g(z)$ in $P \setminus L$ match the zeros of $f(z)$ in $P \setminus L$, with the same multiplicities. Applying the same reasoning to the poles (that is, comparing the zeros of $1/f(z)$ with those of $1/g(z)$) shows that the poles of $g(z)$ in $P \setminus L$ match the poles of $f(z)$ in $P \setminus L$ and their multiplicities. The conclusion is that the function $f(z)/g(z)$ is a meromorphic L-periodic function all of whose zeros and poles are elements of L. However, such a function must be constant: for otherwise, if it had a zero of any order at $z = 0$, then by periodicity it would have a zero at any $\omega \in L$ and therefore no poles, and similarly, if it had a pole at $z = 0$, then it would have a pole at all $\omega \in L$ and therefore no zeros. Since we know that any nonconstant doubly periodic function must have both zeros and poles, neither of those situations can occur.

To summarize, we proved that $f(z)$ coincides with the function $cg(z)$ for some constant c, as claimed. The proof is complete. ☐

Theorem 4.20. *Let $L \subset \mathbb{C}$ be a lattice. The set of meromorphic functions that are periodic with respect to L coincides with the set of functions of the form*

$$f(z) = R(\wp(z), \wp'(z)), \tag{4.28}$$

where $R(\xi, \zeta)$ is a rational function in two variables.

Proof. If f is of the form (4.28), then it is meromorphic and L-periodic. Conversely, given a meromorphic and L-periodic function f, decompose $f(z)$ in the standard way as a sum $f(z) = g(z) + h(z)$ of an even function $g(z)$ and an odd function $h(z)$, where

$$g(z) = \frac{f(z) + f(-z)}{2}, \quad h(z) = \frac{f(z) - f(-z)}{2}.$$

Now note that $g(z)$ is an even L-periodic function and therefore by Theorem 4.19 can be represented as a rational function in $\wp(z)$. Similarly, $h(z)$ is an *odd* L-periodic function, which means that $h(z)/\wp'(z)$ is even and L-periodic. Therefore $h(z)$ can be represented as $\wp'(z)$ times a rational function in $\wp(z)$. Combining the two representations for $g(z)$ and $h(z)$ gives the desired representation for $f(z)$. ☐

Suggested exercises for Section 4.9. 4.8, 4.9.

4.10 $\wp(z)$ as a conformal map for rectangles

Among the remarkable properties of the Weierstrass $\wp$-function, it provides a solution to the natural geometric problem of conformally mapping a rectangle onto a half-plane. This happens in the case where the associated lattice L is a rectangular lattice, that is, when it is of the form $L = \mathbb{Z} + iA\mathbb{Z}$ for a real parameter $A > 0$. The precise result is as follows.

Theorem 4.21. *Let $A > 0$, let $L = \mathbb{Z} + iA\mathbb{Z}$ be a rectangular lattice, and let $\wp(z) = \wp(z; L)$ be the associated Weierstrass elliptic function. The map $\wp(z)$ restricted to the rectangle $R = (0, \frac{1}{2}) \times (0, \frac{1}{2}A)$ is a conformal map from R to the lower half-plane $\{z : \text{Im}(z) < 0\}$.*

Proof. Denote by R' the *closed* rectangle $[0, \frac{1}{2}] \times [0, \frac{1}{2}A]$. First, note that the restriction of $\wp(z)$ to R' is injective. Indeed, $\alpha = 0$ is the unique point in R' that gets mapped to ∞. On the other hand, if $\alpha \in R' \setminus \{0\}$, then the function $\wp(z) - \wp(\alpha)$ has simple zeros at α and $1 + iA - \alpha$ and (since $\wp(z)$ is a doubly periodic function of order 2) at no other points in the fundamental parallelogram $P_0(1, iA)$. When $\alpha = (1 + iA)/2$, those two points coincide, and for any other $\alpha \in R' \setminus \{0\}$, the second zero $1 + iA - \alpha$ is not in R'. This proves the injectivity claim. It follows that $\wp(z)$ maps R conformally to its image $\wp(\Omega)$.

To understand why the image $\wp(\Omega)$ is the lower half-plane, it is helpful to examine the behavior of $\wp(z)$ as one traverses the boundary ∂R of the rectangle in an anticlockwise direction, starting at 0. Denote $e_1 = \wp(1/2)$, $e_2 = \wp(iA/2)$, $e_3 = \wp((1 + iA)/2)$ as in (4.25). We claim that ∂R is mapped under $\wp(z)$ to the real line (including the point at infinity, the image of 0). More specifically, the numbers e_1, e_2, e_3 have the ordering $-\infty < e_2 < e_3 < e_1 < \infty$, and as z moves successively along the four boundary edges $[0, 1/2]$, $[1/2, (1 + iA)/2]$, $[(1 + iA)/2, iA/2]$, and $[iA/2, 0]$,[3] the image $\wp(z)$ descends from $+\infty$ to e_1 (the image of the first boundary edge), then from e_1 to e_3 (second boundary edge image), then from e_3 to e_2 (third boundary edge image), and finally from e_2 to $-\infty$ (fourth boundary edge image).

This geometric picture is easily justified by the following list of simple claims.

1. $\wp(z)$ takes real values on the segment $(0, 1/2)$.

   Proof. This is immediate from (4.11).

2. $\wp(z)$ is decreasing on $(0, 1/2]$.

   Proof. The derivative $\wp'(z)$ is nonzero everywhere in R' except at the three points $1/2, (1+iA)/2$, and $iA/2$. Thus $\wp(t)$ regarded as a function of a real variable $t \in (0, 1/2]$ is monotone. It must be decreasing rather than increasing, since the Laurent expansion (4.10) around $\omega = 0$ implies that

   $$\lim_{t \searrow 0} \wp(t) = +\infty$$

   (in the sense of ordinary real limits from calculus).

3. $\wp(z)$ takes real values on the segment $[1/2, (1 + iA)/2]$.

   Proof. By representation (4.12) for the derivative of $\wp(z)$, we have

   $$\wp'\left(\frac{1}{2} + it\right) = -2 \sum_{m,n \in \mathbb{Z}} \frac{1}{(\frac{1}{2} + it - m - iAn)^3}.$$

---

3 Here we use the notation $[a, b]$ to denote the directed straight line segment connecting a point a to another point b. Similarly, the notations (a, b), $(a, b]$, and $[a, b)$ are further used to denote open and half-open straight line segments, consistently with the usual notation for intervals from real analysis.

$$= -2 \sum_{n \in \mathbb{Z}} \sum_{m=1}^{\infty} \left[\frac{1}{(\frac{1}{2} + it - m - iAn)^3} + \frac{1}{(\frac{1}{2} + it - (1-m) - iAn)^3} \right]$$

$$= -2 \sum_{n \in \mathbb{Z}} \sum_{m=1}^{\infty} \left[\frac{1}{(\frac{1}{2} - m + i(t - An))^3} + \frac{1}{(m - \frac{1}{2} + i(t - An))^3} \right].$$

This represents $\wp'(\frac{1}{2} + it)$ as a sum of terms of the form

$$\frac{1}{(x + iy)^3} + \frac{1}{(-x + iy)^3} = -2i \frac{y(3x^2 - y^2)}{(x^2 + y^2)^3}$$

over pairs $x = 1/2 - m$ and $y = t - An$ (both real numbers if t is assumed real). Thus we see that $\wp'(z)$ takes imaginary values on the segment $[1/2, (1 + iA)/2]$. Since we already know that $\wp(1/2)$ is a real number, we get that

$$\wp(1/2 + it) = \wp(1/2) + \int_{1/2}^{1/2+it} \wp'(z)\, dz$$

is also real for $0 \leq t \leq A$.

4. $\wp(z)$ is decreasing on the segment $[1/2, (1 + iA)/2]$.

   Proof. Again, from the knowledge of where $\wp'(z)$ takes nonzero values we conclude that the function $t \mapsto \wp(1/2 + it)$ is monotone for $0 \leq t \leq A$. Again, it is not only monotone but in fact must be decreasing: if it were *increasing*, then $\wp(1/2 + it)$ for $0 \leq t \leq A$ would be a real number in (e_1, ∞). That is impossible, since as discussed above, $\wp(z)$ is injective on the closed rectangle R', and the real numbers in (e_1, ∞) were already shown to belong to the image of the interval $(0, 1/2)$.

Using similar arguments, it is not difficult to verify the following additional claims:
5. $\wp(z)$ takes real values on the segment $[(1 + iA)/2, iA/2]$.
6. $\wp(z)$ is decreasing on the segment $[(1 + iA)/2, iA/2]$.
7. $\wp(z)$ takes real values on the segment $[iA/2, 0)$.
8. $\wp(z)$ is decreasing on the segment $[iA/2, 0)$.

This completes the explanation about the mapping properties of $\wp(z)$ on the boundary of R. Now since $\wp(z)$ maps the rectangle boundary to the real axis and is injective on R', we see that R itself must get mapped either to the lower half-plane or to the upper half-plane. Appealing again to the Laurent expansion (4.10), we see that for z in R that is close to 0 (for example, z of the form $\epsilon(1 + i)$ where $\epsilon > 0$ is small), $\wp(z)$ lies in the lower half-plane, so $\wp(R)$ is the lower half-plane, as claimed. □

Fig. 4.4 illustrates how the Weierstrass $\wp$-function associated with the square lattice $\mathbb{Z}^2$ can be used to conformally map a square to the unit disc.

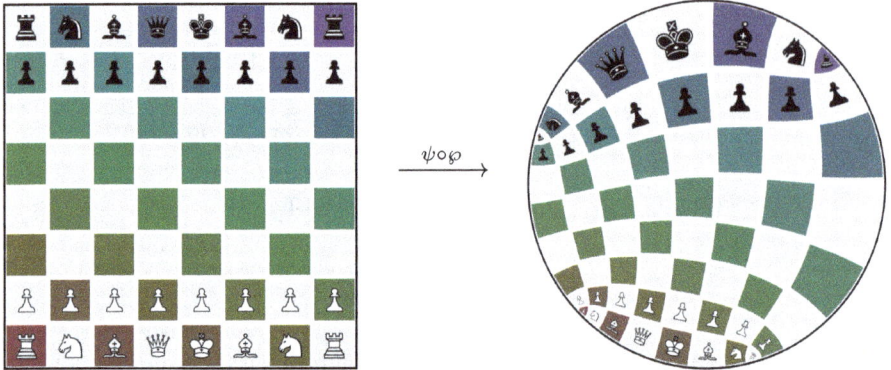

Figure 4.4: For the square lattice $\mathbb{Z}^2 = \mathbb{Z} + i\mathbb{Z}$, if we take ψ to be any conformal map from the lower half-plane to the unit disc, then the map $z \mapsto \psi \circ \wp(z)$ maps the square $(0, \frac{1}{2}) \times (0, \frac{1}{2})$ conformally onto the unit disc. The figure shows the action of the map in the case where $\psi(w) = -\frac{1+i}{\sqrt{2}} \frac{w+4i}{w-4i}$.

4.11 The discriminant of a cubic polynomial

The **discriminant** of a complex polynomial $p(z) = a_n z^n + \cdots + a_1 z + a_0$ of degree $n \geq 1$ is defined by

$$\Delta_p = a_n^{2n-2} \prod_{1 \leq i < j \leq n} (z_i - z_j)^2, \tag{4.29}$$

where $z_1, \ldots, z_n$ denote the roots of $p(z)$, counting multiplicities. Note that this definition does not depend on the ordering of the roots. Trivially, $p(z)$ has multiple (that is, nonsimple) zeros if and only if $\Delta_p = 0$. What in addition makes Δ_p a useful quantity is that it is of the form a_n^{2n-2} multiplied by a symmetric polynomial in the zeros of $p(z)$, and therefore, by a standard result from algebra, it can be expressed as a polynomial in the coefficients of $p(z)$, providing an explicit criterion for checking if a polynomial has multiple zeros. For example, for a quadratic polynomial $p(z) = az^2 + bz + c$, we learn in basic algebra that $\Delta_p = b^2 - 4ac$. The derivation is trivial.

If $p(z) = 4z^3 - az - b$ is a cubic polynomial given in the "reduced" form we are using for our elliptic curves discussion, then the formula expressing the discriminant in terms of the coefficients a, b is less well known, and its derivation is a bit less trivial.

Lemma 4.22. *The discriminant of the cubic $p(z) = 4z^3 - az - b$ is given by*

$$\Delta_p = 16(a^3 - 27b^2). \tag{4.30}$$

We note that in some books, the discriminant of a cubic polynomial $4z^3 - az - b = 4(z - z_1)(z - z_2)(z - z_3)$ is defined as $16(z_1 - z_2)^2(z_1 - z_3)^2(z_2 - z_3)^2$, which differs from our definition (4.29), the usual definition for general degree n polynomials, by a factor of 1/16. For that alternative scaling, the correct formula would be $\Delta_p = a^3 - 27b^2$. See also (4.35).

Proof of Lemma 4.22. Denote the zeros of $p(z)$ by z_1, z_2, z_3. By comparing coefficients of powers of z in the equation

$$p(z) = 4z^3 - az - b = 4(z - z_1)(z - z_2)(z - z_3)$$

we get the relations

$$\mu_1 := z_1 + z_2 + z_3 = 0,$$
$$\mu_2 := z_1 z_2 + z_1 z_3 + z_2 z_3 = -\frac{a}{4},$$
$$\mu_3 := z_1 z_2 z_3 = \frac{b}{4}.$$

Next, differentiate $p(z)$ to get that

$$p'(z) = 4(z - z_1)(z - z_2) + 4(z - z_1)(z - z_3) + 4(z - z_2)(z - z_3),$$

so in particular

$$p'(z_1) = 4(z_1 - z_2)(z_1 - z_3),$$
$$p'(z_2) = 4(z_2 - z_1)(z_2 - z_3),$$
$$p'(z_3) = 4(z_3 - z_1)(z_3 - z_2).$$

Therefore

$$\Delta_p = -4p'(z_1)p'(z_2)p'(z_3).$$

On the other hand, $p'(z) = 12z^2 - a$, so we get that

$$\Delta_p = -4(12z_1^2 - a)(12z_2^2 - a)(12z_3^2 - a)$$
$$= -4\left[12^3 z_1^2 z_2^2 z_3^2 - 12^2 a(z_1^2 z_2^2 + z_1^2 z_3^2 + z_2^2 z_3^2) + 12a^2(z_1^2 + z_2^2 + z_3^2) - a^3\right]. \qquad (4.31)$$

In this expansion, we have that

$$z_1^2 z_2^2 z_3^2 = \mu_3^2 = \frac{b^2}{16}, \qquad (4.32)$$
$$z_1^2 + z_2^2 + z_3^2 = (z_1 + z_2 + z_3)^2 - 2(z_1 z_2 + z_1 z_3 + z_2 z_3) = 0 - 2\mu_2 = \frac{a}{2}. \qquad (4.33)$$

This also gives that

$$\frac{a^2}{4} = 4\mu_2^2 = 4(z_1 z_2 + z_1 z_3 + z_2 z_3)^2$$

$$= 4(z_1^2 z_2^2 + z_1^2 z_3^2 + z_2^2 z_3^2) + 8z_1 z_2 z_3(z_1 + z_2 + z_3),$$

which yields the relation

$$z_1^2 z_2^2 + z_1^2 z_3^2 + z_2^2 z_3^2 = \frac{a^2}{16}. \tag{4.34}$$

Substituting (4.32), (4.33), and (4.34) into representation (4.31) for Δ_p gives finally that

$$\Delta_p = -4\left(12^3 \frac{b^2}{16} - 12^2 \frac{a^3}{16} + 12\frac{a^3}{2} - a^3\right) = 16(a^3 - 27b^2),$$

as claimed. $\square$

4.12 The discriminant of a lattice

Let $L \subset \mathbb{C}$ be a lattice, and let g_2, g_3 be the associated elliptic invariants defined in (4.15). The quantity

$$\Delta = g_2^3 - 27g_3^2 \tag{4.35}$$

is called the **discriminant** of the lattice L. In the context of the theory of modular forms, which is the subject of the next chapter, it is called the **modular discriminant**. Note that, as we see from (4.30), Δ is simply the discriminant of the cubic polynomial $4z^3 - g_2 z - g_3$ (with the different scaling convention mentioned after the statement of Lemma 4.22). By (4.29), (4.30), and Lemma 4.17 it can also be rewritten as

$$\Delta = 16(e_1 - e_2)^2(e_1 - e_3)^2(e_2 - e_3)^2, \tag{4.36}$$

where e_1, e_2, e_3 are given by (4.25). We also get the following conceptually important result.

Corollary 4.23. *The discriminant Δ of a lattice L is always nonzero.*

4.13 The J-invariant of a lattice

Another important parameter associated with a lattice L is known as **Klein's J-invariant**. It is defined by

$$J = \frac{g_2^3}{\Delta} = \frac{g_2^3}{g_2^3 - 27g_3^2},$$

which evaluates to a complex number since Δ is never 0. Klein's J-invariant plays an important role in the theory of modular functions and modular forms, and we will have more to say about it later; see Sections 5.9–5.10.

4.14 The modular variable τ: from elliptic functions to elliptic modular functions

Up until now, we considered the Weierstrass elliptic function associated with a specific fixed lattice L and denoted it by $\wp(z)$, letting the dependence on L remain implicit in the notation. However, it turns out that there is much to gain from considering the lattice itself as another variable the Weierstrass $\wp$-function and other related quantities depend on. Moreover, while a priori it might seem that "functions of a lattice-valued variable" are a cumbersome notion to attempt to study, it turns out that we can encode the dependence on the lattice in a natural way with a single complex variable, called the **modular variable** and denoted τ. From this new point of view, the function $\wp(z)$ (which, as we have also said, can sometimes be denoted $\wp(z; L)$) becomes a function of two complex variables, now denoted $\wp(z; \tau)$. Historically, the functions that we now refer to as elliptic functions were known as **elliptic modular functions** to signify this double dependence on the variable z, with respect to which they are doubly periodic, and the variable τ, the dependence on which has its own interesting flavor, captured by the term "modular." This term seems to be mostly used in older textbooks.

To explain the connection between L and τ, note that our convention to represent lattices as $L = \omega_1 \mathbb{Z} + \omega_2 \mathbb{Z}$ involve certain degrees of freedom that are not interesting in the sense that they can easily be eliminated and play no further role in the analysis. First, the ordering of ω_1, ω_2 is immaterial; that is, the ordered pair ω_1, ω_2 represents the same lattice as ω_2, ω_1. We can get rid of this double representation of lattices by considering the pair ω_1, ω_2 to come ordered in such a way that the parallelogram with vertices $0, \omega_1, \omega_1 + \omega_2, \omega_2$ is "oriented in the positive direction." Equivalently, this means that their quotient ω_2/ω_1 lies in the upper half-plane.

Second, lattices can also be scaled and rotated; that is, a pair ω_1, ω_2 representing the lattice $L = \omega_1 \mathbb{Z} + \omega_2 \mathbb{Z}$ can be replaced by $\omega_1' = \lambda\omega_1$, $\omega_2' = \lambda\omega_2$ for some scalar $\lambda \neq 0$ to obtain the lattice $L' = \omega_1' \mathbb{Z} + \omega_2' \mathbb{Z}$. Although L' are L are technically distinct lattices, from the point of view of complex analysis, they are equivalent in the sense that the Riemann surfaces $\mathbb{C}/L$ and $\mathbb{C}/L'$ are conformally equivalent via the scaling map $z \mapsto \lambda z$; meromorphic functions that are L-periodic are trivially in bijection with those that are L'-periodic; the Weierstrass $\wp$-function associated with L is in a simple relation to the $\wp$-function associated with L'; etc. Formally, we say that lattices L, L' related by $L' = \lambda L$ for some $\lambda \neq 0$ are **homothetic**. The above remarks can be summarized as saying that our main interest is in understanding lattices up to the equivalence relation of homothety.

For this reason, we now define the modular variable

$$\tau = \frac{\omega_2}{\omega_1},$$

a parameter taking values in the upper half-plane $\mathbb{H}$, and which we consider to be canonically associated with the lattice

$$L_\tau = \mathbb{Z} + \tau\mathbb{Z}.$$

As remarked above, this lattice is equivalent via a rescaling operation as described above to the lattice

$$L = \omega_1\mathbb{Z} + \omega_2\mathbb{Z}.$$

With this notation, the original lattice L and fundamental period pair ω_1, ω_2 need not play any further role in the analysis.

As we will see in the next chapter, the transition to the parameterization of lattices using the modular variable τ will reveal many additional layers of depth and beauty to the theory and open up a new complex-analytic area to explore, that of the **modular surface** and various families of meromorphic functions that are associated with it, which are known as **modular functions** and **modular forms**.

4.15 The classification problem for complex tori

You might have noticed by now, or seen it pointed out somewhere, that the doubly periodic functions we have been studying can be naturally identified with functions on a quotient space $\mathbb{C}/L$ in which we consider points z, z' as equivalent if they are congruent modulo the lattice L. This quotient space (which is indeed a quotient *group*) is topologically homeomorphic to the **torus** $\mathbb{T}^2 = S^1 \times S^1$, a compact surface. It also comes naturally equipped with the structure of a Riemann surface, inherited from $\mathbb{C}$ (in this book, we will not discuss the formal details of how this structure is set up, but at an intuitive level, it is not hard to appreciate that quotienting by a discrete subgroup leaves the complex structure "locally" similar to that of a normal complex region $\mathfrak{Q}$), so when thought of in that way, we refer to it as a **complex torus**. The doubly periodic functions that are periodic with respect to the lattice L, which are the meromorphic functions on $\mathbb{C}$ that "respect" the equivalence relation of congruence modulo the lattice, can be seen from this point of view as simply meromorphic functions on the complex torus $\mathbb{C}/L$. So the theory of doubly periodic functions is precisely the study of the complex-analytic structure of complex tori.

This way of thinking takes us back to the discussion of conformal mappings from Chapter 3 and the problem of classifying complex regions, or in the current context Riemann surfaces, up to conformal equivalence. Each lattice L gives rise to its own complex torus, but what can be said about how to decide when one complex torus $\mathbb{C}/L$ is biholomorphic to another complex torus $\mathbb{C}/L'$?[4] (Note that there is no hope for a complex torus to be biholomorphic to anything that is not topologically a torus, such as an ordinary complex region $\Omega \subset \mathbb{C}$, since conformal equivalence is stronger than topological homeomorphism.) Thus we arrive at the **classification problem for complex tori**. This consists broadly of several related questions:

1. First, what are necessary and sufficient conditions that two lattices $L, L' \subset \mathbb{C}$ must satisfy for the biholomorphism relation $\mathbb{C}/L \cong \mathbb{C}/L'$ to hold?
2. Second, can we find a nicely behaved set of representatives covering all conformal equivalence classes for the tori $\mathbb{C}/L$, with each class being covered exactly once?
3. Third, can this set of representatives be parameterized using a canonical "invariant" of some kind to make its description even simpler? (What this means exactly will become clearer later.)

Before you continue reading, pause for a minute to think what you might expect a solution to this classification problem to look like, keeping in mind some of the phenomena we discussed in Chapter 3, such as the Riemann mapping theorem and the classification of annuli and doubly connected regions up to conformal equivalence.

We will have to develop some additional theory to fully answer these questions. As we will see, the answers are related to the theory of modular forms, discussed in the next chapter. For now, we can formulate an initial attempt at a solution that answers the first of the questions formulated above. The remaining questions are answered in Sections 5.5 and 5.11.

Theorem 4.24 (Classification of complex tori: first part). *Let $L, L' \subset \mathbb{C}$ be two lattices in the complex plane.*

(a) *The complex tori $\mathbb{C}/L$ and $\mathbb{C}/L'$ are biholomorphic as Riemann surfaces if and only if the lattices L and L' are homothetic.*

(b) *If L, L' are given explicitly as*

$$L = \omega_1 \mathbb{Z} + \omega_2 \mathbb{Z}, \quad L' = \omega_1' \mathbb{Z} + \omega_2' \mathbb{Z}, \tag{4.37}$$

in terms of respective fundamental period pairs (ω_1, ω_2), (ω_1', ω_2') for the two lattices, then the homothety condition in part (a) is satisfied if and only if

$$\frac{\omega_2'}{\omega_1'} = \frac{a\omega_2 + b\omega_1}{c\omega_2 + d\omega_1}$$

for some $a, b, c, d \in \mathbb{Z}$ such that $ad - bc = \pm 1$.

---

4 When talking about Riemann surfaces, it seems a bit more customary to use the term "biholomorphic" rather than "conformally equivalent", although the two terms are generally regarded as synonymous.

Proof. (b) It is immediate from Lemma 4.1 that L and L' given in (4.37) are homothetic if and only if

$$\omega_1' = \lambda(\alpha\omega_1 + \beta\omega_2),$$
$$\omega_2' = \lambda(\gamma\omega_1 + \delta\omega_2)$$

for some complex number $\lambda \neq 0$ and integers $\alpha, \beta, \gamma, \delta \in \mathbb{Z}$ such that $\alpha\delta - \beta\gamma = \pm 1$. It is easy to see that this is equivalent to the condition described in the theorem.

(a) We start by proving the "if" part of the claim. Assume that $L' = \lambda L$ with $\lambda' \neq 0$. Define the map $f : \mathbb{C}/L \to \mathbb{C}/L'$ by

$$f(z + L) = \lambda z + L',$$

that is, the map taking the coset $z + L$ in the quotient group $\mathbb{C}/L$ to the coset $\lambda z + L'$ in the quotient group $\mathbb{C}/L'$. We claim that f is well-defined (i. e., that the definition is independent of the choice of a member z of the coset). Indeed, if z_1, z_2 are members of the same coset of $\mathbb{C}/L$, that is, $z_1 + L = z_2 + L$, then

$$\lambda z_1 + L' = \lambda z_1 + \lambda L = \lambda(z_1 + L) = \lambda(z_2 + L) = \lambda z_2 + L',$$

so λz_1 and λz_2 are in the same coset of $\mathbb{C}/L'$.

It is easily checked that this map also respects the Riemann surface structure of the quotient groups $\mathbb{C}/L$ and $\mathbb{C}/L'$, that is, that it is holomorphic. Applying the same reasoning with the roles of L and L' swapped, the map $g : \mathbb{C}/L' \to \mathbb{C}/L$ defined by

$$g(w + L') = \lambda^{-1}w + L$$

is a well-defined holomorphic map of $\mathbb{C}/L'$ into $\mathbb{C}/L$, and trivially g and f are inverse to each other, thus the two surfaces are biholomorphic.

Now we prove the "only if" part, which is the less obvious part. Assume that $\mathbb{C}/L \cong \mathbb{C}/L'$ (meaning that the two tori are biholomorphic), and let $f : \mathbb{C}/L \to \mathbb{C}/L'$ be a biholomorphism. We can assume without loss of generality that f maps the zero coset $0 + L$ to the zero coset $0 + L'$ (otherwise, replace f with its composition with a translation map $z + L' \mapsto z + a + L'$ for a suitable a). Motivated by the proof of the "if" part above, it seems natural to ask whether f can be represented as a map of cosets inherited from an "ordinary" complex-valued function of a complex-valued parameter. In other words, we look for an entire function $\tilde{f} : \mathbb{C} \to \mathbb{C}$ for which

$$f(z + L) = \tilde{f}(z) + L' \tag{4.38}$$

for all $z \in \mathbb{C}$. Schematically, it is helpful to think of such $\tilde{f}$ as the "solution" to the problem of completing the dashed line in the commutative diagram

$$
\begin{array}{ccc}
\mathbb{C} & \xrightarrow{\;\;\tilde{f}\;\;} & \mathbb{C} \\
\downarrow{\scriptstyle \varphi_L} & & \downarrow{\scriptstyle \varphi_{L'}} \\
\mathbb{C}/L & \xrightarrow{\;\;f\;\;} & \mathbb{C}/L'
\end{array}
$$

where $\varphi_L : \mathbb{C} \to \mathbb{C}/L$ and $\varphi_{L'} : \mathbb{C} \to \mathbb{C}/L'$ denote the quotient maps associated with the quotient groups $\mathbb{C}/L$ and $\mathbb{C}/L'$, respectively. That is, φ_L and $\varphi_{L'}$ are given by

$$
\varphi_L(z) = z + L, \quad \varphi_{L'}(w) = w + L'.
$$

If you have studied topology or other areas of mathematics where such diagrams appear, you are probably aware that the question of when we can "solve" such an equation in the unknown map is a rather subtle one in general; in our particular situation, it will not be very hard, fortunately. If such $\tilde{f}$ exists, it is often referred to as a **lifting** of f (with respect to the quotienting maps φ_L, $\varphi_{L'}$ that "descend" from the "upstairs" part of the diagram to the "downstairs" part).

Now assume that such $\tilde{f}$ can be shown to exist—we will prove this shortly. Since $f(0 + L) = 0 + L'$, we must have $\tilde{f}(0) \in L'$, and again we may assume without loss of generality that $\tilde{f}(0) = 0$ by replacing $\tilde{f}$ by its composition with translation $w \mapsto w - \tilde{f}(0)$ if necessary.

The function $\tilde{f}$ is entire by assumption. We claim that it is in fact a conformal automorphism of $\mathbb{C}$. The reason is that if $g : \mathbb{C}/L' \to \mathbb{C}/L$ denotes the inverse map to f, then the same assumption we made above about the existence of a lifting for f also implies that there exists a lifting for g, that is, an entire function $\tilde{g} : \mathbb{C} \to \mathbb{C}$ such that $g(w + L') = \tilde{g}(w) + L$ for all $w \in \mathbb{C}$. Then it is easy to see that the fact that f and g are inverse to each other or, in other words, that $f \circ g$ is the identity function, together with the normalization $\tilde{f}(0) = 0 = \tilde{g}(0)$, implies also that the composition $\tilde{f} \circ \tilde{g}$ of the lifted maps coincides with the identity function at least locally in a neighborhood of 0; and similarly for $\tilde{g} \circ \tilde{f}$. Therefore by analytic continuation in fact $f \circ g$ and $g \circ f$ both coincide with the identity function *globally* on all of the complex plane. Thus we see that $\tilde{f}$ and $\tilde{g}$ are inverse maps, and thus $\tilde{f}$ is an automorphism, as claimed.

Now we can apply the classification theorem for automorphisms of the complex plane (Theorem 3.3) and conclude that $\tilde{f}(z)$ is of the form $\tilde{f}(z) = \lambda z + b$ with $\lambda \neq 0$. In our case, $\tilde{f}(0) = 0$, so $b = 0$ and $\tilde{f}(z) = \lambda z$. In that case, for any $\omega \in L$, we have

$$
L' = 0 + L' = f(0 + L) = f(\omega + L) = \tilde{f}(\omega) + L' = \lambda\omega + L',
$$

so $\lambda\omega \in L'$. This proves that $\lambda L \subseteq L'$. Applying the same reasoning to the inverse map $\tilde{g}(w) = \tilde{f}^{-1}(w) = \lambda^{-1}w$ gives the opposite inclusion $\lambda L \supseteq L$, so finally we get that $L' = \lambda L$, as claimed.

It remains to prove the existence of the lifting $\tilde{f}$ of f. The reason why it exists is fundamentally a *topological* one and has to do with the notion of a **covering map**. I will sketch the argument, which is somewhat abstract and uses some background from the theory of Riemann surfaces, and then also provide a self-contained proof that manages to avoid any Riemann surface machinery.

The abstract explanation is as follows. In a general version of this situation, visualized by the diagram

$$
\begin{array}{ccc}
X & \dashrightarrow{\ \widetilde{f}\ } & Y \\
\varphi \downarrow & & \downarrow \psi \\
U & \xrightarrow{\ f\ } & V
\end{array}
$$

in which X, Y, U, V are Riemann surfaces and $\varphi : X \to U$ and $\psi : Y \to V$ are covering maps, a theorem from Riemann surfaces says that the lifting $\widetilde{f}$ is guaranteed to exist if the Riemann surface X at the top-left corner of the diagram is *simply connected*. (In that case, X is called the **universal cover** or **universal covering space** of U.) Fortunately, we are in precisely that scenario. So if you are familiar with that result, then the proof is complete, and no more effort is required.

Now for the self-contained argument: the function $f \circ \varphi_L$ is a holomorphic map from $\mathbb{C}$ to $\mathbb{C}/L'$. Let $z_0 \in \mathbb{C}$. By the definition of the Riemann surface structure on $\mathbb{C}/L'$, in some open disc U_{z_0} centered at z_0, this map is represented by an ordinary holomorphic map $g_{z_0} : U_{z_0} \to \mathbb{C}$ such that $f \circ \varphi_L = \varphi_{L'} \circ g_{z_0}$, that is, $f(z + L) = g_{z_0}(z) + L'$ for all $z \in U_{z_0}$.

It is also easy to see that any other holomorphic map $h : U_{z_0} \to \mathbb{C}$ representing f in such a way will have the form

$$
h(z) = g_{z_0}(z) + \omega' \tag{4.39}
$$

for some $\omega' \in L'$. This is because the assumption on g_{z_0} and h implies that $h(z) - g_{z_0}(z) \in L'$ for any $z \in U_{z_0}$, so (4.39) has to hold for some $\omega' \in L'$ *that might depend on z*; but $z \mapsto h(z) - g_{z_0}(z)$ is a continuous function of z, U_{z_0} is connected, and L' is discrete, so in fact the ω' has to be the same for all $z \in U_{z_0}$.

Observe further that for any h as above, again by (4.39) we have $h' \equiv g'_{z_0}$, that is, the derivative $g'_{z_0}(z)$ is actually independent of the choice of g_{z_0} from the set of possible choices. By similar reasoning it is also easy to check that if $z_0, z_1 \in \mathbb{C}$ have the property that $U_{z_0} \cap U_{z_1} \neq \emptyset$, then g'_{z_0} and g'_{z_1} agree on $U_{z_0} \cap U_{z_1}$. We can therefore define a global (entire) function $H : \mathbb{C} \to \mathbb{C}$ such that $H_{|U_{z_0}} \equiv g'_{z_0}$ for each of the local representation functions g_{z_0}.

Now let $\widetilde{f} : \mathbb{C} \to \mathbb{C}$ be the primitive of H satisfying $\widetilde{f}(0) = 0$ (guaranteed to exist by Corollary 1.25). We claim that $\widetilde{f}$ satisfies the claimed property (4.38) of being a lifting for f. This equation is true for $z = 0$ by definition. Moreover, assume that we already know that $f(z_0 + L) = \widetilde{f}(z_0) + L'$ for some $z_0 \in \mathbb{C}$. We claim that this implies the same property

$f(z + L) = \tilde{f}(z) + L'$ for all $z \in U_{z_0}$ (the open disc centered at z_0 as above). This is because in that disc we can write

$$f(z + L) = (f \circ \varphi_L)(z) = (\varphi_{L'} \circ g_{z_0})(z) = \varphi_{L'}(g_{z_0}(z))$$

$$= \varphi_{L'}\left(g_{z_0}(z_0) + \int_{z_0}^{z} g'_{z_0}(w)\, dw \right)$$

$$= \varphi_{L'}\left(g_{z_0}(z_0) + \int_{z_0}^{z} H(w)\, dw \right)$$

$$= \varphi_{L'}(g_{z_0}(z_0) + \tilde{f}(z) - \tilde{f}(z_0))$$

$$= \varphi_{L'}(g_{z_0}(z_0)) + \varphi_{L'}(\tilde{f}(z)) - \varphi_{L'}(\tilde{f}(z_0))$$

$$= f(\varphi_L(z_0)) + \varphi_{L'}(\tilde{f}(z)) - \varphi_{L'}(\tilde{f}(z_0)) = \varphi_{L'}(\tilde{f}(z)) = \tilde{f}(z) + L',$$

where we use the fact that φ_L is a group homomorphism (and use "+" to denote addition both in $\mathbb{C}$ and in the quotient group $\mathbb{C}/L'$).

The conclusion from the above discussion is that if we define the set

$$E = \{z \in \mathbb{C} : (f \circ \varphi_L)(z) = (\varphi_{L'} \circ \tilde{f})(z)\}$$

(the set of points for which (4.38) holds), then E is nonempty (it contains $z = 0$) and open. Moreover, E is a closed set: if $(z_n)_{n=1}^{\infty}$ is a sequence of points in E and $z_n \to \xi$ as $n \to \infty$, then

$$(f \circ \varphi_L)(\xi) = (f \circ \varphi_L)\left(\lim_{n \to \infty} z_n \right) = \lim_{n \to \infty} (f \circ \varphi_L)(z_n)$$

$$= \lim_{n \to \infty} (\varphi_{L'} \circ \tilde{f})(z_n) = (\varphi_{L'} \circ \tilde{f})\left(\lim_{n \to \infty} z_n \right) = (\varphi_{L'} \circ \tilde{f})(\xi),$$

so ξ is in E as well.

We showed that $E \subset \mathbb{C}$ is closed and open (that is, it is a "clopen" set in topology jargon) and is nonempty. The complex plane $\mathbb{C}$ is connected, which means that its only clopen subsets are itself and the open set. Thus $E = \mathbb{C}$. This establishes the lifting property of $\tilde{f}$ and finishes the proof. ☐

Suggested exercises for Section 4.15. 4.10.

4.16 Equivalence between complex tori and elliptic curves

At the beginning of this chapter, we presented the topic of elliptic curves as motivation for the study of doubly periodic functions, but until now, we have not explained the precise way in which the study of doubly periodic functions is helpful for understanding the structure of elliptic curves. In fact, the connection between the two subjects is very close and can be summarized by the slogan "elliptic curves are equivalent to complex tori." The key lies in the differential equation (4.16) satisfied by $\wp(z)$, which implies that for a given lattice $L \subset \mathbb{C}$ with invariants g_2, g_3, the point $(x, y) = (\wp(z), \wp'(z))$ lies on the elliptic curve $\mathcal{E}$ described in (4.2). Moreover, the map $z \mapsto (\wp(z), \wp'(z))$ is, when properly interpreted, a biholomorphism and an isomorphism of groups between the complex torus $\mathbb{C}/L$ and the elliptic curve $\mathcal{E}$.

The following result gives a fuller description of this intriguing and highly nonobvious correspondence between two classes of objects.

Theorem 4.25 (Equivalence between complex tori and elliptic curves). *Let $L \subset \mathbb{C}$ be a lattice with associated invariants g_2 and g_3. Let $\mathcal{E} = \mathcal{E}(g_2, g_3)$ denote the elliptic curve*

$$\mathcal{E}: \quad y^2 = 4x^3 - g_2 x - g_3$$

over the complex numbers, including the point at ∞. Then:
1. *The elliptic curve $\mathcal{E}$ is nondegenerate and is equipped in a natural way with the structure of a compact Riemann surface.*
2. *The map $\varphi : \mathbb{C}/L \to \mathcal{E}$ defined by*

$$\varphi(z + L) = \begin{cases} (\wp(z), \wp'(z)) & \text{if } z \notin L, \\ \infty & \text{if } z \in L, \end{cases}$$

   is a biholomorphism of Riemann surfaces.
3. *If $\mathcal{E}$ is also regarded as an abelian group with the group law defined as in Section 4.1, and $\mathbb{C}/L$ is viewed as a quotient group of $\mathbb{C}$, then φ is a group isomorphism in addition to being a biholomorphism.*
4. *The association $L \mapsto \mathcal{E}(g_2, g_3)$ is a bijection from the set of lattices onto the set of nondegenerate elliptic curves over $\mathbb{C}$.*

The upshot of this result is the remarkable fact that the study of elliptic curves over $\mathbb{C}$ coincides (albeit in a rather nontrivial way) with the study of complex tori $\mathbb{C}/L$. In particular, we get that any elliptic curve is topologically a torus, which does not seem obvious from the definition. Moreover, the problem of classifying elliptic curves up to biholomorphism reduces to the already-discussed classification problem of complex tori.

The proof of Theorem 4.25 is beyond the scope of this book and requires a more involved discussion of the group structure and Riemann surface structure on elliptic curves. For the details, see [61, Ch. 6].

Exercises for Chapter 4

4.1 Prove that a topologically discrete additive subgroup of $\mathbb{C}$ must be the zero sub-group of the form $\omega\mathbb{Z}$ for some $\omega \in \mathbb{Z}$ or of the form $\omega_1\mathbb{Z} + \omega_2$ with ω_1, ω_2 linearly independent over the real numbers.

4.2 Prove Lemma 4.9.

4.3 Identify the precise region of convergence of the Laurent expansion (4.14) and prove the necessary bounds that justify that in that region the rearrangement in the proof of Theorem 4.11 is valid.

4.4 To practice the technique demonstrated at the beginning of Section 4.7 that led to the Eisenstein series identities (4.19)–(4.21), use your favorite computer algebra system to extract additional Laurent expansion coefficients from the differential equations (4.16) and (4.18) and see what kinds of explicit identities you get.

4.5 Try to apply the method of proof of Proposition 4.14 by equating the coefficients of z^{2n} in the Laurent expansions for both sides of (4.16) instead of (4.18). Do you get any new identities involving the Eisenstein series?

4.6 This exercise explores an alternative and more direct method for proving the recurrence (4.22), which was found by Zagier [74].

   a) To illustrate the idea behind the method in a simple example, consider the bi-variate rational function

$$R(s, t) = \frac{1}{st^3} + \frac{1}{2s^2t^2} + \frac{1}{s^3t}.$$

   Check that $R(s, t)$ satisfies

$$R(s, t) - R(s + t, t) - R(s, s + t) = \frac{1}{s^2t^2}. \tag{4.40}$$

   b) Sum both sides of (4.40) over all integer pairs $s, t \geq 1$ and perform a bit of creative rearrangement of terms to conclude that

$$\zeta(4) = \frac{2}{5}\zeta(2)^2$$

   (where $\zeta(s)$ is the Riemann zeta function). This is a nice identity in that for example it makes it possible to deduce Euler's identity $\zeta(4) = \frac{\pi^4}{90}$ from its easier cousin $\zeta(2) = \frac{\pi^2}{6}$.

   c) Show that if we sum the sides of (4.40) instead over all pairs of *complex numbers* s, t in the "half-lattice"

$$L_+ = \{p\omega_1 + q\omega_2 : p, q \in \mathbb{Z} \text{ with } p \geq 1 \text{ or } [p = 0 \text{ and } q \geq 1]\},$$

   then by an analogous calculation we in fact obtain identity (4.19) relating the Eisenstein series G_4 and G_8, which is the case $k = 4$ of (4.22).

d) Generalizing the idea above, let $k \geq 2$ and define

$$R_k(s, t) = \frac{1}{st^{2k-1}} + \frac{1}{2}\sum_{r=2}^{2k-2}\frac{1}{s^r t^{2k-r}} + \frac{1}{s^{2k-1}t}$$

$$= \frac{1}{st^{2k-1}} + \frac{1}{s^{2k-1}t} + \frac{1}{2s^{2k-2}t^{2k-2}}\cdot\frac{s^{2k-3} - t^{2k-3}}{s - t}.$$

Show that $R_k(s, t)$ satisfies the identity

$$R_k(s, t) - R_k(s + t, t) - R_k(s, s + t) = \sum_{j=1}^{k-1}\frac{1}{s^{2j}t^{2k-2j}}. \tag{4.41}$$

(To practice your computer algebra skills and save yourself a tedious calculation, see if you can get the computer to prove this for you!)

e) Show that summing both sides of (4.41) over all integer pairs $s, t \geq 1$ yields the
recurrence relation

$$\zeta(2k) = \frac{2}{2k + 1}\sum_{j=1}^{k-1}\zeta(2j)\zeta(2k - 2j) \quad (k \geq 2), \tag{4.42}$$

satisfied by the values of the Riemann zeta function at positive even integers.

f) Show that if we assume that $\zeta(2) = \frac{\pi^2}{6}$, then (4.42), together with standard properties of the Bernoulli numbers discussed in Exercise 1.15, can be used to give
a new proof by induction of formula (2.10) from Chapter 2.

g) Finally, show that summing both sides of (4.41) over all complex numbers s, t
in the half-lattice L_+ as in part c) above gives exactly (4.22).

h) The above calculations highlight an interesting connection between the values
$\zeta(2n)$ and the Eisenstein series G_{2n}, wherein the former can be viewed as a
certain limiting case of the latter. Can you make this notion more precise? See
Section 5.7 for additional clues.

4.7 The Eisenstein series are known to satisfy other summation identities. As an example (taken from [57]), by extending Zagier's method described in Exercise 4.6, or in
any other way, prove the identity

$$G_{6n+2} = \frac{1}{6n + 1}\cdot\frac{(4n + 1)!}{((2n)!)^2}\sum_{k=1}^{n}\frac{\binom{2n}{2k-1}}{\binom{6n}{2n+2k-1}}G_{2n+2k}G_{4n-2k+2}.$$

4.8 (a) Prove the following **addition theorems** for the Weierstrass $\wp$-function and its
derivative:

$$\wp(z + w) = \frac{1}{4}\left(\frac{\wp'(z) - \wp'(w)}{\wp(z) - \wp(w)}\right)^2 - \wp(z) - \wp(w), \tag{4.43}$$

$$\wp'(z+w) = -\frac{1}{4}\left(\frac{\wp'(z)-\wp'(w)}{\wp(z)-\wp(w)}\right)^3 + \frac{1}{(\wp(z)-\wp(w))^3}$$
$$\times \left[(\wp(z)^3\wp'(z)-\wp(w)^3\wp'(w))\right.$$
$$-2(\wp(z)^3\wp'(w)-\wp(w)^3\wp'(z))$$
$$\left.+3\wp(z)\wp(w)(\wp(z)\wp'(w)-\wp(w)\wp'(z))\right]. \tag{4.44}$$

Guidance. It may be useful to note that assuming (4.43), the second identity (4.44) is equivalent to the determinantal identity

$$\begin{vmatrix} 1 & 1 & 1 \\ \wp(z) & \wp(w) & \wp(z+w) \\ \wp'(z) & \wp'(w) & -\wp'(z+w) \end{vmatrix} = 0.$$

(b) Use (4.43)–(4.44) together with the fact that the Weierstrass $\wp$-function and its derivative parameterize the elliptic curve (4.2) to prove that formulas (4.3)–(4.4) define a valid group addition law on the elliptic curve $\mathcal{E}$ in (4.2).

4.9 Prove the **duplication formula**

$$\wp(2z) = \frac{1}{4}\left(\frac{12\wp(z)^2 - g_2}{2\wp'(z)}\right)^2 - 2\wp(z).$$

4.10 (a) Given a lattice $L \subset \mathbb{C}$, identify the complex numbers λ for which $\lambda L = L$.

(b) Given a lattice $L \subset \mathbb{C}$, find all the conformal automorphisms of the complex torus $\mathbb{C}/L$.

5 Modular forms

There are five elementary arithmetical operations: addition, subtraction, multiplication, division, and modular forms.

Martin Eichler[1]

5.1 Motivation: functions of lattices

Our investigations of elliptic functions in the previous chapter gave rise to a host of interesting quantities associated with a lattice $L \subset \mathbb{C}$; among them, the Eisenstein series G_{2k}, modular discriminant Δ, and Klein's J-invariant. As we discussed in Section 4.14, these quantities can be viewed as functions of the modular variable τ that we use to parameterize (up to a trivial scaling operation) the space of lattices, associating it canonically with the lattice $L_\tau = \mathbb{Z} + \tau\mathbb{Z}$. Moreover, we saw that these functions satisfy interesting identities, such as the relations $G_8 = \frac{3}{7}G_4^2$, $G_{10} = \frac{5}{11}G_4 G_6$, and the more general recurrence relation (4.22). As we will see a bit later (Section 5.7), these types of complex-analytic identities encode identities of a purely number-theoretic nature; for example, the relation just mentioned between G_8 and G_4^2 is equivalent to the curious identity

$$\sigma_7(n) = \sigma_3(n) + 120 \sum_{k=1}^{n-1} \sigma_3(k)\sigma_3(n-k) \quad (n \geq 1), \tag{5.1}$$

where $\sigma_\alpha(m)$ denotes the **generalized sum-of-divisors function** defined as

$$\sigma_\alpha(m) = \sum_{d|m} d^\alpha \tag{5.2}$$

(the sum of the α-powers of the divisors of m). And this is just beginning to scratch the surface of the wealth of remarkable phenomena these functions are involved in.

From now on we will make the dependence on the modular variable τ more explicit by writing $G_{2k}(\tau)$, $\Delta(\tau)$, and $J(\tau)$ instead of G_{2k}, Δ, and J. At the heart of the phenomena mentioned above is the fact that the functions $G_{2k}(\tau)$, $J(\tau)$, $\Delta(\tau)$ all satisfy interesting "transformation properties," that is, functional equations that relate their value at τ to their value at $\frac{a\tau+b}{c\tau+d}$ for a certain class of Möbius transformations $\tau \mapsto \frac{a\tau+b}{c\tau+d}$. This fact is essentially immediate from the definitions; we record it as a lemma.

Lemma 5.1. *The functions $G_{2k}(\tau)$ ($k \geq 2$), $J(\tau)$, and $\Delta(\tau)$ satisfy the functional equations*

$$G_{2k}\left(\frac{a\tau+b}{c\tau+d}\right) = (c\tau+d)^{2k}G_{2k}(\tau), \tag{5.3}$$

---

1 This quote may be apocryphal; see the discussion in [W19].

$$J\left(\frac{a\tau + b}{c\tau + d}\right) = J(\tau),\tag{5.4}$$

$$\Delta\left(\frac{a\tau + b}{c\tau + d}\right) = (c\tau + d)^{12}G_{2k}(\tau)\tag{5.5}$$

for all $\tau \in \mathbb{H}$ and $a, b, c, d \in \mathbb{Z}$ satisfying $ad - bc = 1$.

Proof. Relations (5.4)–(5.5) follow immediately from (5.3) and the definitions of $J(\tau)$ and $\Delta(\tau)$. To prove (5.3), apply definition (4.13) of G_{2k} to write

$$
\begin{aligned}
G_{2k}\left(\frac{a\tau + b}{c\tau + d}\right) &= \sum_{(m,n)\in\mathbb{Z}^2\setminus(0,0)} \frac{1}{(m + n\frac{a\tau+b}{c\tau+d})^{2k}} \\
&= (c\tau + d)^{2k} \sum_{(m,n)\in\mathbb{Z}^2\setminus(0,0)} (m(c\tau + d) + n(a\tau + b))^{-2k} \\
&= (c\tau + d)^{2k} \sum_{(m,n)\in\mathbb{Z}^2\setminus(0,0)} ((dm + bn) + (cm + an)\tau)^{-2k}.
\end{aligned}\tag{5.6}
$$

Denoting new summation indices $p = dm + bn$ and $q = cm + an$ or, in matrix notation,

$$\begin{pmatrix} q \\ p \end{pmatrix} = \begin{pmatrix} a & c \\ b & d \end{pmatrix}\begin{pmatrix} n \\ m \end{pmatrix},$$

we can rewrite the last expression in (5.6) as

$$(c\tau + d)^{2k} \sum \frac{1}{(p + q\tau)^{2k}},\tag{5.7}$$

where the summation ranges over the possible pairs (p, q) associated with $(m, n) \in \mathbb{Z}^2 \setminus \{(0,0)\}$ through the above linear transformation. However, the assumptions on a, b, c, d imply that the matrix $\left(\begin{smallmatrix} a & c \\ b & d \end{smallmatrix}\right)$ maps $\mathbb{Z}^2 \setminus \{(0,0)\}$ bijectively onto itself, so the summation range is exactly $\mathbb{Z}^2 \setminus \{(0,0)\}$, and we see that (5.7) is precisely $(c\tau + d)^{2k}G_{2k}(\tau)$. □

Coneptually, the transformation properties (5.3)–(5.5) can be regarded as a kind of family of internal symmetries of the functions $G_{2k}(\tau)$, $J(\tau)$, and $\Delta(\tau)$. As the easy calculation above shows, these symmetries are simply a manifestation of the fact that the functions were originally defined in terms of infinite summations over a lattice, and so they must transform in a specific way when we switch from one fundamental period pair ω_1, ω_2 generating the lattice to another. However, it turns out that functions with similar internal symmetries arise in many other places where the reason for the symmetry holding is not nearly as self-evident (we will see examples of this later; see Section 5.13). The systematic study of functions with these types of symmetries, which we now undertake, is the beginning of the theory of **modular forms**, a rich subbranch of complex analysis that has strong connections to elliptic functions, number theory, and numerous other topics in mathematics.

Suggested exercises for Section 5.1. 5.1.

5.2 The modular group $\Gamma = \mathrm{PSL}(2, \mathbb{Z})$

Lemma 4.1 and Theorem 4.24(b) give conditions for two lattices to be equal and homothetic, respectively. It is convenient to think about these types of equivalences in terms of group actions. The condition for the equivalence of two lattices $\omega_1\mathbb{Z}+\omega_2\mathbb{Z}$ and $\omega_1'\mathbb{Z}+\omega_2'\mathbb{Z}$ given in Lemma 4.1 can be interpreted as the statement that (ω_1, ω_2) and (ω_1', ω_2') are in the same orbit under the action of the **general linear group of order** 2 **over** $\mathbb{Z}$ defined by

$$\mathrm{GL}(2, \mathbb{Z}) = \left\{ \begin{pmatrix} a & b \\ c & d \end{pmatrix} : a, b, c, d \in \mathbb{Z}, \ ad - bc = \pm 1 \right\}.$$

Our interest is mainly in describing lattices up to homothety, which means that we can consider the action of a smaller group. Let

$$\mathrm{SL}(2, \mathbb{Z}) = \left\{ \begin{pmatrix} a & b \\ c & d \end{pmatrix} : a, b, c, d \in \mathbb{Z}, \ ad - bc = 1 \right\}$$

be the **special linear group of order** 2 **over** $\mathbb{Z}$. Note that $\mathrm{SL}(2, \mathbb{Z})$ has a normal subgroup $\{\pm I\}$ of order 2 comprising the identity matrix I and its negative. We define the group Γ as the quotient group

$$\Gamma = \mathrm{SL}(2, \mathbb{Z})/\{\pm I\}.$$

This group is known as the **modular group** (or in certain contexts as the **projective special linear group of order** 2 **over** $\mathbb{Z}$). The notation Γ is in common use in the theory of modular forms. The alternative notation $\mathrm{PSL}(2, \mathbb{Z})$ is also sometimes used to denote the same group.

It turns out that Γ is the "correct" group to work with for our complex-analytic purposes, since it measures the precise extent of nonuniqueness when studying lattices up to homothety and parameterizing them using the modular variable τ as discussed in Section 4.14. This will be explained in Sections 5.3–5.4. We start however by thinking about Γ from a more abstract group-theoretic point of view.

Working with quotient groups is a bit cumbersome, and in the case of Γ the quotienting is quite minimal, involving the identification of pairs $\pm A$ of matrices. It is therefore common to abuse notation slightly and still denote elements of Γ as 2×2 matrices with the understanding that both such a matrix A and its negation $-A$ represent the same element of Γ and that all matrix equations written in this context are only assumed to hold modulo the subgroup $\{\pm I\}$.[2]

---

2 Note that we can get away with this without running into trouble as long as we only *multiply* matrices, as opposed to adding them or performing other operations that do not behave well under the quotienting homomorphism.

Three elements of Γ that play a special role in its analysis are the matrices

$$S = \begin{pmatrix} 0 & 1 \\ -1 & 0 \end{pmatrix}, \quad T = \begin{pmatrix} 1 & 1 \\ 0 & 1 \end{pmatrix}, \quad U = ST = \begin{pmatrix} 0 & 1 \\ -1 & -1 \end{pmatrix}. \tag{5.8}$$

Note that $S^2 = I$ (in the sense of the abuse of notation mentioned above), $U^3 = I$, and $T^k = \left(\begin{smallmatrix} 1 & k \\ 0 & 1 \end{smallmatrix}\right)$, so that S, U, and T generate cyclic subgroups of Γ of orders 2, 3, and ∞, respectively.

Theorem 5.2. *The group Γ is generated by the elements T, S.*

Proof. Let $A = \left(\begin{smallmatrix} a & b \\ c & d \end{smallmatrix}\right) \in \Gamma$. We may assume that $c \geq 0$; otherwise, replace A by $-A$ (recall that the two are equal as elements of Γ). We prove by induction on c that A can be represented as a product of elements of the form S and T^k, $k \in \mathbb{Z}$. In the case $c = 0$, A is of the form $\left(\begin{smallmatrix} a & b \\ 0 & d \end{smallmatrix}\right)$, and since $\det A = ad = 1$ and the entries are integers, actually

$$A = \begin{pmatrix} 1 & b \\ 0 & 1 \end{pmatrix} = T^b \quad \text{or} \quad A = \begin{pmatrix} -1 & b \\ 0 & -1 \end{pmatrix} = \begin{pmatrix} 1 & -b \\ 0 & 1 \end{pmatrix} = T^{-b},$$

both of which are of the required form.

For the inductive step, we assume that the claim has been proved in the case where the entry in the south-west corner of the matrix is strictly less than c. Dividing d by c with remainder, we let $q, r \geq 0$ denote the integers for which

$$d = qc + r, \quad 0 \leq r < c.$$

Then

$$AT^{-q} = \begin{pmatrix} a & b \\ c & d \end{pmatrix}\begin{pmatrix} 1 & -q \\ 0 & 1 \end{pmatrix} = \begin{pmatrix} a & -aq + b \\ c & r \end{pmatrix},$$

and therefore

$$AT^{-q}S = \begin{pmatrix} aq - b & a \\ -r & c \end{pmatrix} = \begin{pmatrix} -aq + b & -a \\ r & -c \end{pmatrix} =: M.$$

Applying the inductive hypothesis to the matrix M on the right-hand side, we see that it can be expressed as a product of group elements involving appearances of S and powers (negative or positive) of T. Therefore $A = MST^q$ can also be expressed in such a way, and we are done. □

5.3 The modular group as a group of Möbius transformations

In Section 4.14, we introduced the point of view whereby the space of lattices up to homothety is parameterized in terms of the modular variable τ taking values in the upper half-plane. The following lemma adapts the statement of Theorem 4.24(b) to that new point of view.

Lemma 5.3. *Let $\tau, \tau' \in \mathbb{H}$. The lattices $L_1 = \mathbb{Z} + \tau\mathbb{Z}$ and $L_2 = \mathbb{Z} + \tau'\mathbb{Z}$ are homothetic if and only if τ' is related to τ via*

$$\tau' = \frac{a\tau + b}{c\tau + d} \quad \text{for some } a, b, c, d \in \mathbb{Z}, \ ad - bc = 1. \tag{5.9}$$

Proof. Exercise 5.2. □

From Lemma 5.3 we see that the study of lattices up to homothety can be regarded as the study of the set of points in $\mathbb{H}$ quotiented out by the action of a group of Möbius transformations of the form (5.9). In fact, this group is canonically isomorphic to the modular group Γ with the isomorphism sending the element $\pm\left(\begin{smallmatrix} a & b \\ c & d \end{smallmatrix}\right)$ of Γ to the Möbius transformation $\tau \mapsto \frac{a\tau+b}{c\tau+d}$. In another small abuse of notation that is standard practice in the field, we still use the same letter Γ to denote this group and still refer to it as the modular group. That is, we write

$$\Gamma = SL(2, \mathbb{Z})/\{\pm I\} = \left\{ \tau \mapsto \frac{a\tau + b}{c\tau + d} : a, b, c, d \in \mathbb{Z}, \ ad - bc = 1 \right\}$$

with the convention that the map $\tau \mapsto \frac{a\tau+b}{c\tau+d}$ is simply another way to represent the group element $\pm\left(\begin{smallmatrix} a & b \\ c & d \end{smallmatrix}\right)$ of Γ. When referring to group elements, we will often use the same letter to denote an element of Γ thought of either as a matrix (with a $\pm$ sign ambiguity) or as a Möbius transformation. In particular, the group elements S, T, and U defined in (5.8) have the expressions

$$S(\tau) = \frac{-1}{\tau}, \quad T(\tau) = \tau + 1, \quad U(\tau) = \frac{-1}{\tau + 1}$$

in their interpretation as Möbius transformations.

Being able to switch at will between the two alternative points of view of working with matrices on the one hand and Möbius transformations on the other is convenient, since some arguments become simpler when considered from one of the points of view, and others are easier to understand from the alternative one.

Suggested exercises for Section 5.3. 5.2, 5.3.

5.4 The fundamental domain and the modular surface $\mathbb{H}/\Gamma$

Having identified the modular group as capturing the notion of the equivalence of two modular parameters τ, τ' that represent the same lattice, it is natural to ask for a complete set of equivalence class representatives, that is, a set of values of τ such that each point in the upper half-plane is equivalent to precisely one. (This question is precisely analogous to the idea that led us to the notion of a fundamental parallelogram in the study of elliptic functions.) The identification of such a set is one of the famous results of the field. It is given by

$$\mathcal{D} = \left\{ \tau \in \mathbb{H} : -\frac{1}{2} \le \mathrm{Re}(\tau) < \frac{1}{2} \right.$$

$$\left. \text{and} \left[|\tau| > 1 \text{ or } |\tau| = 1, \frac{\pi}{2} \le \arg \tau \le \frac{2\pi}{3} \right] \right\}.$$

We call $\mathcal{D}$ the **fundamental domain** under the action of Γ; see Fig. 5.1.

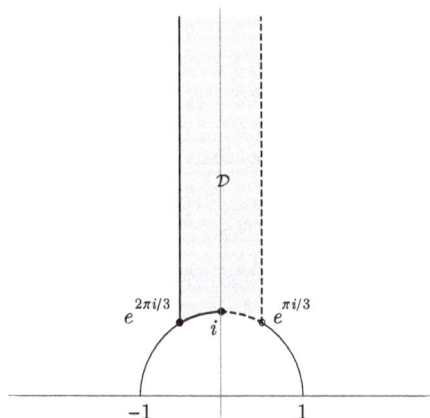

Figure 5.1: The fundamental domain $\mathcal{D}$.

Theorem 5.4. *The translates $A(\mathcal{D})$, $A \in \Gamma$ of the fundamental domain $\mathcal{D}$ under the elements of Γ tile the upper half-plane without overlap, except for specific exceptions given below. More precisely, each $\tau \in \mathbb{H}$ has a representation of the form*

$$\tau = A(\tau_0) \tag{5.10}$$

for some $A \in \Gamma$ and $\tau_0 \in \mathcal{D}$. The point τ_0 is unique. The Möbius transformation A is also unique if $\tau_0 \ne i, e^{2\pi i/3}$. If $\tau_0 = i$, then there are precisely two distinct representations

$$\tau = A_1(i) = A_2(i)$$

where $A_1, A_2 \in \Gamma$ are related by $A_2 = A_1 S$. If $\tau_0 = e^{2\pi i/3}$, then there are precisely three distinct representations

$$\tau = A_1(e^{2\pi i/3}) = A_2(e^{2\pi i/3}) = A_3(e^{2\pi i/3})$$

where $A_1, A_2, A_3 \in \Gamma$ are related by $A_2 = A_1 U$ and $A_3 = A_1 U^2$.

Proof. Let $\tau \in \mathbb{H}$. We prove the existence of A and τ_0 satisfying (5.10). Recall from (3.11) that for $a, b, c, d \in \mathbb{R}$, we have the formula

$$\text{Im}\left(\frac{a\tau + b}{c\tau + d}\right) = \frac{ad - bc}{|c\tau + d|^2} \, \text{Im}(\tau). \tag{5.11}$$

In particular, $\text{Im}(A(\tau)) = \frac{\text{Im}(\tau)}{|c\tau+d|^2}$ for $A = \left(\begin{smallmatrix} a & b \\ c & d \end{smallmatrix}\right) \in \Gamma$. Now the set of points

$$\{c\tau + d \, : \, (c, d) \in \mathbb{Z}^2 \setminus (0, 0)\}$$

is discrete and in particular disjoint from a neighborhood of 0; hence there exists some point of the form $c_0\tau + d_0$ in this set for which $|c\tau + d|$ is minimal. It is clear that for this c_0, d_0 we must have $\gcd(c_0, d_0) = 1$ (otherwise, divide each of c_0 and d_0 by their g. c. d. to get a pair with a smaller value of $|c\tau + d|$). This in turn implies that there exist integers a_0 and b_0 for which $a_0 c_0 + b_0 d_0 = 1$, in other words, such that the matrix $A_0 = \left(\begin{smallmatrix} b_0 & -a_0 \\ c_0 & d_0 \end{smallmatrix}\right)$ is an element of Γ. By the construction this A_0 has the property that $\text{Im}(A_0(\tau))$ is maximal over all $A \in \Gamma$. By replacing A_0 by $T^k A_0$ for a suitable $k \in \mathbb{Z}$ (thus replacing $A_0(\tau)$ with $A_0(\tau) + k$, which does not affect the imaginary value) we can also assume without loss of generality that $-\frac{1}{2} \le \text{Re}(A_0\tau) < \frac{1}{2}$, still retaining the maximality property.

Having chosen A_0, denote $\tau' = A_0(\tau)$. We claim that $|\tau'| \ge 1$. To see this, assume by contradiction that $|\tau'| < 1$. Then letting $B = SA_0$, we have

$$|\text{Im}(B(\tau))| = |\text{Im}(S\tau')| = |\text{Im}(-1/\tau')| > |\text{Im}(\tau')| = |\text{Im}(A_0\tau)|,$$

contradicting the maximality property of A_0.

Now if $\tau' \in \mathcal{D}$, then we can denote $A = A_0^{-1}$, $\tau_0 = \tau'$, and get that (5.10) holds, so we are done with the proof of the existence claim. Otherwise, we must have $|\tau'| = 1$ and $\frac{\pi}{3} \le \arg(\tau') < \frac{\pi}{2}$. In that case, let $\tau_0 = S\tau' = -1/\tau'$ and note that $\tau_0 \in \mathcal{D}$, so that if we define $A = (SA_0)^{-1}$, then (5.10) again holds. Thus the existence of the representation has been proved.

Now assume that τ has two *distinct* representations $\tau = A\tau_0 = A'\tau_0'$ with $\tau_0, \tau_0' \in \mathcal{D}$ and $A, A' \in \Gamma$. Our goal is to show that this can only happen in the specific situations listed in the theorem.

Assume without loss of generality that $\text{Im}(\tau_0') \ge \text{Im}(\tau_0)$ (otherwise, switch their labels). Denote $B = (A')^{-1}A$. Then $\tau_0' = B\tau_0 = \frac{a\tau_0+b}{c\tau_0+d}$, where a, b, c, d denote the entries of B. Then by (5.11) we get that

$$|c\tau_0 + d| \le 1. \tag{5.12}$$

Since $\tau_0 \in \mathcal{D}$ and c, d are integers and $\tau_0 \in \mathcal{D}$, there are not too many ways this inequality can hold. First, we could have $c = 0$ and $d = \pm 1$. In that case, we must have $a = d$, and therefore B is of the form

$$B = \begin{pmatrix} \pm 1 & b \\ 0 & \pm 1 \end{pmatrix} = \begin{pmatrix} 1 & \pm b \\ 0 & 1 \end{pmatrix} = T^{\pm b}.$$

Then $\tau_0' = B\tau_0 = \tau_0 \pm b$. The conditions $-\frac{1}{2} \le \mathrm{Re}(\tau_0) < \frac{1}{2}$ then guarantee that $b = 0$, so B is the identity map, $A' = A$, and $\tau_0' = \tau_0$, so that this is the case where the two representations $\tau = A\tau_0 = A'\tau_0'$ are the same, which is not relevant to the current discussion.

A second possibility for (5.12) to hold is that $d = 0$, $c = \pm 1$, and $|\tau_0| = 1$. In that case, B is of the form

$$B = \begin{pmatrix} a & \mp 1 \\ \pm 1 & 0 \end{pmatrix} = \begin{pmatrix} \pm a & -1 \\ 1 & 0 \end{pmatrix} = T^{\pm a} S.$$

Therefore $\tau_0' = -\frac{1}{\tau_0} \pm a$, or alternatively, if we write $\tau_0 = e^{i\theta}$ and $\alpha = \pm a$, then

$$\tau_0' = e^{i(\pi - \theta)} + \alpha.$$

For this to hold with τ_0, τ_0' elements of $\mathcal{D}$ and α an integer, we must have that either

$$\alpha = 0, \ B = S, \ \text{and} \ \tau_0' = \tau_0 = i, \tag{5.13}$$

or

$$\alpha = -1, \ B = T^{-1}S, \ \text{and} \ \tau_0 = \tau_0' = e^{2\pi i/3}. \tag{5.14}$$

In the first subcase (5.13), the two representations for τ become

$$\tau = A(i) = AS(i). \tag{5.15}$$

In the second subcase (5.14), we get that $B^{-1} = U$, so the two representations are

$$\tau = A(e^{2\pi i/3}) = AU(e^{2\pi i/3}). \tag{5.16}$$

The third and final possibility for (5.12) to hold is that $c = d = \pm 1$ and $\tau_0 = \tau_0' = e^{2\pi i/3}$. Assume without loss of generality that $c = d = 1$ (in the other case, replace a, b, c, d with the numbers $-a, -b, -c, -d$, respectively, which represent the same element of Γ). In that case the condition $ad - bc = 1$ forces $a = b + 1$, and we see that B is of the form

$$B = \begin{pmatrix} b+1 & b \\ 1 & 1 \end{pmatrix} = T^b S U^{-1}.$$

Then

$$\tau_0' = B\tau_0 = \frac{(b+1)\tau_0 + b}{\tau_0 + 1} = b + \frac{\tau_0}{\tau_0 + 1} = b + 1 - \frac{1}{\tau_0 + 1}$$

$$= b + 1 - \frac{1}{e^{\pi i/3}} = b + 1 + e^{2\pi i/3} = b + 1 + \tau_0'.$$

Thus we must have $b = -1$, and therefore $B = \left(\begin{smallmatrix} 0 & -1 \\ 1 & 1 \end{smallmatrix}\right) = U$, $B^{-1} = U^{-1} = U^2$, and we get that the two representations for τ are

$$\tau = A\left(e^{2\pi i/3}\right) = AU^2\left(e^{2\pi i/3}\right). \tag{5.17}$$

Summarizing, we showed that representation (5.10) is unique except for the three possible exceptions we identified, which are given by (5.15), (5.16), and (5.17). Those were precisely the exceptions listed in the theorem. This finishes the proof. □

The fundamental domain $\mathcal{D}$ can be thought of as the "arena" where modular functions and modular forms "live." We will do all our analysis in reference to this arena. This is mostly straightforward, except for some technical subtleties that will arise when functions have zeros or poles on the boundary of $\mathcal{D}$. (This is analogous to the issue that led us to consider fundamental parallelograms of the form $P_{z_0}(\omega_1, \omega_2)$ with an arbitrary origin point z_0 in Chapter 4 as a way to avoid having to worry about doubly periodic functions that have zeros or poles on the boundary of the parallelogram. In the case of modular forms, this issue is harder to work around using a simple translation trick of that type.)

We mention in passing that there is a more advanced, but conceptually clearer, point of view, in which the correct object to regard as the arena on which modular forms and functions are defined is the quotient space $\mathbb{H}/\Gamma$, that is, the space of orbits of $\mathbb{H}$ under the action of Γ. This quotient space is equipped in a natural way with the structure of a Riemann surface and is called the **modular surface**. The fundamental domain $\mathcal{D}$ is just one particular coordinate chart (in the sense of being an element of the atlas of charts a Riemann surface and other manifold-like objects come equipped with) that is used to perform calculations on it. Understanding this point of view will make various arguments and calculations in some of the proofs in this chapter appear more intuitive and motivated but is not strictly necessary from a formal point of view, so we will not discuss the details of how such arguments can be presented from the point of view of Riemann surfaces.

Suggested exercises for Section 5.4. 5.4.

5.5 The classification problem for complex tori, part II

We now return to the classification problem for complex tori discussed in Section 4.15. Previously we solved the first part of the problem when we gave a necessary and sufficient condition for two tori $\mathbb{C}/L$ and $\mathbb{C}/L'$ to be biholomorphic. Now we can use the results of the previous section to give a solution to the second part, namely finding a canonical system of representatives under this equivalence relation on the family of lattices.

Theorem 5.5 (Classification of complex tori; second part). *The family of complex tori*

$$\{\mathbb{C}/L_\tau : \tau \in \mathcal{D}\} \tag{5.18}$$

(where $L_\tau = \mathbb{Z} + \tau\mathbb{Z}$ as before) forms a complete set of biholomorphism representatives of the complex tori $\mathbb{C}/L$, that is, each complex torus $\mathbb{C}/L$ is biholomorphic to $\mathbb{C}/L_{\tau_0}$ for precisely one $\tau_0 \in \mathcal{D}$. If L is given explicitly as $L = \omega_1\mathbb{Z} + \omega_2\mathbb{Z}$ with $\omega_2/\omega_1 \in \mathbb{H}$, then τ_0 is the unique element of $\mathcal{D}$ related to $\tau = \omega_2/\omega_1$ via (5.10) for some $A \in \Gamma$, with the biholomorphism being the homothety $z \mapsto \omega_1 z$ (more precisely: the map of Riemann surfaces whose lifting is the homothety map, in the sense discussed in the proof of Theorem 4.24).

Proof. First, we show that no two elements of the family (5.18) are biholomorphic. Assume that $\tau_1, \tau_2 \in \mathcal{D}$ where $\mathbb{C}/L_{\tau_1}$ and $\mathbb{C}/L_{\tau_2}$ are biholomorphic. Then by Theorem 4.24(a), L_{τ_1} and L_{τ_2} are homothetic. By Lemma 5.3 we have

$$\tau_2 = \frac{a\tau_1 + b}{c\tau_1 + d} = A(\tau_1) \quad \text{for some } A = \begin{pmatrix} a & b \\ c & d \end{pmatrix} \in \Gamma.$$

Of course, τ_2 can also be represented as $I(\tau_2)$, where I is the identity element of Γ, so since $\tau_1, \tau_2 \in \mathcal{D}$, the uniqueness claim in Theorem 5.4 implies that $\tau_1 = \tau_2$.

For the remaining claim that the tori (5.18) include a representative of all biholomorphism classes of complex tori, let $L = \omega_1\mathbb{Z} + \omega_2\mathbb{Z}$ be a lattice, where the ordering of ω_1, ω_2 is chosen such that $\tau := \omega_2/\omega_1 \in \mathbb{H}$. Let $\tau_0 \in \mathcal{D}$ be the unique point in the fundamental domain, guaranteed to exist by Theorem 5.4, such that

$$\tau = A(\tau_0) = \frac{a\tau_0 + b}{c\tau_0 + d} \quad \text{for some } A = \begin{pmatrix} a & b \\ c & d \end{pmatrix} \in \Gamma.$$

By Lemma 5.3 the lattices $\mathbb{Z} + \tau\mathbb{Z}$ and $\mathbb{Z} + \tau_0\mathbb{Z}$ are homothetic, that is, we have

$$\mathbb{Z} + \tau\mathbb{Z} = \lambda(\mathbb{Z} + \tau_0\mathbb{Z})$$

for some $\lambda \neq 0$. It then follows that

$$L = \omega_1\mathbb{Z} + \omega_2\mathbb{Z} = \omega_1(\mathbb{Z} + \tau\mathbb{Z}) = \omega_1\lambda(\mathbb{Z} + \tau_0\mathbb{Z}) = \omega_1\lambda L_{\tau_0}.$$

Thus L and L_{τ_0} are also homothetic, and by Theorem 4.24(a), $\mathbb{C}/L$ is biholomorphic to $\mathbb{C}/L_{\tau_0}$, as claimed. □

5.6 The point at $i\infty$, premodular forms, and their Fourier expansions

In the sections below, we will start defining certain classes of functions that generalize properties (5.3)–(5.5) of the explicit functions we constructed. All of them will share one particular property that will be useful to name: we say that a function $f : \mathbb{H} \to \mathbb{C}$ is a **premodular form**[3] if it is

1. holomorphic;
2. periodic with period 1, that is, satisfies $f(\tau + 1) = f(\tau)$ for all $\tau \in \mathbb{H}$; and
3. for some constant $C \in \mathbb{R}, f(\tau)$ satisfies the asymptotic bound

$$\left| f(\tau) \right| = O\!\left(e^{C \operatorname{Im}(\tau)} \right) \quad \text{as } \operatorname{Im}(\tau) \to \infty, \text{ uniformly in } \operatorname{Re}(\tau). \tag{5.19}$$

We say that a function $f : \mathbb{H} \to \mathbb{C}$ is a **weak premodular form** if it satisfies the same conditions as for a premodular form, but with the first condition being relaxed to that of f being *meromorphic*.

Proposition 5.6. *Let $f : \mathbb{H} \to \mathbb{C}$. Then $f(\tau)$ is a premodular form if and only if it has an expansion of the form*

$$f(\tau) = \sum_{n=-m}^{\infty} a(n) e^{2\pi i n \tau} \quad (\tau \in \mathbb{H}), \tag{5.20}$$

*which converges absolutely, uniformly on compacts in $\mathbb{H}$, and where $m \geq 0$ is an integer. We refer to expansion (5.20) as the **Fourier expansion** of f. The coefficients $a(n)$ are called the **Fourier coefficients** of f and can be recovered as*

$$a(n) = \int_{-1/2}^{1/2} f(x + iy) e^{-2\pi i n(x+iy)} \, dx \quad (y > 0 \text{ arbitrary}). \tag{5.21}$$

Proof. The change of variables $q = e^{2\pi i \tau}$ defines the bijective correspondence

$$f(\tau) \longleftrightarrow g(q)$$

defined via the relation

$$f(\tau) = g\!\left(e^{2\pi i \tau} \right) \tag{5.22}$$

between holomorphic functions $f : \mathbb{H} \to \mathbb{C}$ that are periodic with period 1 and holomorphic functions $g : \mathbb{D} \setminus \{0\} \to \mathbb{C}$ on the punctured unit disc. If we add the assumption that $f(\tau)$ satisfies a bound of the form (5.19), then that translates to the condition that $g(q)$ must satisfy a bound of the form

$$\left| g(q) \right| = O(|q|^{M}), \quad q \to 0,$$

---

3 Note that this term is not standard in the literature.

for some constant $M \in \mathbb{R}$. Of course, this asymptotic bound is nothing particularly exotic; it is easily seen to be equivalent to the statement that $g(q)$ has either a removable singularity or a pole at $q = 0$. From Section 1.18 and Exercise 1.43 we know that any such function has a Laurent series of the form

$$g(q) = \sum_{n=-m}^{\infty} a(n)q^n,$$

which converges (absolutely and uniformly on compacts) for q in the punctured disc. Translating this back to the language of $f(\tau)$, this shows exactly that the condition of f being a premodular form is equivalent to it having the Fourier expansion (5.20) with the appropriate convergence. Finally, the coefficients can be extracted in the usual way as an integral on the circular contour $\{|q| = r\}$, $0 < r < 1$, using the residue theorem. Specifically, if we denote for convenience $r = e^{-2\pi y}$, $y > 0$, then we have

$$a(n) = \frac{1}{2\pi i} \oint_{|q|=r} \frac{g(q)}{q^{n+1}} \, dq = \frac{1}{2\pi i} \int_{-1/2}^{1/2} \frac{g(re^{2\pi ix})}{r^{n+1}e^{2\pi i(n+1)x}} (2\pi i)re^{2\pi ix} \, dx$$

$$= \int_{-1/2}^{1/2} g\big(e^{2\pi i(x+iy)}\big)e^{-2\pi in(x+iy)} \, dx = \int_{-1/2}^{1/2} f(x+iy)e^{-2\pi in(x+iy)} \, dx,$$

which is exactly (5.21). □

As we see from the proof above, the growth restriction on $|f(\tau)|$ as $\text{Im}(\tau) \to \infty$ for premodular forms is equivalent to the statement that under the change of variables $q = e^{2\pi i\tau}$, such a function expressed as a function of q is a holomorphic function on the punctured unit disc with a pole or removable singularity at $q = 0$. This suggests introducing the notion of "the point at $i\infty$" as a way of discussing the behavior of premodular forms near $q = 0$ while still thinking in terms of the variable τ. We will use the notation $\widehat{\mathcal{D}} = \mathcal{D} \cup \{i\infty\}$ to denote the fundamental domain with this point at $i\infty$ added. We will refer to $\widehat{\mathcal{D}}$ as the **extended fundamental domain.** We also introduce the following bit of terminology to describe the behavior of $f(\tau)$ near the point $i\infty$: if the function $g(q)$ associated with $f(\tau)$ as in (5.22) has a pole of some order $k \geq 1$, then we say that $f(\tau)$ **has a pole of order k at $i\infty$.** If $g(q)$ has a zero of order $k \geq 1$ at $q = 0$, we say that $f(\tau)$ **has a zero of order k at $i\infty$.** As usual, we can unify those two concepts and regard both zeros and poles as two aspects of the same thing by declaring that $f(\tau)$ **has a (generalized) zero of order k at $\tau = i\infty$** if $g(q)$ has a generalized zero of order k at $q = 0$ in the sense of having an ordinary zero of order k if $k \geq 1$, a pole of order $-k$ if $k < 0$, or neither a zero nor a pole if $k = 0$. (Refer to the parallel discussion on this terminology in Section 1.10.)

5.7 Fourier expansions and number-theoretic identities

The functions $G_{2k}(\tau)$, $\Delta(\tau)$, and $J(\tau)$ are all periodic functions with period 1 and, as we will see shortly, satisfy the growth condition (5.19) for being a premodular form. It turns out that their associated Fourier expansions, which we will now derive, are extremely interesting and lead to identities of a purely arithmetic nature.

Theorem 5.7 (Fourier expansion of the Eisenstein series). *For $k \geq 2$, the Eisenstein series $G_{2k}(\tau)$ is a premodular form and has the Fourier expansion*

$$G_{2k}(\tau) = 2\zeta(2k) + 2\frac{(2\pi i)^{2k}}{(2k-1)!} \sum_{n=1}^{\infty} \sigma_{2k-1}(n)q^n \quad (q = e^{2\pi i \tau}, \tau \in \mathbb{H}), \tag{5.23}$$

where $\sigma_{2k-1}(n)$ is the generalized sum-of-divisors function defined in (5.2).

Proof. Start with the partial fraction expansion of the cotangent function

$$\pi \cot(\pi z) = \frac{1}{z} + \sum_{\substack{n \in \mathbb{Z} \\ n \neq 0}} \left(\frac{1}{z+n} - \frac{1}{n} \right) \tag{5.24}$$

(see (1.73)). Differentiating this expansion p times gives

$$\frac{d^p}{dz^p}(\pi \cot(\pi z)) = (-1)^p p! \sum_{n=-\infty}^{\infty} \frac{1}{(z+n)^{p+1}} \quad (p \geq 1). \tag{5.25}$$

On the other hand, note that for $z \in \mathbb{H}$,

$$\pi \cot \pi z = \pi \frac{\cos \pi z}{\sin \pi z} = \pi i \left(1 - \frac{2}{1 - e^{2\pi i z}} \right) = -\pi i \left(1 + 2 \sum_{\ell=1}^{\infty} e^{2\pi i \ell z} \right),$$

and therefore also

$$\frac{d^p}{dz^p}(\pi \cot(\pi z)) = -(2\pi i)^{p+1} \sum_{\ell=1}^{\infty} \ell^p e^{2\pi i \ell z} \quad (p \geq 1). \tag{5.26}$$

Now

$$G_{2k}(\tau) = \sum_{(m,n) \neq (0,0)} \frac{1}{(m\tau + n)^{2k}} = \sum_{n \neq 0} \frac{1}{n^{2k}} + \sum_{m \neq 0} \sum_{n=-\infty}^{\infty} \frac{1}{(m\tau + n)^{2k}}$$

$$= 2\zeta(2k) + 2 \sum_{m=1}^{\infty} \sum_{n=-\infty}^{\infty} \frac{1}{(m\tau + n)^{2k}}$$

$$= 2\zeta(2k) + 2 \sum_{m=1}^{\infty} \frac{(-1)^{2k-1}}{(2k-1)!} \cdot \frac{1}{m^{2k-1}} \cdot \frac{d^{2k-1}}{d\tau^{2k-1}}(\pi \cot(\pi m \tau))$$

$$= 2\zeta(2k) + 2 \sum_{m=1}^{\infty} \frac{(-1)^{2k}(2\pi i)^{2k}}{(2k-1)!} \cdot \sum_{\ell=1}^{\infty} \ell^{2k-1} e^{2\pi i \ell m \tau}$$

$$= 2\zeta(2k) + \frac{2(2\pi i)^{2k}}{(2k-1)!} \sum_{n=1}^{\infty} \left(\sum_{\substack{\ell,m \geq 1 \\ \ell m = n}} \ell^{2k-1} \right) e^{2\pi i n \tau}$$

$$= 2\zeta(2k) + \frac{2(2\pi i)^{2k}}{(2k-1)!} \sum_{n=1}^{\infty} \sigma_{2k-1}(n) e^{2\pi i n \tau}, \tag{5.27}$$

which is the claimed expansion. Since $\sigma_{2k-1}(n)$ is bounded by a polynomial in n, expansion (5.23) clearly converges absolutely and uniformly in a neighborhood of $q = 0$ and defines a holomorphic function there. As we remarked in the previous section, this implies that $G_{2k}(\tau)$ is a premodular form. ☐

Theorem 5.8 (Fourier expansion of the modular discriminant). *The modular discriminant* $\Delta(\tau)$ *is a premodular form. Its Fourier expansion is given by*

$$\Delta(\tau) = (2\pi)^{12}(q - 24q^2 + 252q^3 - 1\,472q^4 + \cdots)$$

$$= (2\pi)^{12} \sum_{n=1}^{\infty} \tau(n) q^n \quad (q = e^{2\pi i \tau}, \tau \in \mathbb{H}). \tag{5.28}$$

Here the normalized coefficients $(\tau(n))_{n=1}^{\infty}$ *are a sequence of integers, which are given explicitly by*

$$\tau(n) = 8\,000 \sum_{\substack{j,k \geq 0 \\ j+k \leq n}} \sigma_3(j)\sigma_3(k)\sigma_3(n-j-k) - 147 \sum_{j=0}^{n} \sigma_5(j)\sigma_5(n-j) \tag{5.29}$$

for all $n \geq 1$, *where* σ_3 *and* σ_5 *denote the generalized sum-of-divisors functions as before with the additional convention that* $\sigma_3(0) = \frac{1}{240}$ *and* $\sigma_5(0) = -\frac{1}{504}$.

Proof. We have $\Delta(\tau) = 60^3 G_4(\tau)^3 - 27 \cdot 140^2 G_6(\tau)^2$, so $\Delta(\tau)$ trivially inherits the property of being a premodular form from $G_4(\tau)$ and $G_6(\tau)$. To get its Fourier expansion, note that, by (5.23) and (1.95),

$$60^3 G_4(\tau)^3 = 60^3 \left(\frac{\pi^4}{45} + 2\frac{(2\pi)^4}{6} \sum_{n=1}^{\infty} \sigma_3(n) q^n \right)^3$$

$$= (2\pi)^{12} \frac{8 \cdot 60^3}{6^3} \left(\sum_{n=0}^{\infty} \sigma_3(n) q^n \right)^3$$

$$= (2\pi)^{12} \cdot 8\,000 \sum_{n=0}^{\infty} \left(\sum_{\substack{j,k \geq 0 \\ j+k \leq n}} \sigma_3(j)\sigma_3(k)\sigma_3(n-j-k) \right) q^n$$

$$= (2\pi)^{12} \left(\frac{1}{1\,728} + 8\,000 \sum_{n=1}^{\infty} \left(\sum_{\substack{j,k \geq 0 \\ j+k \leq n}} \sigma_3(j)\sigma_3(k)\sigma_3(n-j-k) \right) q^n \right)$$

and, similarly,

$$27 \cdot 140^2 G_6(\tau)^2 = 27 \cdot 140^2 \left(\frac{2\pi^6}{945} - 2\frac{(2\pi)^6}{120} \sum_{n=1}^{\infty} \sigma_5(n)q^n \right)^2$$

$$= (2\pi)^{12} \cdot 4 \cdot 27 \cdot \frac{140^2}{120^2} \left(-\sum_{n=0}^{\infty} \sigma_5(n)q^n \right)^2$$

$$= (2\pi)^{12} \cdot 147 \sum_{n=0}^{\infty} \left(\sum_{j=0}^{n} \sigma_5(j)\sigma_5(n-j) \right) q^n$$

$$= (2\pi)^{12} \left(\frac{1}{1\,728} + 147 \sum_{n=1}^{\infty} \left(\sum_{j=0}^{n} \sigma_5(j)\sigma_5(n-j) \right) q^n \right).$$

Subtracting these two expressions leads to (5.28)–(5.29).

It remains to show that $\tau(n)$ is an integer. Observe that in representation (5.29), all the summands are integers, except possibly those for which one or both of the summation indices j, k are equal to 0. The total contribution of these exceptional summands to $\tau(n)$ can be expressed as

$$3 \times 8\,000\sigma_3(0)^2\sigma_3(n) + 3 \times 8\,000\sigma_3(0) \sum_{k=1}^{n-1} \sigma_3(k)\sigma_3(n-k) + 2 \times 147\sigma_5(0)\sigma_5(n)$$

$$= \frac{5}{12}\sigma_3(n) + 100 \sum_{k=1}^{n-1} \sigma_3(k)\sigma_3(n-k) + \frac{7}{12}\sigma_5(n)$$

$$= 100 \sum_{k=1}^{n-1} \sigma_3(k)\sigma_3(n-k) + \sum_{d\,|\,n} \frac{5d^3 + 7d^5}{12}.$$

This is in fact an integer, since it is easy to check that $5d^3 + 7d^5$ is divisible by 12 for any integer d. (Another famous formula for $\Delta(\tau)$ that we will prove later makes it immediate to see that the $\tau(n)$ are integers; see Theorem 5.31 in Section 5.14.) $\qquad\square$

The sequence of normalized Fourier coefficients

$$(\tau(n))_{n=1}^{\infty} = 1, -24, 252, -1\,472, 4\,830, -6\,048, \ldots$$

of the modular discriminant is called **Ramanujan's tau function.**[4] It is a celebrated mathematical object with many remarkable properties. To name one example, one of the surprising results of the theory of modular forms, which we will not prove here, is the following property, conjectured by Ramanujan in 1916 and proved shortly afterward by Mordell.

---

[4] Beware the small notational quirk of the theory wherein the letter τ is used to denote both the sequence $\tau(n)$ and the modular variable τ.

Theorem 5.9 (Multiplicativity of Ramanujan's tau function). *If* $m, n \geq 1$ *are relatively prime integers, then* $\tau(mn) = \tau(m)\tau(n)$.

For the proof, see [5, Ch. 6].

Theorem 5.10 (Fourier expansion of Klein's J-invariant). *Klein's J-invariant is a premodular form and has the Fourier expansion*

$$J(\tau) = \frac{1}{1728}\left(\frac{1}{q} + 744 + 196\,884q + 21\,493\,760q^2 + \cdots\right)$$

$$= \frac{1}{1728}\left(\frac{1}{q} + \sum_{n=0}^{\infty} c(n)q^n\right) \quad (q = e^{2\pi i\tau}, \tau \in \mathbb{H}).$$

The coefficients $c(n)$ are all positive integers.

Proof. Exercise 5.5. □

The coefficients $c(n)$ are also a much-studied sequence of numbers. In the late 1970s, they were found to be related to dimensions of the irreducible representations of the so-called **monster group**, a connection that was developed into a deep mathematical theory and is sometimes referred to as **monstrous moonshine**. The story of this discovery and some of the amazing mathematical ideas it led to is told in [29].

More mundane, but still interesting, is a result due to Petersson from 1932, which states that the asymptotic rate of growth of the coefficients $c(n)$ is given by

$$c(n) \sim \frac{1}{\sqrt{2}n^{3/4}}e^{4\pi\sqrt{n}} \quad \text{as } n \to \infty. \tag{5.30}$$

This result is conceptually related to another famous result, the **Hardy–Ramanujan formula** for the asymptotic rate of growth of the number $p(n)$ of integer partitions of n. That formula states that

$$p(n) \sim \frac{1}{4\sqrt{3}n}e^{\pi\sqrt{2n/3}} \quad \text{as } n \to \infty. \tag{5.31}$$

Both (5.30) and (5.31) can be proved using complex analysis; see [22], [66, Appendix A].

The Fourier expansions (5.23) and (5.28) make it possible to translate various identities involving the functions G_{2k} and Δ into number-theoretic identities.

Theorem 5.11. *We have the following number-theoretic identities for all $n \geq 1$:*

$$\sigma_7(n) = \sigma_3(n) + 120 \sum_{k=1}^{n-1} \sigma_3(k)\sigma_3(n-k), \tag{5.32}$$

$$\sigma_9(n) = \frac{1}{11}\left(21\sigma_5(n) - 10\sigma_3(n) + 5\,040 \sum_{k=1}^{n-1} \sigma_3(k)\sigma_5(n-k)\right), \tag{5.33}$$

$$\sigma_{13}(n) = 11\sigma_9(n) - 10\sigma_3(n) + 2\,640 \sum_{k=1}^{n-1} \sigma_3(k)\sigma_9(n-k), \tag{5.34}$$

$$\tau(n) = \frac{65}{756}\sigma_{11}(n) + \frac{691}{756}\sigma_5(n) - \frac{691}{3}\sum_{k=1}^{n-1}\sigma_5(k)\sigma_5(n-k). \tag{5.35}$$

Proof of (5.32) *and* (5.33). We consider the Fourier expansions of both sides of the identity $G_8 = \frac{3}{7}G_4^2$ from Section 4.7. By (5.23) the left-hand side is

$$G_8(\tau) = 2\zeta(8) + 2\frac{(2\pi i)^8}{7!}\sum_{n=1}^{\infty}\sigma_7(n)q^n = 2\frac{(2\pi)^8}{7!}\left(\frac{1}{480} + \sum_{n=1}^{\infty}\sigma_7(n)q^n\right).$$

The right-hand side is

$$\frac{3}{7}\left(2\zeta(4) + 2\frac{(2\pi i)^4}{3!}\sum_{n=1}^{\infty}\sigma_3(n)q^n\right)^2$$

$$= \frac{\pi^8}{4\,725} + \frac{32\pi^8}{315}\sum_{n=1}^{\infty}\sigma_3(n)q^n + \frac{256\pi^8}{21}\sum_{n=1}^{\infty}\left(\sum_{k=1}^{n-1}\sigma_3(k)\sigma_3(n-k)\right)q^n.$$

Equating the coefficients at q^n in the above expressions gives identity (5.32).

Identity (5.33) follows similarly from the Eisenstein series identity $G_{10} = \frac{5}{11}G_4G_6$, which we also discussed in Section 4.7. We omit the details of this simple calculation. □

The principle behind identities (5.34) and (5.35) is similar. They follow by equating the Fourier coefficients in the Eisenstein series identities

$$G_{14} = \frac{6}{13}G_4G_{10}, \tag{5.36}$$

$$\Delta = 1\,200(1\,430G_{12} - 691G_6^2), \tag{5.37}$$

respectively. These are not identities that we have previously derived, but they are conceptually similar to (4.19)–(4.21) and can be proved without great effort using the results of Section 4.7. However, rather than pursue this method, we will instead show in Section 5.12 a more elegant way of obtaining them (and similar identities) by applying more general ideas we will develop about modular forms.

Many more identities with a similar flavor to (5.32)–(5.35) are known to exist and can be proved using modular form techniques (or, through a much more painstaking analysis, using manipulations of a purely elementary nature [63]). As an example of a more sophisticated identity whose proof requires additional background, we mention the following identity due to Niebur [50]:

$$\tau(n) = n^4\sigma_1(n) - 24\sum_{k=1}^{n-1}(35k^4 - 52k^3n + 18k^2n^2)\sigma_1(k)\sigma_1(n-k).$$

The above discussion gives a glimpse into some of the close connections that exist between arithmetic and modular forms. As seen in the examples above, one way in which these connections manifest themselves is that the Fourier coefficients of naturally occurring modular forms (or mildly renormalized versions of them) are often integers with interesting arithmetic properties.

Suggested exercises for Section 5.7. 5.5.

5.8 Modular functions

A meromorphic function $f : \mathbb{H} \to \mathbb{C}$ is called a **modular function** if it is a weak premodular form and satisfies

$$f\left(\frac{a\tau + b}{c\tau + d}\right) = f(\tau) \tag{5.38}$$

for all $\tau \in \mathbb{H}$ and $A = \left(\begin{smallmatrix} a & b \\ c & d \end{smallmatrix}\right) \in \Gamma$. That is, a modular function is a true meromorphic function on the modular surface (including also the point $i\infty$). Note that since Γ is generated by the elements T, S, to verify the modular invariance property (5.38), it suffices to check that $f(\tau)$ satisfies

$$f(\tau + 1) = f(\tau), \quad f(-1/\tau) = f(\tau), \quad (\tau \in \mathbb{H}). \tag{5.39}$$

(The first of these two equations is already guaranteed by the condition that $f(\tau)$ is a weak premodular form.)

A modular function $f(\tau)$ that is not the zero function has only finitely many zeros and poles in $\mathcal{D}$: indeed, the zeros and poles cannot have $i\infty$ as an accumulation point (otherwise, $i\infty$ would be an essential singularity rather than a pole or removable singularity), which means that all the zeros and poles of $f(\tau)$ in the *closure* $\mathrm{cl}(\mathcal{D})$ are concentrated in the intersection of the closure with the strip $\{0 < \mathrm{Im}(\tau) \le M\}$ for some $M > 0$. This intersection is compact, so if there were an infinite sequence of zeros or poles of $f(\tau)$ in it, it would have an accumulation point, so it would be identically zero or have an essential singularity in $\mathbb{H}$, which is not allowed.

An essential property of modular functions is analogous to Proposition 4.6 we encountered in our discussion of elliptic functions in Chapter 4; loosely speaking, it states that the total number of zeros of a modular functions in the fundamental domain is equal to its total number of poles (as usual, counted with multiplicities, and the point $i\infty$ needs to be included in the count as well). An additional caveat in the current setting is that the "numbers" being referred to are actually *weighted* counts of points with respect to a certain weight function. We define the **weight** $w(\tau)$ of a point $\tau \in \widehat{\mathcal{D}}$ by

$$w(\tau) = \begin{cases} \frac{1}{2} & \text{if } \tau = i, \\ \frac{1}{3} & \text{if } \tau = e^{\pi i/3}, \\ 1 & \text{if } \tau = i\infty, \\ 1 & \text{otherwise.} \end{cases}$$

Theorem 5.12 (The weight formula for modular functions). *Let $f : \mathbb{H} \to \mathbb{C}$ be a modular function other than the zero function. Then*

$$\sum_{f(\xi)=0} w(\xi) = \sum_{f(\zeta)=\infty} w(\zeta). \tag{5.40}$$

Here the summation on the left-hand side ranges over zeros ξ of $f(\tau)$ in $\widehat{\mathcal{D}}$, counted with multiplicities, and the summation on the right-hand side ranges over poles ζ of $f(\tau)$ in $\widehat{\mathcal{D}}$, counted with multiplicities. In both summations, we include the point $i\infty$ with appropriate multiplicity if $f(\tau)$ has a zero or a pole there.

Proof. Consider a contour integral of the form

$$\oint_{\mathcal{G}} \frac{f'(\tau)}{f(\tau)} \, d\tau \tag{5.41}$$

around a suitable contour $\mathcal{G}$ that, as a first approximation, hugs the boundary of the fundamental domain $\mathcal{D}$ up to some vertical level M in the imaginary direction (Fig. 5.2(a)). The parameter $M > 0$ is chosen larger than the imaginary values of any of the zeros or poles of $f(\tau)$, other possibly than the point $i\infty$ (we discussed earlier why such an M exists). Now the general idea of the proof is to evaluate the contour integral in two ways. This is not conceptually difficult, but involves some technicalities of a somewhat tedious nature (which are nonetheless essential to check carefully), so to make things clearer pedagogically, we build up the calculation in several successive versions, each improving on the previous one.

First version. In the first version of the proof, we assume for simplicity that $f(\tau)$ does not have any zeros or poles on the boundary of $\mathcal{D}$. The integration contour in that case takes the form shown in Fig. 5.2(a) and decomposes as a sum of five subcontours

$$\mathcal{G} = \gamma_1 + \gamma_2 + \gamma_3 + \gamma_4 + \gamma_5.$$

Denote by X the total number of zeros of $f(\tau)$ in $\mathcal{D}$ and by Y the total number of poles, counted with multiplicities. Then the integral (5.41) is equal to $2\pi i(X - Y)$.

On the other hand, denote by Z the order of the zero $f(\tau)$ has at $i\infty$ (with the usual convention that Z is taken negative if $f(\tau)$ has a pole there). Breaking up the inte-

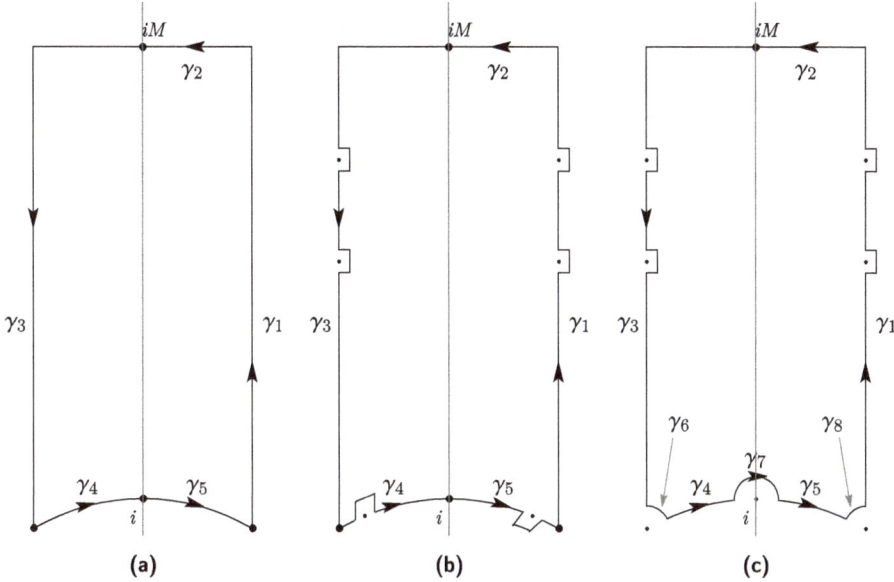

Figure 5.2: (a) The integration contour $\mathcal{G}$ used in the first version of the proof. (b) The modified version of $\mathcal{G}$ with detours added around zeros and poles on the boundary of $\mathcal{D}$. The detours on γ_5 are images of the detours on γ_4 under the inversion map $\tau \mapsto -1/\tau$. (c) The third version in which detours (labeled $\gamma_6, \gamma_7, \gamma_8$) are also added around i, $e^{\pi i/3}$, and $e^{2\pi i/3}$.

gral (5.41) into the integrals over the five subcontours γ_j, $1 \leq j \leq 5$, we wish to show that it is equal to $2\pi i Z$. This will give the equation

$$X + Z = Y \quad \text{or, equivalently,} \quad X = Y - Z,$$

and one or both of these two equations (depending on whether $Z > 0$, $Z = 0$, or $Z < 0$; that is, whether $i\infty$ is a zero, a pole, or neither a zero nor a pole) is what (5.40) claims under our simplified assumptions on $f(\tau)$.

Start with the contributions to (5.41) from the subcontours γ_1 and γ_3. Those are trivially seen to cancel each out, summing up to 0 because of the periodicity property $f(\tau + 1) = f(\tau)$.

Second, we show that the contributions from the subcontours γ_4 and γ_5 likewise cancel each other out. This follows by making the change of variables $\rho = -1/\tau$ in the integral over γ_5, which maps γ_5 to $-\gamma_4$ and therefore gives that

$$\int_{\gamma_5} \frac{f'(\tau)}{f(\tau)} \, d\tau = \int_{-\gamma_4} \frac{f'(-1/\rho)}{f(-1/\rho)} \frac{d\rho}{\rho^2} = -\int_{\gamma_4} \frac{f'(\rho)}{f(\rho)} \, d\rho. \tag{5.42}$$

(The last equality follows from the relation $\tau^{-2} f'(-1/\tau) = f'(\tau)$, obtained by differentiating the second identity in (5.39).)

The contribution from the integral over the remaining subcontour γ_2 can be evaluated by once again using the change of variables $q = e^{2\pi i \tau}$, which transforms this subcontour into a circle of radius $e^{-2\pi M}$ around $q = 0$ in the q-plane. Denote $\phi(q) = f(\tau)$; as discussed in the proof of Proposition 5.6, $\phi(q)$ is a holomorphic function on the punctured unit disc because $f(\tau)$ is periodic and has a zero of order Z (or a pole of order $-Z$) at the origin. Under this change of variables, we have

$$f'(\tau) = \phi'(q)q'(\tau),$$

and therefore

$$\frac{f'(\tau)}{f(\tau)} d\tau = \frac{\phi'(q)}{\phi(q)} q'(\tau) d\tau = \frac{\phi'(q)}{\phi(q)} dq,$$

which then implies that

$$\int_{\gamma_2} \frac{f'(\tau)}{f(\tau)} d\tau = - \oint_{|q|=e^{-2\pi M}} \frac{\phi'(q)}{\phi(q)} dq,$$

nicely mapping the integral to a similar-looking one in the variable q, except that the integral in the q variable is over a closed curve (and receives a minus sign since the change of variables maps the subcontour γ_2 to a circle oriented in the *negative* (clockwise) direction around $q = 0$). By the argument principle (Theorem 1.48) this last integral is equal to $2\pi i$ times the number of zeros minus the number of poles of $\phi(q)$ inside the circle $|q| = e^{-2\pi M}$. Since M was taken large enough so that $f(\tau)$ has no zeros or poles with imaginary value greater than M, its value is equal to $2\pi i Z$.

Putting the above results together, we have shown that

$$X - Y = \frac{1}{2\pi i} \oint_{\mathcal{G}} \frac{f'(\tau)}{f(\tau)} d\tau$$

$$= \frac{1}{2\pi i} \left[\left(\int_{\gamma_1} \frac{f'(\tau)}{f(\tau)} d\tau + \int_{\gamma_3} \frac{f'(\tau)}{f(\tau)} d\tau \right) \right.$$

$$\left. + \left(\int_{\gamma_4} \frac{f'(\tau)}{f(\tau)} d\tau + \int_{\gamma_5} \frac{f'(\tau)}{f(\tau)} d\tau \right) + \int_{\gamma_2} \frac{f'(\tau)}{f(\tau)} d\tau \right] = 0 + 0 - Z.$$

As we pointed out earlier, this was exactly the equality needed to balance the books and conclude that (5.40) holds.

Second version. For the next iteration of the proof, consider a situation in which $f(\tau)$ might now have zeros or poles on the boundary of $\mathcal{D}$ but assume that it does not have poles at $\tau = i$ or $\tau = e^{\pi i/3}$. The above proof can then be amended by modifying the integration contour $\mathcal{G}$ to add small "detours" bypassing each of the boundary zeros and poles, as shown in Fig. 5.2(b). The requirements for the detours are as follows: first, the detours on the subcontour y_3 of $\mathcal{G}$ dip into $\mathcal{D}$, are matched by detours of the same shape moving *away* from $\mathcal{D}$ along the subcontour y_1, and are small enough so that each detour goes around exactly one of the distinct zeros and poles. In this way, the contributions to the integral (5.41) from the subcontours y_1 and y_3 still trivially cancel each other out as before.

Second, for the detours along the two circular arc segments y_4 and y_5, they also move away from the unit circle in opposite directions, with the detours on y_4 dipping into the fundamental domain, and those in y_5 moving away from it. The precise shapes of these detours are not important; it is important to make them small enough (so that each detour only goes around a single pole or zero), and the shape of each detour around a zero or pole τ_0 along y_5 should be associated with the shape of the detour around the "reflected" point $-1/\tau_0$ lying along y_4 in such a way that y_5 coincides with the image of y_4 under the map $\tau \mapsto -1/\tau$.

Again, because the contours y_4 and y_5 have been matched to each other as we described, we will still have cancelation of the contributions to the integral (5.41) from y_4 and y_5 (since the first equality in (5.42) remains valid).

Now, with the modifications to the contour $\mathcal{G}$ described above, you can easily convince yourself that the integral (5.41) is still equal to $2\pi i(X - Y)$, where X and Y denote the same quantities as before. As a result, all the arguments from the first version of the proof remain valid, and we conclude that (5.40) holds in the same way as before.

Third version. So far we have avoided thinking about zeros and poles at $\tau = i$ and $\tau = e^{2\pi i/3}$, so we did not really have to grapple with the question of where the weights $1/2$ and $1/3$ in the definition of the weight function $w(\tau)$ come from. We now prove the theorem in its full generality, in the setting where $f(\tau)$ is allowed to have zeros and poles on the boundary of $\mathcal{D}$, including possibly at $\tau = i$ and $\tau = e^{\pi i/3}$. Let X, Y, Z be as before, except that we now define X and Y more carefully as the respective numbers of zeros and poles of f that are in $\mathcal{D}$ *other than the points* $\tau = i$ *and* $\tau = e^{2\pi i/3}$. Now denote additionally by Q and R the orders of the zeros of $f(\tau)$ at $\tau = i$ and $\tau = e^{2\pi i/3}$, respectively (again with the convention that they are negative if we have poles instead of zeros). In this setting, we modify the contour again, introducing additional detours around $\tau = i$, $\tau = e^{\pi i/3}$, and $\tau = e^{2\pi i/3}$, as shown in Fig. 5.2(c). These detours are taken as circular arcs of some radius r, chosen small enough so that no other zeros or poles of $f(\tau)$ lie within distance r of the special points $\tau = i, e^{\pi i/3}$, and $e^{2\pi i/3}$.

With this notation, the decomposition of $\mathcal{G}$ into segments now has the form

$$\mathcal{G} = y_1 + y_2 + y_3 + y_4 + y_5 + y_6(r) + y_7(r) + y_8(r),$$

where $y_6 = y_6(r)$, $y_7 = y_7(r)$, and $y_8 = y_8(r)$ denote the three added circular arcs; we emphasize their dependence on r in our notation for reasons that will become clear shortly.

Now retracing the reasoning in the previous version of the proof, we see that the conclusion $X - Y = -Z$ is now modified to

$$X - Y = -Z + \frac{1}{2\pi i} \int_{\gamma_6(r)} \frac{f'(\tau)}{f(\tau)} d\tau + \frac{1}{2\pi i} \int_{\gamma_7(r)} \frac{f'(\tau)}{f(\tau)} d\tau + \frac{1}{2\pi i} \int_{\gamma_8(r)} \frac{f'(\tau)}{f(\tau)} d\tau. \tag{5.43}$$

To understand the contribution from the new integrals over $\gamma_6(r)$, $\gamma_7(r)$, and $\gamma_8(r)$, consider the local behavior of $f(\tau)$ near $\tau = i, e^{2\pi i/3}$. Using our notation Q, R for the orders of the zeros at these exceptional points, we can factor $f(\tau)$ as

$$f(\tau) = (\tau - i)^Q g(\tau)$$

for τ in some neighborhood V of i, with $g(\tau)$ being holomorphic and nonzero in that neighborhood. Therefore the integral over $\gamma_7(r)$ can be evaluated (assuming that r is small enough so that the disc of radius r around i is contained in V) as

$$\int_{\gamma_7(r)} \frac{f'(\tau)}{f(\tau)} d\tau = Q \int_{\gamma_7(r)} \frac{1}{\tau - i} d\tau + \int_{\gamma_7(r)} \frac{g'(\tau)}{g(\tau)} d\tau.$$

Denote by θ_r the angle subtended by the circular arc $\gamma_7(r)$ (relative to the center point i of the circle of which that arc forms a part). Then by explicit parameterization of the integral of $1/(\tau - i)$ above, it is easily seen that that integral (without the constant Q in front of it) is equal to $-\theta_r$. For the second integral involving $g'(\tau)/g(\tau)$, we can bound it as

$$\left| \int_{\gamma_7(r)} \frac{g'(\tau)}{g(\tau)} d\tau \right| \le 2\pi M r,$$

where M is a positive constant such that $|g'(\tau)/g(\tau)| \le M$ for $\tau \in V$. Thus we have shown that

$$\int_{\gamma_7(r)} \frac{f'(\tau)}{f(\tau)} d\tau = -Q\theta_r + O(r) \tag{5.44}$$

for small r. Furthermore, it is geometrically obvious (and trivial to show formally if desired) that $\theta_r \to \pi$ as $r \to 0$.

Similarly, the integrals over $\gamma_6(r)$ and $\gamma_8(r)$ can be understood by writing a factorization for $f(\tau)$ of the form

$$f(\tau) = \left(\tau - e^{2\pi i/3}\right)^R h(\tau), \tag{5.45}$$

valid in a neighborhood of $e^{2\pi i/3}$, with $h(\tau)$ holomorphic and nonzero in that neighborhood. By the periodicity of f this also implies that for τ in a neighborhood of $e^{\pi i/3} = e^{2\pi i/3} + 1$, we have a similar factorization

$$f(\tau) = f(\tau - 1) = \left(\tau - e^{\pi i/3}\right)^R h(\tau - 1), \tag{5.46}$$

where again $\tau \mapsto h(\tau - 1)$ is holomorphic and nonzero in the neighborhood of $e^{\pi i/3}$. From representations (5.45)–(5.46) by a similar calculation as the one that led us to (5.44) we get that

$$\int_{\gamma_6(r)} \frac{f'(\tau)}{f(\tau)}\, d\tau = -R\phi_r + O(r) \quad \text{and} \quad \int_{\gamma_8(r)} \frac{f'(\tau)}{f(\tau)}\, d\tau = -R\phi_r + O(r)$$

for r near 0, where ϕ_r denotes the angle subtended by each of the circular arcs $\gamma_6(r)$ and $\gamma_8(r)$ relative to the center points $e^{2\pi i/3}$ and $e^{\pi i/3}$ of the circles of which these arcs are a part. It is easy to see that $\phi_r \to \pi/3$ as $r \to 0$.

Combining (5.43) and the other results noted above, we have shown that

$$X - Y = -Z - \frac{\theta_r}{2\pi}Q - \frac{\phi_r}{\pi}R + O(r)$$

for r near 0. Passing to the limit as $r \to 0$, this becomes

$$X - Y = -Z - \frac{1}{2}Q - \frac{1}{3}R,$$

which, as we see upon inspection, is simply another way of writing (5.40). □

Corollary 5.13. *Let $f : \mathbb{H} \to \mathbb{C}$ be a nonconstant modular function. Then f takes on any value an equal number of times in $\widehat{\mathcal{D}}$; that is, the weighted number of zeros of $f(\tau) - \alpha$ in $\widehat{\mathcal{D}}$ calculated in the sense of the left-hand side of (5.40) is the same for any $\alpha \in \mathbb{C}$.*

Proof. The right-hand side of (5.12) remains the same when we replace $f(\tau)$ by $f(\tau) - \alpha$. □

Corollary 5.14. *A modular function without poles in $\widehat{\mathcal{D}}$ is a constant.*

Proof. If f is a modular function without poles in $\widehat{\mathcal{D}}$, then $f(\tau)$ must in fact be bounded, since f is bounded in a neighborhood of $i\infty$ (that is, a half-plane of the form $\{\mathrm{Im}(\tau) \geq M\}$), and separately from that, it is bounded in the (compact) intersection of $\{0 < \mathrm{Im}(\tau) \leq M\}$ with the closure $\mathrm{cl}(\mathcal{D})$ of the fundamental domain.

Now since f is bounded, that means that for some $\alpha \in \mathbb{C}$, the equation $f(\tau) = \alpha$ has no solutions. By Corollary 5.13, if f were nonconstant, then the equation $f(\tau) = \alpha$ would have no solutions for *all* $\alpha \in \mathbb{C}$, which is obviously impossible. Therefore f is a constant. □

5.9 Klein's J-invariant

Let us now apply some of the understanding we developed on modular functions to Klein's function $J(\tau)$.

Lemma 5.15. *The Eisenstein series* $G_{2k}(\tau)$ *satisfy*

$$G_{2k}(i) = 0 \quad \text{if } k \text{ is odd, and}$$
$$G_{2k}(e^{2\pi i/3}) = 0 \quad \text{is } k \text{ is not divisible by 3.}$$

Proof. We have

$$G_{2k}(i) = \sum_{(m,n)\in\mathbb{Z}^2\backslash(0,0)} \frac{1}{(m+ni)^{2k}} = \sum_{(m,n)\in\mathbb{Z}^2\backslash(0,0)} \frac{1}{i^{2k}(-im+n)^{2k}}$$

$$= (-1)^k \sum_{(m,n)\in\mathbb{Z}^2\backslash(0,0)} \frac{1}{(-im+n)^{2k}} = (-1)^k G_{2k}(i),$$

which implies that $G_{2k}(i) = 0$ if k is odd. Similarly, we write

$$G_{2k}(e^{2\pi i/3}) = \sum_{(m,n)\in\mathbb{Z}^2\backslash(0,0)} \frac{1}{(m+ne^{2\pi i/3})^{2k}}$$

$$= \sum_{(m,n)\in\mathbb{Z}^2\backslash(0,0)} \frac{1}{e^{2(2k)\pi i/3}(me^{-2\pi i/3}+n)^{2k}}$$

$$= e^{-4k\pi i/3} \sum_{(m,n)\in\mathbb{Z}^2\backslash(0,0)} \frac{1}{(m(-e^{2\pi i/3}-1)+n)^{2k}}$$

$$= e^{-4k\pi i/3} \sum_{(m,n)\in\mathbb{Z}^2\backslash(0,0)} \frac{1}{((n-m)+(-m)e^{2\pi i/3})^{2k}}$$

$$= e^{-4k\pi i/3} \sum_{(p,q)\in\mathbb{Z}^2\backslash(0,0)} \frac{1}{(p+qe^{2\pi i/3})^{2k}} = e^{-4k\pi i/3} G_{2k}(e^{2\pi i/3}).$$

Since $e^{-4k\pi i/3} \neq 1$ if k is not divisible by 3, the desired conclusion follows. $\square$

Proposition 5.16. *The function* $J(\tau)$ *is a modular function. At the special points* $\tau = i$, $\tau = e^{2\pi i/3}$, *and* $\tau = i\infty$, *it takes the values*

$$J(e^{2\pi i/3}) = 0, \quad J(i) = 1, \quad J(i\infty) = \infty.$$

The zero at $e^{2\pi i/3}$ *is of order 3, the zero of* $J(\tau) - 1$ *at* $\tau = i$ *is of order 2, and the pole at* $i\infty$ *is simple.*

Proof. We know from Lemma 4.23 that $\Delta(\tau)$ is never zero for $\tau \in \mathbb{H}$, and from the Fourier expansion (5.28) we see that $\Delta(\tau)$ has a simple zero at $\tau = i\infty$. Therefore $J(\tau)$ has a simple pole at $i\infty$ and no other poles. We can also see using Lemma 5.15 that

$$J(i) = \frac{g_2(i)^3}{g_2(i)^3 - 27g_3(i)^2} = 1,$$

since $g_3(i) = 140G_6(i) = 0$. Similarly, $g_2(e^{2\pi i/3}) = 60G_4(e^{2\pi i/3}) = 0$, so

$$J(e^{2\pi i/3}) = \frac{g_2(e^{2\pi i/3})^3}{\Delta(e^{2\pi i/3})} = 0.$$

Now the zero of $J(\tau)$ at $e^{2\pi i/3}$ must be of an order that balances out the simple pole at $i\infty$ in accordance with (5.40). This implies that it is a zero of order 3. Applying the same reasoning to the zero of $J(\tau) - 1$ at $\tau = i$ shows that that zero is of order 2. □

Corollary 5.17. *The function $J(\tau)$ takes on any value in $\widehat{\mathcal{D}}$ exactly once; that is, the weighted number of zeros of $J(\tau) - a$ in $\widehat{\mathcal{D}}$ calculated in the sense of the left-hand side of (5.12) is equal to 1 for any $a \in \mathbb{C}$.*

Proof. By Proposition 5.16 the right-hand side of the sum in (5.40) for the case $f = J$ is equal to 1. The claim therefore follows from Corollary 5.13. □

We now show that $J(\tau)$ gives rise to all possible modular functions, as the next result explains.

Theorem 5.18. *A meromorphic function $f : \mathbb{H} \to \mathbb{C}$ is a modular function if and only if it is of the form*

$$f(\tau) = R(J(\tau))$$

for some rational function $R(w)$.

Proof. The "if" part is obvious; for the "only if," let $f(\tau)$ be a modular function that is not identically zero. Denote by $\mu_{i\infty}$ the order of the zero of $f(\tau)$ at $i\infty$. Denote by μ_i the order of the zero of $f(\tau)$ at i. Denote by μ_ρ the order of the zero of $f(\tau)$ at $\rho := e^{2\pi i/3}$. In these definitions, we use the usual convention that μ_a (for $a = i, \rho, i\infty$) is negative and equal to minus the order of the pole at a if there is a pole at that point instead of a zero.

Denote the zeros of f in the fundamental domain, counted with multiplicities but excluding the points $i, \rho, i\infty$, by $z_1, \ldots, z_n$. Denote the poles of f in the fundamental domain, with the same exclusions, by $w_1, \ldots, w_k$.

Relation (5.40) translates to the concrete statement that

$$n + \mu_{i\infty} + \frac{1}{2}\mu_i + \frac{1}{3}\mu_\rho = k. \tag{5.47}$$

Since n, k, and $\mu_{i\infty}$ are integers, we see that μ_i must be even, and μ_ρ must be a multiple of 3.

Now define the function

$$g(\tau) = (J(\tau) - 1)^{\mu_i/2} J(\tau)^{\mu_\rho/3} \frac{\prod_{j=1}^{n}(J(\tau) - J(z_j))}{\prod_{j=1}^{k}(J(\tau) - J(w_j))}. \tag{5.48}$$

Let $h(\tau) = f(\tau)/g(\tau)$. This is a modular function; let us examine where it has zeros and poles. By Corollary 5.17 each of the factors $J(\tau) - J(a)$ participating in the product in (5.48) (where $a = z_j$ or $a = w_j$ for some j) has a simple zero at $z = a$ and no other zeros. Therefore the zeros of f at $z_1, \ldots, z_n$ and the poles of $f(\tau)$ at $w_1, \ldots, w_k$ are precisely canceled out by the factors $J(\tau) - J(z_j)$ and $(J(\tau) - J(w_j))^{-1}$ in g, so the points $z_1, \ldots, z_n$, $w_1, \ldots, w_k$ are not zeros or poles of h. No other zeros or poles at any other points of the fundamental domain that are not the special points $i\infty, i, \rho$ are contributed by any multiplicand. Thus h may have zeros or poles at the three special points but nowhere else.

In fact, there are no zeros or poles at the special points either, since by Proposition 5.16 the factor $(J(\tau) - 1)^{\mu_i/2}$ has a zero of order μ_i at i, which cancels out the zero of order μ_i of $f(\tau)$ at i; similarly, the factor $J(\tau)^{\mu_\rho/3}$ has a zero of order μ_ρ at ρ, canceling out the zero of the same order of $f(\tau)$ at ρ; and, finally, the order of the zero of $h(\tau)$ at $i\infty$ is

$$\mu_{i\infty} + \frac{\mu_i}{2} + \frac{\mu_\rho}{3} + n - k = 0$$

by (5.47).

The conclusion is that $h(\tau)$ is a modular function with no poles or zeros and is therefore a constant by Corollary 5.14, that is, $h \equiv c$ with $c \in \mathbb{C}$. We have therefore shown that $f(\tau) = cg(\tau)$, which is a rational function in $J(\tau)$, as claimed. $\qquad\square$

5.10 The J-invariant as a conformal map

Another thing that makes $J(\tau)$ a natural function is that it is a conformal map and elucidates the structure of the modular surface H/Γ as a Riemann surface.

Theorem 5.19. *The function $J(\tau)$ is a biholomorphism between the modular surface $\mathbb{H}/\Gamma$ and the Riemann sphere $\widehat{\mathbb{C}}$.*

Sketch of proof. $J(\tau)$ maps $\mathbb{H}$ to $\mathbb{C}$ but respects the equivalence relation induced by the action of the modular group Γ. Thus it induces a function (which, abusing notation slightly, we also denote by J) $J : \mathbb{H}/\Gamma \to \mathbb{C}$. Adding the point $i\infty$, which gets mapped by J to the point ∞ on the Riemann sphere (this is just the fancy Riemann surface way of saying $J(\tau)$ has a simple pole at $i\infty$, as we stated in Proposition 5.16), turns J into a function from the full modular surface to the Riemann sphere. This function is holomorphic: this is reasonably obvious at a generic point of $\mathbb{H}/\Gamma$ but requires an explanation in terms of the Riemann surface structure of $\mathbb{H}/\Gamma$ at the special points $\tau = i, e^{2\pi i/3}, i\infty$. To avoid an involved digression into Riemann surfaces, we omit the details.

Moreover, we claim that the induced function is in fact a bijection and therefore a biholomorphism of Riemann surfaces. Indeed, Corollary 5.17 states that $J(\tau)$ takes on any value α exactly once on $\widehat{\mathcal{D}}$ (or, equivalently, on $\mathbb{H}/\Gamma$) in the sense of the weighted sum (5.40) over solutions of $f(\xi) = \alpha$. For $\alpha = 0$, this corresponds to the triple zero at

$\tau = e^{2\pi i/3}$ (which is the only zero; otherwise, the weighted sum would be greater than 1); for $\alpha = 1$, this corresponds to the double zero of $J(\tau) - 1$ at $\tau = i$, which again must be the only solution to the equation $J(\tau) = 1$; for $\alpha = \infty$, this corresponds to the simple pole at $\tau = i\infty$. For any other $\alpha \in \mathbb{C}$, the equation $J(\tau) = \alpha$ must have at least one solution $\tau \in \widehat{\mathcal{D}}$, and since τ is not one of the special points $i, e^{2\pi i/3}, i\infty$, it has weight $w(\tau) = 1$, and therefore (5.40) guarantees that it is the *only* solution. Thus J is a bijection. □

5.11 The classification problem for complex tori, part III

We saw in Section 5.5 that the fundamental domain $\mathcal{D}$ is a natural index set for the family of biholomorphism classes of complex tori $\mathbb{C}/L$. While this is satisfying at one level, it still leaves some room to complain that the fundamental domain is an oddly shaped region, with various identifications along its boundary induced by the action of Γ making its structure odder and still more mysterious. However, the result of the previous section clarifies things by showing that this structure is in fact simply that of the set of complex numbers, with the J-invariant acting as a conformal map translating between the two sets. Thus we arrive at the following result, which complements the results of Sections 4.15 and 5.5 and completes our solution of the classification problem for complex tori.

Theorem 5.20 (Classification of complex tori; third part). *The conformal map J^{-1} parameterizes the biholomorphism classes of complex tori $\mathbb{C}/L$ in the following precise sense: for any $z \in \mathbb{C}$, denote by $\tau_0(z)$ the point in the fundamental domain $\mathcal{D}$ for which $J(\tau_0) = z$. (Theorem 5.19 guarantees that $\tau(z)$ exists and is unique.) Then the map $z \mapsto L_{\tau_0(z)}$ is a bijection between $\mathbb{C}$ and the biholomorphism classes of complex tori.*

Proof. Immediate from Theorem 5.5. □

Recall also that Theorem 4.25 established a bijection between the family of complex tori $\mathbb{C}/L$ and the family of elliptic curves $\mathcal{E}(g_2, g_3)$. Thus Theorems 4.24, 5.5, and 5.20, which together formed our solution to the classification problem for complex tori, when combined with Theorem 4.25, also give a complete solution to the analogous classification problem for elliptic curves.

5.12 Modular forms

As Theorem 5.18 makes evident, the property of being a modular function is such a strong one that we end up with a fairly small collection of functions, the rational functions in $J(\tau)$, which, moreover, does not include most of the interesting functions we already encountered and which served as motivation for the much of the theory we developed so far in this chapter, such as the Eisenstein series and the modular discriminant.

Fortunately, the true richness and beauty of the theory starts to emerge once we expand our notion of modularity from modular functions to the more general concept of modular *forms*. For an integer $\ell \geq 0$, we say that a function $f : \mathbb{H} \to \mathbb{C}$ is an **entire modular form of weight** ℓ if it is a pre-modular form, is holomorphic at $i\infty$ (that is, the Fourier expansion (5.20) contains no terms with $n < 0$), and satisfies the condition

$$f\left(\frac{a\tau + b}{c\tau + d}\right) = (c\tau + d)^\ell f(\tau) \quad \text{for all } \tau \in \mathbb{H}, \ \begin{pmatrix} a & b \\ c & d \end{pmatrix} \in \Gamma. \tag{5.49}$$

We say that $f : \mathbb{H} \to \mathbb{C}$ is a **weak modular form of weight** ℓ if it is a weak premodular form and satisfies (5.49). Note that the notion of modular functions coincides with that of weak modular forms of weight $\ell = 0$.[5]

In practice, to check that a function is a modular form, it is sufficient and necessary to check that it is periodic and transforms in a certain way under the map $\tau \mapsto -1/\tau$, as the next lemma explains.

Lemma 5.21. *A function $f : \mathbb{H} \to \mathbb{C}$ satisfies (5.49) if and only if it satisfies the functional equations*

$$f(\tau + 1) = f(\tau), \quad f(-1/\tau) = \tau^\ell f(\tau). \tag{5.50}$$

Proof. Exercise 5.6. □

Another simple observation is that if f is a nonzero (weak or entire) modular form of weight ℓ, the weight must be an even integer. This is necessary for the condition (5.49) to be self-consistent, since we can apply this relation with the group element $\begin{pmatrix} a & b \\ c & d \end{pmatrix}$ of Γ being equal to either $\begin{pmatrix} 0 & -1 \\ 1 & 0 \end{pmatrix}$ or $\begin{pmatrix} 0 & 1 \\ -1 & 0 \end{pmatrix}$ (both representing the same Möbius transformation $\tau \mapsto -1/\tau$), to get that

$$\tau^\ell f(\tau) = f\left(\frac{-1}{\tau}\right) = f\left(\frac{1}{-\tau}\right) = (-\tau)^\ell f(\tau) = (-1)^\ell \tau^\ell f(\tau),$$

implying that either f is identically zero or ℓ is even.

The following result is an analogue of Theorem 5.12 for modular forms and is of fundamental importance.

Theorem 5.22 (The weight formula for modular forms). *Let $f : \mathbb{H} \to \mathbb{C}$ be an entire modular form of weight ℓ that is not the zero function. Then*

$$12 \sum_{f(\xi)=0} w(\xi) = \ell. \tag{5.51}$$

Here the summation extends over all zeros ξ of $f(\tau)$ counted with multiplicities, including the point $i\infty$ if it is a zero.

---

5 The logic behind not attaching the label "weak" to modular *functions* is that, as Corollary 5.14 shows, there is no useful notion of a "strong" or "entire" modular function. Nonetheless, this terminology is a bit inconsistent and a possible source of confusion to be aware of.

Proof. The proof involves a repetition of the calculation used in the proof of Theorem 5.12, where we consider the same contour integral (5.41) as we did in that proof. In the current setting, the modular transformation property (5.49) that generalizes the simple notion of modular invariance associated with a modular function will affect the calculation in a specific way, which needs to be carefully examined. We will not go over the full calculation again, but simply point out where the change happens, which is in the consideration of the contour integrals of $f'(\tau)/f(\tau)$ over the subcontours γ_4 and γ_5 of the overall integration contour $\mathcal{G}$ (refer to the proof of Theorem 5.12 for the definitions). Where previously we saw in (5.42) that the two integrals cancel led each other out, now there will be a residual effect from the factor τ^ℓ appearing in the transformation property (5.50). Specifically, the version of (5.42) updated for the current situation is

$$
\int_{\gamma_5} \frac{f'(\tau)}{f(\tau)}\, d\tau = \int_{-\gamma_4} \frac{f'(-1/\rho)}{f(-1/\rho)} \frac{d\rho}{\rho^2} = -\int_{\gamma_4} \frac{f'(\rho)}{f(\rho)}\, d\rho - \int_{\gamma_4} \frac{\ell}{\rho}\, d\rho
$$

$$
= -\int_{\gamma_4} \frac{f'(\rho)}{f(\rho)}\, d\rho + \ell \int_i^{e^{2\pi i/3}} \frac{d\rho}{\rho} = -\int_{\gamma_4} \frac{f'(\rho)}{f(\rho)}\, d\rho + \frac{\pi i}{6}\ell.
$$

We leave to the reader to check that when the reasoning of the proof of Theorem 5.12 is carried out again but with the new term $\pi i \ell/6$ included, the result is precisely (5.51). (Note that another difference from the case of modular functions is that in the current setting, poles are not allowed, which means that when repeating the calculation from the proof of Theorem 5.12, all the terms associated with counting poles can be set to 0.) □

Theorem 5.22 gives us a powerful tool for understanding what sort of functions can be entire modular forms of different weights. We now aim to use it to classify the modular forms of even weight $\ell = 2k$ for any $k \geq 0$. We start by answering this question for small values of k.

Proposition 5.23. *Let $f(\tau)$ be a modular form of even weight $\ell = 2k \leq 10$. Then:*
(a) *If $\ell = 0$, then f is a constant.*
(b) *If $\ell = 2$, then f is the zero function.*
(c) *If $\ell \in \{4, 6, 8, 10\}$, then f is a constant multiple of G_{2k}.*

Proof. The case $\ell = 0$ is the case of modular functions without poles. In this case, we already saw in Corollary 5.14 that the only functions with these properties are the constant functions.

For the case $\ell = 2$, note that by the definition of the weight function $w(\xi)$ formula (5.22) cannot be satisfied with any possible (multi)set of zeros, as the smallest positive contribution on the left-hand side can be 4, so f must be the zero function.

Similarly, for other values $\ell \in \{4, 6, 8, 10\}$, formula (5.22) can be satisfied but only in very limited ways. Specifically, it is impossible to have any zeros at points other than $\tau = i, e^{2\pi i/3}$, since for such zeros, we have $12w(\xi) = 12$, which is too large. So we need to consider for each value of ℓ different solutions in nonnegative integers a, b of the equation

$$\ell = 4a + 6b.$$

Here a and b denote the orders of the zero of $f(\tau)$ at $\tau = e^{2\pi i/3}$ and $\tau = i$, respectively.

In the case $\ell = 4$ the only solution is $a = 1$, $b = 0$, that is, $f(\tau)$ must have a simple zero at $\tau = e^{2\pi i/3}$ and no other zeros. By the same reasoning applied to the Eisenstein series G_4 instead of to f, G_4 as well must have a simple zero at $\tau = e^{2\pi i/3}$ and no other zeros. Therefore $f(\tau)/G_4(\tau)$ is an entire modular form of weight 0 and hence a constant by part (a).

In the case $\ell = 6$, we get that $a = 0$ and $b = 1$, so $f(\tau)$ must have a simple zero at $\tau = i$ and no other zeros. Again, the same conclusion must also apply to G_6, so $f(\tau)/G_6(\tau)$ is an entire modular form of weight 0 and hence a constant.

In the case $\ell = 8$ the unique solution is $a = 2$, $b = 0$, so $f(\tau)$ must have a zero of order 2 at $\tau = e^{2\pi i/3}$ and no other zeros. Therefore $f(\tau)/G_4(\tau)^2$ is an entire modular form of weight 0 and hence a constant. Since G_4^2 is proportional to G_8 (see (4.19)), the claim follows in this case.

Finally, in the case $\ell = 10$, we get that $a = b = 1$, so $f(\tau)$ has a simple zero at $\tau = i$, a simple zero at $\tau = e^{2\pi i/3}$, and no other zeros. Therefore $f(\tau)/(G_4(\tau)G_6(\tau))$ is an entire modular form of weight 0 and hence a constant. Since we know from (4.20) that $G_4 G_6$ is proportional to G_{10}, this case is also proved. □

The next result characterizes *all* entire modular forms of an arbitrary even weight. This is best stated in terms of linear algebra. For $k \geq 0$, we define the vector space M_{2k} (over the field of complex numbers, naturally) as the space consisting of all entire modular forms of weight $2k$.

Theorem 5.24. (a) *The vector spaces M_{2k} are finite-dimensional. Their dimensions are equal to*

$$\dim M_{2k} = \begin{cases} \frac{2k-2}{12} & \text{if } 2k \equiv 2 \ (\mathrm{mod} \ 12), \\ \lfloor \frac{2k}{12} \rfloor + 1 & \text{otherwise.} \end{cases}$$

(b) *A linear basis for M_{2k} is the set*

$$\mathcal{A}_k = \{G_4(\tau)^a G_6(\tau)^b \ : \ a, b \in \mathbb{Z}, \ a, b \geq 0, \ 4a + 6b = 2k\}. \tag{5.52}$$

(c) *Another linear basis for M_{2k} is the set*

$$\mathcal{B}_k = \left\{ G_{2k-12a}(\tau)\Delta(\tau)^a \ : \ a \in \mathbb{Z}, \ 0 \leq a \leq \left\lfloor \frac{2k}{12} \right\rfloor, \ 12a \neq 2k - 2 \right\} \tag{5.53}$$

with the notational convention that $G_0 = 1$.

Proof. We prove part (c) (which also trivially implies part (a)) by induction on k. The base cases $2k = 0, 2, 4, 6, 8, 10$ form precisely the content of Proposition 5.23. For the inductive step, let $2k \geq 12$. We claim that M_{2k} is spanned by the set $\mathcal{B}_k$. To show this, let $f \in M_{2k}$. Let $\alpha = f(i\infty)$ be the constant coefficient in the Fourier expansion for f. Then $g(\tau) = f(\tau) - \alpha \frac{G_{2k}(\tau)}{G_{2k}(i\infty)}$ is an entire modular form of weight $2k$ and satisfies $g(i\infty) = 0$. Therefore the function $g(\tau)/\Delta(\tau)$ is an entire modular form of weight $2k - 12$, that is, an element of the space M_{2k-12}. By the inductive hypothesis it can be represented as a linear combination of the form

$$\frac{g(\tau)}{\Delta(\tau)} = \sum_a c_a G_{2k-12-12a}(\tau)\Delta(\tau)^a$$

for some coefficients c_a, where the sum ranges over all $a \geq 0$ for which $2k - 12 - 12a \geq 0$ and $2k - 12 - 12a \neq 2$. In terms of the original modular form f, this means that we have represented f in the form

$$f(\tau) = \frac{\alpha}{G_{2k}(i\infty)} G_{2k}(\tau) + \sum_a c_a G_{2k-12(a+1)}(\tau)\Delta(\tau)^{a+1},$$

which is a linear combination of elements of $\mathcal{B}_k$. This proves that $\mathcal{B}_k$ spans M_{2k}.

To establish linear independence, assume that we have a linear relation of the form

$$\sum_a c_a G_{2k-12a}(\tau)\Delta(\tau)^a = 0$$

over the appropriate range of indices a. In particular, for $\tau = i\infty$, this implies that $c_0 = 0$, since $G_{2k}(i\infty) = 2\zeta(2k) \neq 0$ (recall (5.23)). The remaining expression can be factored as

$$\Delta(\tau) \sum_{a \geq 1} c_a G_{2k-12a}(\tau)\Delta(\tau)^{a-1} = 0,$$

that is,

$$\sum_{a \geq 1} c_a G_{2k-12-12(a-1)}(\tau)\Delta(\tau)^{a-1} = 0,$$

so by the inductive hypothesis, $c_a = 0$ for all a. The proof by induction is complete.

Finally, to prove part (b), since we already showed that $\mathcal{B}_k$ is a linear basis, it is sufficient to show that any element of $\mathcal{B}_k$ can be represented as a linear combination of elements of $\mathcal{A}_k$ and that $\mathcal{A}_k$ and $\mathcal{B}_k$ have the same cardinality. The second claim is left as an exercise (Exercise 5.7). For the first claim, use (4.22) and an induction to show that for any $j \geq 2$, G_{2j} can be represented as a linear combination of terms of the form $G_4^p G_6^q$, where $p, q \geq 0$ are integers satisfying $4p + 6q = 2j$ (this is a slightly more precise version of Corollary 4.15). Then, taking $2j = 2k - 2a$ and using the fact that Δ is a linear combination of G_4^3 and G_6^2, we see that $G_{2k-12a}\Delta^a$ can similarly be expressed as a linear combination of monomials $G_4^a G_6^b$ with $4a + 6b = 2k$, as claimed. □

Corollary 5.25. *Any entire modular form can be expressed as a polynomial in G_4 and G_6.*

Proof of (5.36) *and* (5.37). We now revisit our earlier discussion about the Eisenstein series identities (5.36)–(5.37) and the number-theoretic identities (5.34)–(5.35) they imply. The main thing to observe is that Theorem 5.24 reduces these identities and similar ones to essentially a triviality, since it represents an equality between elements of a finite-dimensional (indeed, very low-dimensional in the situation at hand) vector space, whose existence can be guessed based on simple linear-algebraic considerations, and whose precise form can be derived mechanically.

The verification in the case of (5.36) is as follows: since the space M_{14} of modular forms of weight 14 is of dimension 1 and contains both G_{14} and $G_4 G_{10}$, there must be a linear dependence between these two modular forms, that is, a relation of the form $G_{14} = c G_4 G_{10}$ for some constant c. The value of the constant c can now be found simply by comparing the zeroth Fourier coefficient of the two sides of the relation. (You can check that this leads to $c = 6/13$.)

The verification of (5.37) is similar but involves the *two-dimensional* space M_{12}, which contains the modular forms Δ, G_{12}, and G_6^2 as elements. Again, because of our knowledge of the dimension of the space, we can deduce the existence of a linear dependence relation of the form $\Delta = a G_{12} + b G_6^2$ for some unknown constants a, b. Looking at the first *two* Fourier coefficients gives two linear equations for the coefficients a, b, which (again, you are encouraged to check) are easily solved to give the values $a = 1\,200 \times 1\,430$ and $b = -1\,200 \times 691$. □

Suggested exercises for Section 5.12. 5.6, 5.7, 5.8.

5.13 Examples of modular forms

We have already encountered some of the most important examples of modular forms, namely:

1. The Eisenstein series G_{2k}, $k \geq 2$, is a modular form of weight $2k$.
2. The modular discriminant $\Delta = g_2^3 - 27 g_3^2$ is a modular form of weight 12.
3. Klein's J-function $J = g_2^3/\Delta$ is a modular function and a weak modular form of weight 0.

Although Corollary 5.25 guarantees that *all* modular forms can in fact be represented in terms of these known, "obvious" examples, other examples of modular forms sometimes appear "in the wild," arising out of formulas that do not make it at all obvious that these functions are either modular forms or related toEisenstein series.[6] Below we survey a few important examples.

---

6 Moreover, many more examples come up in more advanced parts of the theory when we broaden the notion of what a modular form is to allow for functions that have nice transformation properties

5.13.1 Theta functions

In our study of the Riemann zeta function in Chapter 2, we encountered the function

$$\theta(t) = \sum_{n=-\infty}^{\infty} e^{-\pi n^2 t}$$

(see (2.15)) whose functional equation $\theta(1/t) = \sqrt{t}\theta(t)$ (Theorem 2.7) provides one of the standard ways of analytically continuing $\zeta(s)$ to a meromorphic function on $\mathbb{C}$ and proving its functional equation. This function is in fact a mildly disguised modular form (although of weight half, and under the action of a subgroup of Γ rather than the full modular group) and belongs to a much larger family of functions known as **theta functions**. Switching to the notation more customary to use in the theory of modular forms, we define functions

$$\theta_2(\tau) = \sum_{n=-\infty}^{\infty} e^{\pi i (n+1/2)^2 \tau}, \tag{5.54}$$

$$\theta_3(\tau) = \sum_{n=-\infty}^{\infty} e^{\pi i n^2 \tau}, \tag{5.55}$$

$$\theta_4(\tau) = \sum_{n=-\infty}^{\infty} (-1)^n e^{\pi i n^2 \tau}. \tag{5.56}$$

We will refer to them as the **Jacobi thetanull functions**.[7]

Theorem 5.26. *The functions $\theta_j(\tau)$ satisfy the following transformation properties under the generators T, S of the modular group Γ:*

$$\theta_2(\tau+1) = e^{\pi i/4}\theta_2(\tau), \quad \theta_2(-1/\tau) = \sqrt{-i\tau}\,\theta_4(\tau), \tag{5.57}$$

$$\theta_3(\tau+1) = \theta_4(\tau), \quad \theta_3(-1/\tau) = \sqrt{-i\tau}\,\theta_3(\tau), \tag{5.58}$$

$$\theta_4(\tau+1) = \theta_3(\tau), \quad \theta_4(-1/\tau) = \sqrt{-i\tau}\,\theta_2(\tau). \tag{5.59}$$

Proof. This is Exercise 5.9. Note that the relations involving $\theta_j(\tau+1)$ are immediate from the definitions; the relation between $\theta_3(-1/\tau)$ and $\theta_3(\tau)$ is the same as the transformation property $\theta(1/x) = \sqrt{x}\theta(x)$ discussed above from the theory of the Riemann zeta function; and the remaining relations involving $\theta_j(-1/\tau)$ for $j = 2, 4$ are proved using an argument

---

with respect to only a *subgroup* of the full modular group or otherwise relax or generalize the various conditions a modular form is expected to satisfy. Here we focus mostly on the forms that are modular under the full action of Γ.

7 The functions $\theta_j(\tau)$ are also sometimes referred to as **Jacobi theta constants** or **Jacobi theta functions**. The term "Jacobi theta function" also denotes a more general function of two complex variables z and τ, which specializes to our θ_j under certain substitutions of the "elliptic" variable z.

involving the Poisson summation formula similarly to that used to prove the functional equation for $\theta_3(-1/\tau)$. □

Theorem 5.27. *We have the following identities:*

$$G_4 = \frac{\pi^4}{90}(\theta_2^8 + \theta_3^8 + \theta_4^8), \tag{5.60}$$

$$G_6 = \frac{\pi^6}{945}(\theta_3^{12} + \theta_4^{12} - 3\theta_2^8(\theta_3^4 + \theta_4^4)) \tag{5.61}$$

$$= \frac{\pi^6}{\sqrt{2}\cdot 945}((\theta_2^8 + \theta_3^8 + \theta_4^8)^3 - 54(\theta_2\theta_3\theta_4)^8)^{1/2}, \tag{5.62}$$

$$\Delta = 16\pi^{12}(\theta_2\theta_3\theta_4)^8. \tag{5.63}$$

Proof. Exercise 5.10. □

5.13.2 The modular lambda function

Define the function $\lambda : \mathbb{H} \to \mathbb{C}$ by

$$\lambda(\tau) = \frac{e_3(\tau) - e_2(\tau)}{e_1(\tau) - e_2(\tau)}, \tag{5.64}$$

where $e_1(\tau), e_2(\tau), e_3(\tau)$ are the quantities derived from the Weierstrass $\wp$-function associated with the lattice $L = \mathbb{Z} + \tau\mathbb{Z}$ according to (4.25). The function $\lambda(\tau)$ is known as the **modular lambda function**. It is a modular form, although not quite of the ordinary kind we are used to work with. The next result adds more details.

Theorem 5.28. (a) $\lambda(\tau)$ *is a modular function under the action of the congruence group* $\Gamma(2)$ *discussed in Exercise 5.4, that is,* $\lambda(\tau)$ *satisfies*

$$\lambda\left(\frac{a\tau + b}{c\tau + d}\right) = \lambda(\tau) \quad \text{for all} \quad \begin{pmatrix} a & b \\ c & d \end{pmatrix} \in \Gamma(2).$$

(b) *Klein's J-invariant can be expressed in terms of* $\lambda(\tau)$ *as*

$$J = 256\frac{(1 - \lambda + \lambda^2)^3}{\lambda^2(1 - \lambda)^2}.$$

Proof. Exercise 5.11. □

The modular λ function has interesting applications to parts of complex analysis that seem a priori unrelated to modular forms. The most well-known such application is its use in giving a slick proof of a deep result known as **Picard's theorem**.

Theorem 5.29 (Picard's theorem). *Let $f : \mathbb{C} \to \mathbb{C}$ be an entire function such that two distinct complex numbers a, b are not in the image of f. Then f is a constant.*

The proof, although conceptually simple, involves a use of the monodromy theorem, which is outside the scope of this book. See [1, Ch. 8] for the details.

Another appearance of the modular lambda function is in connection with a maximization problem in the theory of conformal mapping of doubly connected regions. This is discussed in [2, Sec. 4.12].

5.13.3 The zeros of $\wp(z)$ and their modular properties

Fix a lattice $L = \omega_1 \mathbb{Z} + \omega_2 \mathbb{Z}$. In our discussion of doubly periodic functions in Chapter 4, we saw that both $\wp(z)$ and its derivative $\wp'(z)$ have their poles at the points of L and that $\wp'(z)$ has its zeros at the half-periods $\frac{1}{2}\omega_1, \frac{1}{2}\omega_2, \frac{1}{2}(\omega_1 + \omega_2)$. We also discussed that $\wp(z)$ takes every value twice in any fundamental parallelogram as a doubly periodic function of order 2. It might therefore seem like a curious omission that we never discussed the question of where the *zeros* of $\wp(z)$ are located. In fact, the question of the location of the zeros as a function of the lattice L turns out to be quite nontrivial and gives rise to an interesting modular form.

Let us denote the location of one of the zeros of $\wp(z)$ by Z. This is a function of the lattice L, so we can write $Z = Z(L)$ or

$$Z = Z(\tau)$$

if we switch to the notation involving the modular variable τ taking values in the upper half-plane and representing the "canonical" lattice $L_\tau = \mathbb{Z} + \tau\mathbb{Z}$, that is, the defining equation of $Z(\tau)$ is

$$\wp(Z(\tau); \tau) \equiv 0 \quad (\tau \in \mathbb{H}).$$

It is natural to think of $Z(\tau)$ as a multivalued function of τ in the sense that— similarly to the logarithm and kth root functions, we are familiar with from basic complex analysis—it takes its values in the quotient of the complex plane by some discrete group of symmetries. In our case the set of zeros of $\wp(z)$ has two obvious symmetries: it is L-periodic, and (since $\wp(z)$ is even) it is invariant under reflection $z \mapsto -z$. Thus $Z(\tau)$ can be thought of as a function of τ that is well-defined up to a translation by an arbitrary element of L and a sign change. Moreover, the location of any one zero of $\wp(z)$ determines the location of all of its zeros, since if Z lies in some fundamental parallelogram, then either Z is a half-period and then must be of order 2 (in which case there are no other zeros in the parallelogram), or Z is not a half-period, is a simple zero, and is matched by another zero at the unique point in the parallelogram that is congruent to −Z modulo L. That is, geometrically, the zeros come in pairs of points that are reflections of each other around the center of the parallelogram.

It is worth keeping in mind that when we discuss multivalued functions, we are really talking about functions taking values in a certain Riemann surface. We will not explore this point of view in depth, but if you find it interesting, then try to think what the Riemann surface is in this case.

The question of understanding the behavior of $Z(\tau)$ seems to have been addressed for the first time in a 1982 paper by Eichler and Zagier [26], who derived a formula for this function. A more explicit formula was found in 2008 by Duke and Imamoğlu [24]. It seems possible that the last word has not yet been said on this interesting and quite nontrivial problem.

We present below without proof Eichler and Zagier's result, which ties in a nice way to our current discussion of modular forms.

Theorem 5.30. (a) *The function $Z(\tau)$ is holomorphic.*

(b) *The function $Z''(\tau)^2$ is a single-valued function of τ, that is, an ordinary holomorphic function on $\mathbb{H}$.*

(c) *The function $Z''(\tau)^2$ is a weak modular form of weight 6 for the modular group Γ. It is given explicitly by*

$$Z''(\tau)^2 = -124\,416\pi^2 \frac{\Delta(\tau)^2}{E_6(\tau)^3}.$$

(d) *$Z(\tau)$ can be expressed explicitly as*

$$Z(\tau) = \mathbb{Z} + \tau\mathbb{Z} + \frac{1}{2} \pm \left(\frac{\log(5 + 2\sqrt{6})}{2\pi i} + 144\pi i\sqrt{6} \int_\tau^{i\infty} (\rho - \tau)\frac{\Delta(\rho)}{E_6(\rho)^{3/2}}\, d\rho \right).$$

5.13.4 Infinite products

Modular forms often arise in applications in the form of certain types of infinite products, where, again, the fact that the function expressed in such a way is a modular form is not easily apparent. This is the subject of the next section.

Suggested exercises for Section 5.13. 5.9, 5.10, 5.11.

5.14 Infinite products for modular forms

One additional beautiful and somewhat mysterious aspect of the theory of modular forms is the fact that many modular forms that are commonly encountered in the theory have elegant representations as infinite products. It is not clear whether there is a good conceptual explanation for why this happens so frequently [W20], or whether instead it is yet another vivid illustration of John von Neumann's famous quip that "in mathematics you don't understand things. You just get used to them."[8] Our goal in this section is to prove a few of the most well-known identities of this type.

5.14.1 The modular discriminant

The following result is one of the famous identities of modular form theory.

Theorem 5.31. *The modular discriminant Δ has the infinite product representation*

$$\Delta(\tau) = (2\pi)^{12} q \prod_{n=1}^{\infty} (1 - q^n)^{24} \quad (q = e^{2\pi i \tau}, \ \tau \in \mathbb{H}), \tag{5.65}$$

One reason why identity (5.65) is interesting is that it highlights an unexpected connection between the modular discriminant and integer partitions, since the function on the right-hand side of (5.65) is, up to trivial factors, the generating function of integer partitions raised to the power -24. The connection between modular forms and integer partitions goes much further than this single identity and has far-reaching consequences that go quite deep into the theory; you can learn about it in more specialized books, such as [5].

The existence of identity (5.65) is closely tied to yet another intriguing object, which we will now study, the **weight 2 Eisenstein series** G_2. One motivation for introducing G_2 is that Theorem 5.24 suggests an annoying gap in the dimensions of the vector spaces $M_{2k}(\Gamma)$. Noticing this, we might wonder whether the definition of a modular form of weight 2 can be modified somehow to lead to some useful family of functions rather than the empty set and—which is related—whether formula (4.13) defining the Eisenstein series can be made to make sense for the exponent 2 through some simple modification. The answer to both these questions is "yes"; in fact, the modification to (4.13) is the most obvious one that one can think of and consists of replacing an absolutely convergent series by a conditionally convergent one. The next result explains what happens when such a modification is carried out.

Theorem 5.32. *Define the weight 2 Eisenstein series G_2 by*

$$G_2(\tau) = \sum_{m \in \mathbb{Z}}^{\infty} \left[\sum_{\substack{n \in \mathbb{Z} \\ (m,n) \neq (0,0)}} \frac{1}{(m\tau + n)^2} \right]. \tag{5.66}$$

---

8 Von Neumann said this in response to a complaint from a colleague that he did not understand the method of characteristics [75, p. 208].

(a) *Expression (5.66) defines a meromorphic function $G_2(\tau)$ on the upper half-plane.*

(b) *G_2 transforms under the actions of the generators T, S of Γ as*

$$G_2(\tau + 1) = G_2(\tau),$$ (5.67)

$$G_2(-1/\tau) = \tau^2 G_2(\tau) - 2\pi i \tau.$$ (5.68)

(c) *G_2 is a premodular form with the Fourier expansion*

$$G_2(\tau) = \frac{\pi^2}{3}\left(1 - 24\sum_{n=1}^{\infty}\sigma_1(n)q^n\right) \quad (q = e^{2\pi i \tau}, \tau \in \mathbb{H}).$$ (5.69)

Note that (5.69) is the case $2k = 2$ of the Fourier expansion (5.23). Thus we see yet another way in which G_2 can be thought of as extending the definition of the original Eisenstein series G_{2k}, $k \geq 2$, in the most natural way possible by a kind of "analytic continuation" (very loosely speaking), that is, by taking one of the formulas that represent those series and simply observing that it continues to represent a well-defined object even in the case $2k = 2$.

Proof. (a) For $m = 0$, the inner sum in (5.66) is equal to $2\zeta(2) = \pi^2/3$. For $m \neq 0$, this inner sum can be summed using (5.25) as

$$\sum_{n=-\infty}^{\infty}\frac{1}{(m\tau + n)^2} = -\frac{1}{m}\cdot\frac{d}{d\tau}(\pi\cot(\pi m\tau)) = \frac{\pi^2}{\sin^2(\pi m\tau)}.$$

It is now easy to see that the infinite series $\sum_{m\neq 0}\sin^{-2}(\pi m\tau)$ converges absolutely uniformly on compacts in $\mathbb{H}$ (since $|\sin(z)|$ grows exponentially in $|\operatorname{Im}(z)|$). Thus $G_2(z)$ is well-defined and holomorphic on $\mathbb{H}$.

(c) The calculation is essentially a repetition of (5.27): again using (5.25) and also (5.26), we have

$$\sum_{m\in\mathbb{Z}}\left[\sum_{\substack{n\in\mathbb{Z}\\(m,n)\neq(0,0)}}\frac{1}{(m\tau + n)^2}\right]$$

$$= \frac{\pi^2}{3} + \sum_{m\neq 0}\sum_{n=-\infty}^{\infty}\frac{1}{(m\tau + n)^2} = \frac{\pi^2}{3} + 2\sum_{m=1}^{\infty}\sum_{n=-\infty}^{\infty}\frac{1}{(m\tau + n)^2}$$

$$= \frac{\pi^2}{3} - 8\pi^2\sum_{m=1}^{\infty}\sum_{\ell=1}^{\infty}\ell e^{2\pi i\ell m\tau} = \frac{\pi^2}{3} - 8\pi^2\sum_{n=1}^{\infty}\left(\sum_{\substack{\ell,m\geq 1\\\ell m=n}}\ell\right)e^{2\pi i n\tau}$$

$$= \frac{\pi^2}{3}\left(1 - 24\sum_{n=1}^{\infty}\sigma_1(n)e^{2\pi i n\tau}\right),$$

as claimed.

(b) The first relation (5.67) is obvious from (5.69) and also easy to check directly from the definition of G_2. Thus the main challenge is to prove (5.68). As in the proof of (c) above, we start by attempting to replicate the calculation that we used to prove the analogous property (5.3) (in the particular case of the transformation $S(z) = -1/z$ for the "proper" Eisenstein series. However, in this case, we are in for a surprise. Specifically, multiplying the left-hand side of (5.68) by τ^{-2} gives

$$\tau^{-2} G_2(-1/\tau) = \sum_{m \neq 0} \left[\sum_{n \in \mathbb{Z}} \frac{1}{(m(-1/\tau) + n)^2} \right] \tau^{-2} + 2\zeta(2)\tau^{-2}$$

$$= \frac{\pi^2}{3\tau^2} + \sum_{m \neq 0} \left[\sum_{n \in \mathbb{Z}} \frac{1}{(n\tau - m)^2} \right] = \frac{\pi^2}{3\tau^2} + \sum_{m \neq 0} \left[\frac{1}{m^2} + \sum_{n \neq 0} \frac{1}{(n\tau - m)^2} \right]$$

$$= \frac{\pi^2}{3\tau^2} + \frac{\pi^2}{3} + \sum_{m \neq 0} \left[\sum_{n \neq 0} \frac{1}{(n\tau - m)^2} \right]$$

$$= \frac{\pi^2}{3} + \sum_{n \neq 0} \frac{1}{(n\tau)^2} + \sum_{m \neq 0} \left[\sum_{n \neq 0} \frac{1}{(n\tau - m)^2} \right]$$

$$= \frac{\pi^2}{3} + \sum_{m} \left[\sum_{n \neq 0} \frac{1}{(n\tau - m)^2} \right] = \frac{\pi^2}{3} + \sum_{n} \left[\sum_{m \neq 0} \frac{1}{(m\tau + n)^2} \right].$$

Comparing this to (5.66) and (5.68), we see that the proof of (5.68) reduces to showing the following curious rearrangement identity:

$$\sum_{m \neq 0} \left(\sum_{n} \frac{1}{(m\tau + n)^2} \right) - \sum_{n} \left(\sum_{m \neq 0} \frac{1}{(m\tau + n)^2} \right) = \frac{2\pi i}{\tau}. \tag{5.70}$$

In other words, what we have here is a naturally occurring example of a conditionally convergent double summation for which changing the order of summation not only changes the value of the series (which can happen, as we know from calculus), but changes it in a predictable and rather interesting way.[9] It is precisely this change that accounts for G_2 satisfying the "exotic" transformation property (5.68) (sometimes described as a "quasimodular" relation) rather than the more standard modular transformation relation satisfied by the other, absolutely convergent G_{2k}.

Denote the first double sum on the left-hand side of (5.70) by X and the second by Y. We have

$$X = \sum_{m \neq 0} \left[\sum_{n} \left(\frac{1}{(m\tau + n)^2(m\tau + n + 1)} + \frac{1}{m\tau + n} - \frac{1}{m\tau + n + 1} \right) \right]$$

$$= \sum_{m \neq 0} \left[\sum_{n} \frac{1}{(m\tau + n)^2(m\tau + n + 1)} \right] + \sum_{m \neq 0} \left[\sum_{n} \left(\frac{1}{m\tau + n} - \frac{1}{m\tau + n + 1} \right) \right].$$

---

9 The examples illustrating this sort of order-dependence phenomenon in calculus textbooks often have a rather contrived feel to them. This one seems more natural.

A key observation here is that the first of these two double summations is *absolutely convergent*. Similarly,

$$Y = \sum_n \left[\sum_{m \neq 0} \left(\frac{1}{(m\tau + n)^2(m\tau + n + 1)} + \frac{1}{m\tau + n} - \frac{1}{m\tau + n + 1} \right) \right]$$

$$= \sum_n \left[\sum_{m \neq 0} \frac{1}{(m\tau + n)^2(m\tau + n + 1)} \right] + \sum_n \left[\sum_{m \neq 0} \left(\frac{1}{m\tau + n} - \frac{1}{m\tau + n + 1} \right) \right],$$

and thus, by absolute convergence, the difference $X - Y$ is now seen to be equal to

$$\sum_{m \neq 0} \left[\sum_n \left(\frac{1}{m\tau + n} - \frac{1}{m\tau + n + 1} \right) \right] - \sum_n \left[\sum_{m \neq 0} \left(\frac{1}{m\tau + n} - \frac{1}{m\tau + n + 1} \right) \right]. \tag{5.71}$$

The first of these new double series is trivial to evaluate, since the internal summation is telescoping: we have

$$\sum_{m \neq 0} \left[\sum_n \left(\frac{1}{m\tau + n} - \frac{1}{m\tau + n + 1} \right) \right]$$

$$= \sum_{m \neq 0} \left[\lim_{N \to \infty} \sum_{n=-N}^{N} \left(\frac{1}{m\tau + n} - \frac{1}{m\tau + n + 1} \right) \right]$$

$$= \sum_{m \neq 0} \left[\lim_{N \to \infty} \left(\frac{1}{m\tau - N} - \frac{1}{m\tau + N + 1} \right) \right] = \sum_{m \neq 0} 0 = 0.$$

The second double series in (5.71) is only slightly more challenging. Write

$$\sum_n \left[\sum_{m \neq 0} \left(\frac{1}{m\tau + n} - \frac{1}{m\tau + n + 1} \right) \right]$$

$$= \lim_{N \to \infty} \sum_{n=-N}^{N-1} \left[\sum_{m \neq 0} \left(\frac{1}{m\tau + n} - \frac{1}{m\tau + n + 1} \right) \right]$$

$$= \lim_{N \to \infty} \sum_{m \neq 0} \left[\sum_{n=-N}^{N-1} \left(\frac{1}{m\tau + n} - \frac{1}{m\tau + n + 1} \right) \right]$$

$$= \lim_{N \to \infty} \sum_{m \neq 0} \left(\frac{1}{m\tau - N} - \frac{1}{m\tau + N} \right)$$

$$= \lim_{N \to \infty} \sum_{m=1}^{\infty} \left(\frac{1}{m\tau - N} - \frac{1}{m\tau + N} + \frac{1}{-m\tau - N} - \frac{1}{-m\tau + N} \right)$$

$$= \lim_{N \to \infty} 2 \sum_{m=1}^{\infty} \left(\frac{1}{m\tau - N} - \frac{1}{m\tau + N} \right)$$

$$= -\frac{2}{\tau} \lim_{N \to \infty} \sum_{m=1}^{\infty} \left(\frac{1}{\frac{N}{\tau} + m} + \frac{1}{\frac{N}{\tau} - m} \right). \tag{5.72}$$

Appealing to the partial fraction expansion of the cotangent function in one of its variant forms,

$$\pi \cot(\pi z) = \frac{1}{z} + \sum_{m=1}^{\infty}\left(\frac{1}{z+m} + \frac{1}{z-m}\right)$$

(a trivial recasting of (5.24)), we see that the last expression in (5.72) is equal to

$$-\frac{2}{\tau}\lim_{N\to\infty}\left(\pi\cot\left(\frac{\pi N}{\tau}\right) - \frac{\tau}{N}\right) = -\frac{2\pi}{\tau}\lim_{N\to\infty}\cot\left(\frac{\pi N}{\tau}\right).$$

By a straightforward calculation (Exercise 5.12) this is equal to $-2\pi i/\tau$, and therefore (5.71) is equal to $2\pi i/\tau$, which is what we have reduced our claim to. The proof is complete. □

Proof of Theorem 5.31. Define $\tilde{\Delta} : \mathbb{H} \to \mathbb{C}$ by

$$\tilde{\Delta}(\tau) = (2\pi)^{12} q \prod_{n=1}^{\infty}(1 - q^n)^{24}.$$

By Proposition 1.60 the product converges uniformly on compacts in $\mathbb{H}$ and defines a holomorphic function with no zeros, which, since it can be expanded as a series in powers of q with good convergence properties, is a premodular form. Our goal is to prove that $\Delta(\tau) \equiv \tilde{\Delta}(\tau)$, and this will pass through a curious relationship to the Eisenstein series G_2. Namely, the logarithmic derivative of $\tilde{\Delta}(\tau)$ is given by

$$\frac{\tilde{\Delta}'(\tau)}{\tilde{\Delta}(\tau)} = 2\pi i - 24 \sum_{n=1}^{\infty} \frac{2\pi i n\, e^{2\pi i n\tau}}{1 - e^{2\pi i n\tau}} = 2\pi i\left(1 - 24 \sum_{n=1}^{\infty} n \sum_{\ell=1}^{\infty} e^{2\pi i n\ell\tau}\right)$$

$$= 2\pi i\left(1 - 24 \sum_{m=1}^{\infty}\left(\sum_{\substack{n,\ell\geq 1\\ n\ell=m}} n\right)e^{2\pi i m\tau}\right)$$

$$= 2\pi i\left(1 - 24 \sum_{m=1}^{\infty} \sigma_1(m)e^{2\pi i m\tau}\right) = \frac{6i}{\pi}G_2(\tau). \tag{5.73}$$

We claim that this connection implies that $\tilde{\Delta}(\tau)$ is a modular form of weight 12. By Lemma 5.21 it suffices to prove that $\tilde{\Delta}$ satisfies

$$\tilde{\Delta}(-1/\tau) = \tau^{12}\tilde{\Delta}(\tau) \quad (\tau \in \mathbb{H}). \tag{5.74}$$

However, the logarithmic derivative of the left-hand side is equal to

$$\frac{\frac{d}{d\tau}(\tilde{\Delta}(-1/\tau))}{\tilde{\Delta}(-1/\tau)} = \frac{1}{\tau^2}\frac{\tilde{\Delta}'(-1/\tau)}{\tilde{\Delta}(-1/\tau)} = \frac{6i}{\pi\tau^2}G_2(-1/\tau)$$

$$= \frac{6i}{\pi\tau^2}(\tau^2 G_2(\tau) - 2\pi i\tau) = \frac{6i}{\pi}G_2(\tau) + \frac{12}{\tau},$$

which is the same as the logarithmic derivative of the right-hand side of (5.74). Therefore if we recall the trivial fact that if f, g are two meromorphic functions that are not identically zero for which $f'/f \equiv g'/g$, then $f \equiv cg$ for some constant c, then, together with the fact that (5.74) is satisfied for $\tau = i$, we deduce that (5.74) holds for general $\tau \in \mathbb{H}$. Thus $\tilde{\Delta}(\tau)$ is a modular form of weight 12. As such, it is an element of the vector space $M_{12}(\Gamma)$, which we know (Theorem 5.24) is of dimension 2 and spanned by the original modular discriminant $\Delta(\tau)$ and the Eisenstein series G_{12}. It follows that $\tilde{\Delta} \equiv \alpha\Delta + \beta G_{12}$ for some $\alpha, \beta \in \mathbb{C}$. Comparing the constant and linear terms in the Fourier expansions of Δ, G_{12}, and $\tilde{\Delta}$ shows that $\alpha = 1$ and $\beta = 0$ and finishes the proof.[10] □

The relation between Δ and G_2 that was obtained as part of the proof is of independent interest, so we note it as a corollary.

Corollary 5.33. *The functions $\Delta(\tau)$ and $G_2(\tau)$ are related to each other via*

$$\frac{\Delta'(\tau)}{\Delta(\tau)} = \frac{6i}{\pi}G_2(\tau) \quad (\tau \in \mathbb{H}).$$

5.14.2 The modular lambda function

In this and next subsections, we denote $Q = e^{\pi i\tau}$. (This is the square root of the parameter $q = e^{2\pi i\tau}$ we have been using throughout much of the discussions in this chapter and is more convenient for some expansions discussed below.[11])

Theorem 5.34. *The modular lambda function $\lambda(\tau)$ defined in (5.64) and the complementary function $1 - \lambda(\tau)$ have the infinite product representations*

$$\lambda(\tau) = 16Q \prod_{n=1}^{\infty}\left(\frac{1 + Q^{2n}}{1 + Q^{2n-1}}\right)^8, \tag{5.75}$$

$$1 - \lambda(\tau) = \prod_{n=1}^{\infty}\left(\frac{1 - Q^{2n-1}}{1 + Q^{2n-1}}\right)^8 \tag{5.76}$$

with $Q = e^{\pi i\tau}$, $\tau \in \mathbb{H}$.

---

10 In fact, the constant coefficients of Δ and $\tilde{\Delta}$ are 0, which means they both belong to the codimension-1 subspace of $M_{12}(\Gamma)$ of forms with a constant coefficient 0. Such forms are known as **cusp forms**. So an alternative way of phrasing the argument above without mentioning G_{12} is by saying that $\tilde{\Delta}$ must be proportional to Δ, since they are both cusp forms of weight 12, and since the space of such forms is one-dimensional and spanned by Δ.

11 In many textbooks on modular forms, the letter q may be used alternately for either $e^{\pi i\tau}$ or $e^{2\pi i\tau}$ depending on the context, so pay close attention to the definitions when you read the literature.

Proof of (5.75). Fix $\tau \in \mathbb{H}$, and let $L_\tau = \mathbb{Z} + \tau\mathbb{Z}$ denote as usual the associated lattice with fundamental period pair $\omega_1 = 1$, $\omega_2 = \tau$, and let $\wp(z) = \wp(z; L_\tau)$ be the Weierstrass function of L_τ. Define the meromorphic function $F : \mathbb{C} \to \mathbb{C}$ by the expression

$$F(z) = \prod_{n=-\infty}^{\infty} \frac{(1 + e^{\pi i(2n-1)\tau - 2\pi i z})(1 + e^{\pi i(2n+1)\tau - 2\pi i z})}{(1 + e^{2\pi i n\tau - 2\pi i z})^2}. \tag{5.77}$$

Denote the nth factor in this two-sided infinite product by $\zeta_n = \zeta_n(z; \tau)$. The product of ζ_n over positive values of n clearly converges absolutely, uniformly as z ranges over compacts in $\mathbb{C}$ away from any poles of individual factors, due to the exponential decay of $|e^{2\pi i n\tau}|$. Moreover, ζ_n has the symmetry $\zeta_{-n}(-z; \tau) = \zeta_n(z; \tau)$, which is easy to check, implying the same convergence also for the product over negative n. Thus $F(z)$ is well-defined and is a meromorphic function with poles only at places where one of the individual factors ζ_n has a pole (more on that below).

The usefulness of $F(z)$ is related to the fact, which we now observe, that it is a doubly periodic function with period lattice L_τ. This is easy to see: the relation $F(z + 1) = F(z)$ holds trivially, and to show that $F(z + \tau) = F(z)$, observe that the substitution $z \mapsto z + \tau$ maps each factor ζ_n to its predecessor ζ_{n-1}, that is, we have the relation $\zeta_n(z + \tau; \tau) = \zeta_{n-1}(z; \tau)$.

Next, an examination of the factors involved in the definition of ζ_n and their zeros (as a function of z with fixed τ) reveals that $F(z)$ has double poles at the half-period $z = v_1 = 1/2$ (in the notation of (4.23)) and all its L_τ-translates, and double zeros at the half-period $z = v_3 = (1 + \tau)/2$ and all its L_τ-translates. There are no other zeros or poles. This means that in fact $F(z)$ has the same zeros and poles as the doubly periodic function $\frac{\wp(z)-e_3}{\wp(z)-e_1}$. Therefore the quotient $F(z)/\frac{\wp(z)-e_3}{\wp(z)-e_1}$ is a doubly periodic function with no poles and so must be a constant. Taking $z = 0$ shows that the constant is equal to $F(0)$ (the limit of $\frac{\wp(z)-e_3}{\wp(z)-e_1}$ as $z \to 0$ is 1 because of cancelation of the principal parts of the poles of the numerator and denominator at $z = 0$; refer to (4.10)). So we have shown that

$$F(z) = F(0)\frac{\wp(z) - e_3}{\wp(z) - e_1}.$$

Now set $z = v_2 = \tau/2$ in this identity to get that

$$F(\tau/2) = F(v_2) = F(0)\frac{\wp(v_2) - e_3}{\wp(v_2) - e_1} = F(0)\frac{e_2 - e_3}{e_2 - e_1} = F(0)\lambda(\tau).$$

In other words, we have shown that the lambda function can be represented in terms of $F(z)$ as $\lambda(\tau) = F(\tau/2)/F(0)$. Making the relevant substitutions into (5.77), we see that

$$F(0) = \prod_{n=-\infty}^{\infty} \frac{(1 + e^{\pi i(2n-1)\tau})(1 + e^{\pi i(2n+1)\tau})}{(1 + e^{2\pi i n\tau})^2}$$

$$= \frac{(1 + Q^{-1})(1 + Q)}{4}$$

$$\times \prod_{n=1}^{\infty} \frac{(1 + Q^{2n-1})(1 + Q^{2n+1})(1 + Q^{-(2n-1)})(1 + Q^{-(2n+1)})}{(1 + Q^{2n})^2(1 + Q^{-2n})^2}$$

$$= \frac{(1+Q)^2}{4Q} \prod_{n=1}^{\infty} \frac{(1 + Q^{2n-1})^2(1 + Q^{2n+1})^2}{(1 + Q^{2n})^4} = \frac{1}{4Q} \prod_{n=1}^{\infty} \frac{(1 + Q^{2n-1})^4}{(1 + Q^{2n})^4}.$$

Similarly,

$$F(\tau/2) = \prod_{n=-\infty}^{\infty} \frac{(1 + e^{\pi i(2n-2)\tau})(1 + e^{\pi i(2n)\tau})}{(1 + e^{\pi i(2n-1)\tau})^2}$$

$$= \prod_{n=1}^{\infty} \frac{(1 + Q^{2n-2})(1 + Q^{2n})(1 + Q^{-(2n-2)})(1 + Q^{-2n})}{(1 + Q^{2n-1})^2(1 + Q^{-(2n-1)})^2}.$$

$$= \prod_{n=1}^{\infty} \frac{(1 + Q^{2n-2})^2(1 + Q^{2n})^2}{(1 + Q^{2n-1})^4} = 4 \prod_{n=1}^{\infty} \frac{(1 + Q^{2n})^4}{(1 + Q^{2n-1})^4}.$$

Combining the above results yields precisely the infinite product formula (5.75). □

Proof of (5.76). Exercise 5.15. □

5.14.3 The Jacobi thetanull functions

Our final result on infinite product expansions concerns the Jacobi thetanull functions.

Theorem 5.35. *The Jacobi thetanull functions have the infinite product representations*

$$\theta_2(\tau) = 2Q^{1/4} \prod_{n=1}^{\infty} (1 - Q^{2n})(1 + Q^{2n})^2, \tag{5.78}$$

$$\theta_3(\tau) = \prod_{n=1}^{\infty} (1 - Q^{2n})(1 + Q^{2n-1})^2, \tag{5.79}$$

$$\theta_4(\tau) = \prod_{n=1}^{\infty} (1 - Q^{2n})(1 - Q^{2n-1})^2, \tag{5.80}$$

with the usual notation $Q = e^{\pi i \tau}$, $\tau \in \mathbb{H}$.

As a corollary of (5.75), (5.76), and (5.78)–(5.80), we obtain two additional remarkable identities relating $\lambda(\tau)$ to the Jacobi thetanull functions.

Corollary 5.36. *The modular lambda function* $\lambda(\tau)$ *satisfies the relations*

$$\lambda(\tau) = \left(\frac{\theta_2(\tau)}{\theta_3(\tau)}\right)^4, \quad 1 - \lambda(\tau) = \left(\frac{\theta_4(\tau)}{\theta_3(\tau)}\right)^4. \tag{5.81}$$

Additional interesting corollary worth noting is the following.

Corollary 5.37. *The Jacobi thetanull functions satisfy the identity*

$$\theta_2(\tau)^4 + \theta_4(\tau)^4 = \theta_3(\tau)^4. \tag{5.82}$$

The infinite products (5.78)–(5.80) are particular cases of a more general product identity for the full Jacobi theta function (involving two variables z and τ), known as the **Jacobi triple product identity**.

Theorem 5.38 (Jacobi triple product identity). *We have the identity*

$$\sum_{n=-\infty}^{\infty} \exp(\pi i n^2 \tau + 2\pi i n z)$$
$$= \prod_{n=1}^{\infty}(1 - e^{2n\pi i \tau})(1 + e^{(2n-1)\pi i \tau + 2\pi i z})(1 + e^{(2n-1)\pi i \tau - 2\pi i z}) \tag{5.83}$$

for $\tau \in \mathbb{H}$ and $z \in \mathbb{C}$.

For a complex-analytic proof of identity (5.83) using techniques of a flavor similar to those used in the proof of Theorem 5.34; see [66, Ch. 10]. An alternative approach proceeds by rewriting (5.83) as

$$\sum_{n=-\infty}^{\infty} x^{n^2} y^n = \prod_{n=1}^{\infty}(1 - x^{2n})(1 + yx^{2n-1})(1 + y^{-1}x^{2n-1})$$

(by making the substitutions $x = e^{\pi i \tau}, y = e^{2\pi i z}$) or, equivalently,

$$\prod_{n=1}^{\infty} \frac{1}{1 - x^{2n}} \sum_{n=-\infty}^{\infty} x^{n^2} y^n = \prod_{n=1}^{\infty}(1 + yx^{2n-1})(1 + y^{-1}x^{2n-1}).$$

This can be given a combinatorial proof by interpreting both sides as bivariate generating functions for certain classes of objects associated with integer partitions. These classes are then shown to be in explicit bijection with each other, implying the equality of the coefficients at $x^j y^k$ on both sides of the equation for all j, k. See [54, Sec. 6] for details.

Proof of Theorem 5.35. Exercise 5.17. □

Suggested exercises for Section 5.14. 5.12, 5.13, 5.14, 5.15, 5.16, 5.17, 5.18, 5.19, 5.20, 5.21.

Exercises for Chapter 5

5.1 Show that the Weierstrass $\wp$-function, regarded as a function $\wp(z;\tau)$ of both the "elliptic" variable z and the modular variable τ, satisfies the transformation relation

$$\wp\left(\frac{z}{c\tau+d};\frac{a\tau+b}{c\tau+d}\right) = (c\tau+d)^2\wp(z,\tau) \quad (z\in\mathbb{C},\ \tau\in\mathbb{H}) \tag{5.84}$$

for all $a,b,c,d\in\mathbb{Z}$ for which $ad-bc=1$.

5.2 Prove Lemma 5.3. (Hint: reminding yourself of the statement of Theorem 3.13 from Chapter 3 might be helpful.)

5.3 **Structure of the modular group.** Prove that the algebraic structure of the modular group Γ can be expressed succinctly by the relation

$$\Gamma \cong Z_2 * Z_3.$$

In words, this says that Γ is isomorphic to the free product of the cyclic groups of orders 2 and 3. More precisely, show that it is freely generated by the elements S,U, that is, that if the standard cyclic groups Z_2 and Z_3 have respective generators denoted γ_2 and γ_3, then the map

$$\varphi : Z_2 * Z_3 \to \Gamma$$

(where $Z_2 * Z_3$ denotes the free product of those groups) defined by

$$\varphi(\gamma_2) = S, \quad \varphi(\gamma_3) = U,$$

and extended in the obvious way to a group homomorphism is a group *isomorphism*. (Note: this is a well-known result. A simple proof is given in [4].)

5.4 **The congruence subgroup $\Gamma(2)$.** Let

$$\Gamma(2) = \left\{A = \begin{pmatrix} a & b \\ c & d \end{pmatrix} \in \Gamma : a,d \text{ are odd, } b,c \text{ are even}\right\}.$$

It is easy to see that $\Gamma(2)$ is a subgroup of the modular group Γ either through direct verification or by noting that $\Gamma(2)$ is the kernel of the homomorphism that sends any matrix A in Γ to its reduction mod 2, an element of the matrix group $SL(2,Z_2)$. The group $\Gamma(2)$ belongs to the class of subgroups of Γ known as the **congruence subgroups**.

(a) Prove that the two matrices

$$A = \begin{pmatrix} 1 & 2 \\ 0 & 1 \end{pmatrix} \quad \text{and} \quad B = \begin{pmatrix} 1 & 0 \\ 2 & 1 \end{pmatrix} = A^{\mathsf{T}} \tag{5.85}$$

generate $\Gamma(2)$.

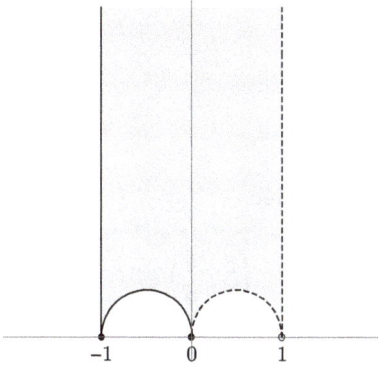

Figure 5.3: The fundamental domain $\mathcal{G}$ for the congruence subgroup $\Gamma(2)$.

(b) Prove that $\Gamma(2)$ is *freely* generated by A and B. That is, the only products we can form from A, A^{-1}, B, and B^{-1} that give the identity element are those that reduce to the identity element by successively canceling out the appearances of $AA^{-1}, A^{-1}A, BB^{-1}$, and $B^{-1}B$.

(c) Prove that the set

$$\mathcal{G} = \left\{ z \in \mathbb{H} : -1 \le |\mathrm{Re}(z)| < 1, \left| z - \frac{1}{2} \right| > 1, \left| z + \frac{1}{2} \right| \ge 1 \right\} \cup \{0\}$$

(Fig. 5.3) is a fundamental domain under the action of $\Gamma(2)$ in a sense that you should formulate precisely as an analogue of the statement of Theorem 5.4.

(d) Find the index $[\Gamma : \Gamma(2)]$.

5.5 Prove Theorem 5.10.

5.6 Prove Lemma 5.21.

5.7 Fill in the missing detail in the proof of Theorem 5.24 by proving that $|\mathcal{A}_k| = |\mathcal{B}_k|$ for all $k \ge 0$, where $\mathcal{A}_k$ and $\mathcal{B}_k$ are defined by (5.52)–(5.53).

5.8 Write a computer program to generate the change of basis matrices (in both directions) between the two linear bases $\mathcal{A}_k$ and $\mathcal{B}_k$ for the vector space M_{2k} described in Theorem 5.24. Investigate these matrices for small values of k and see if you can work out a formula for them that is valid in the general case, or find other interesting patterns.

5.9 Prove the transformation properties (5.57)–(5.59).

5.10 Prove Theorem 5.27. The idea is to show that each of the functions on the right-hand sides of (5.60)–(5.63) has the right structural properties that make it an element of the space M_{2k} for an appropriate value of k, then conclude that it is a constant multiple of the function on the left-hand side, and finally find a way to determine the value of the constant.

5.11 Prove Theorem 5.28.

5.12 Prove that if $z \in \mathbb{C} \setminus \mathbb{R}$, then

$$\lim_{N \to \infty} \cot(Nz) = \begin{cases} -i & \text{if } \mathrm{Im}(z) > 0, \\ i & \text{if } \mathrm{Im}(z) < 0. \end{cases}$$

5.13 Prove that G_2 satisfies the general transformation relation

$$G_2\left(\frac{a\tau + b}{c\tau + d}\right) = (c\tau + d)^2 G_2(\tau) - 2\pi i c(c\tau + d), \quad \begin{pmatrix} a & b \\ c & d \end{pmatrix} \in \Gamma,$$

under the action of the modular group.

5.14 Prove that $G_2(i) = \pi$.

5.15 Prove the infinite product formula (5.76) by applying a similar technique to that used in the proof of (5.75).

5.16 (a) Enter a truncated version of the infinite product formula (5.75) into a computer algebra system of your choice, to obtain the first 10 coefficients in the Q-series expansion of the modular λ function.

(b) Enter the first few coefficients into the search box on the On-Line Encyclopedia of Integer Sequences [W21]. If you have the correct coefficients, then the search results will show you a lot of additional information and references on this sequence of numbers and on the modular lambda function. (You can also try doing the same with the Fourier coefficients for Δ, the Eisenstein series, or other sequences of integers that you encounter in modular forms or any other area of mathematics.)

5.17 Show how to derive formulas (5.78)–(5.80) from (5.83).

In the exercises below, we define renormalized versions of the Eisenstein series G_2, G_4, G_6 by

$$E_2(\tau) = \frac{3}{\pi^2} G_2(\tau) = 1 - 24 \sum_{n=1}^{\infty} \sigma_1(n) q^n, \tag{5.86}$$

$$E_4(\tau) = \frac{45}{\pi^4} G_4(\tau) = 1 + 240 \sum_{n=1}^{\infty} \sigma_3(n) q^n, \tag{5.87}$$

$$E_6(\tau) = \frac{945}{2\pi^6} G_6(\tau) = 1 - 504 \sum_{n=1}^{\infty} \sigma_5(n) q^n. \tag{5.88}$$

These versions of the Eisenstein series are often used in the literature in connection with number-theoretic applications.

5.18 (a) Prove that E_2, E_4, and E_6 satisfy the following system of differential equations, known as **Ramanujan's identities:**

$$\frac{1}{2\pi i} E_2'(\tau) = \frac{1}{12}\left(E_2(\tau)^2 - E_4(\tau)\right), \tag{5.89}$$

$$\frac{1}{2\pi i}E_4'(\tau) = \frac{1}{3}(E_2(\tau)E_4(\tau) - E_6(\tau)), \tag{5.90}$$

$$\frac{1}{2\pi i}E_6'(\tau) = \frac{1}{2}(E_2(\tau)E_6(\tau) - E_4(\tau)^2). \tag{5.91}$$

(b) For each of identities (5.89)–(5.91), find the Fourier expansions of both sides and compare the coefficients to obtain interesting number-theoretic identities.

5.19 Prove the identities

$$E_2(\tau) = \frac{1}{6}\left(4E_2(2\tau) + E_2\left(\frac{\tau}{2}\right) + E_2\left(\frac{\tau + 1}{2}\right) \right),$$

$$E_4(\tau) = \frac{1}{18}\left(16E_4(2\tau) + E_4\left(\frac{\tau}{2}\right) + E_4\left(\frac{\tau + 1}{2}\right) \right).$$

5.20 Prove the identities

$$\theta_2(\tau)^4 = \frac{1}{3}\left(E_2\left(\frac{\tau + 1}{2}\right) - E_2\left(\frac{\tau}{2}\right) \right), \tag{5.92}$$

$$\theta_3(\tau)^4 = \frac{1}{3}\left(4E_2(2\tau) - E_2\left(\frac{\tau}{2}\right) \right), \tag{5.93}$$

$$\theta_4(\tau)^4 = \frac{1}{3}\left(4E_2(2\tau) - E_2\left(\frac{\tau + 1}{2}\right) \right), \tag{5.94}$$

$$\theta_2(\tau)^8 = \frac{16}{15}(E_4(\tau) - E_4(2\tau)), \tag{5.95}$$

$$\theta_3(\tau)^8 = \frac{1}{15}\left(16E_4(\tau) - E_4\left(\frac{\tau + 1}{2}\right) \right), \tag{5.96}$$

$$\theta_4(\tau)^8 = \frac{1}{15}\left(16E_4(\tau) - E_4\left(\frac{\tau}{2}\right) \right). \tag{5.97}$$

Guidance for proving (5.92)–(5.94). Define the functions

$$R_2(\tau) = \frac{E_2(\frac{\tau+1}{2}) - E_2(\frac{\tau}{2})}{3\theta_2(\tau)^4},$$

$$R_3(\tau) = \frac{4E_2(2\tau) - E_2(\frac{\tau}{2})}{3\theta_3(\tau)^4},$$

$$R_4(\tau) = \frac{4E_2(2\tau) - E_2(\frac{\tau+1}{2})}{3\theta_4(\tau)^4},$$

$$\phi_1 = R_2 + R_3 + R_4,$$

$$\phi_2 = R_2R_3 + R_2R_4 + R_3R_4,$$

$$\phi_3 = R_2R_3R_4.$$

Show that ϕ_1, ϕ_2, ϕ_3 are entire modular forms of weight 0 and use this to show that $\phi_1 \equiv 3, \phi_2 \equiv 3, \phi_3 \equiv 1$. Deduce from this that $R_2 \equiv R_3 \equiv R_4 \equiv 1$.

5.21 Show that by expanding both sides of (5.93) and (5.96) as Fourier series and comparing the coefficients, we will obtain interesting number-theoretic identities related to counting the number of ways in which integers can be represented as a sum of squares. Specifically, let $r_4(n)$ and $r_8(n)$ denote the numbers of ways to represent an integer n as a sum of 4 squares and as a sum of 8 squares, respectively. Prove the following identities, due to Jacobi:

$$r_4(n) = 8 \sum_{d \mid n,\, 4 \nmid d} d,$$

$$r_8(n) = 16(-1)^n \sum_{d \mid n} (-1)^d d^3.$$

(In particular, we get the fact that every integer can be expressed as a sum of four squares, a famous result in number theory proved by Lagrange in 1770.)

5.22 Use (5.92)–(5.93) to prove that

$$\frac{2\theta_3(\tau)^4 - \theta_2(\tau)^4}{2} = 1 + 24 \sum_{n=1}^{\infty} \sigma_{\mathrm{odd}}(n) e^{2\pi i n \tau}, \tag{5.98}$$

where $\sigma_{\mathrm{odd}}(n)$, the **odd divisor function**, is defined by

$$\sigma_{\mathrm{odd}}(n) = \sum_{\substack{d \mid n \\ d\ \mathrm{odd}}} d \quad (n \geq 1).$$

5.23 Use the Jacobi triple product identity (5.83) to derive the following identity, known as the **Euler pentagonal number theorem**:

$$\prod_{n=1}^{\infty} (1 - x^n) = \sum_{k=-\infty}^{\infty} (-1)^k x^{k(3k-1)/2} \quad (|x| < 1).$$

6 Sphere packing in 8 dimensions

> Discovery in mathematics is not a matter of logic. It is rather the result of mysterious powers which no one understands, and in which unconscious recognition of beauty must play an important part. Out of an infinity of designs, a mathematician chooses one pattern for beauty's sake and pulls it down to earth, no one knows how.
>
> Marston Morse, "Mathematics and the arts" (1959)

6.1 Motivation: the sphere packing problem in d dimensions

In 1611, two years after publishing the first two of his famous laws of planetary motion, the astronomer Johannes Kepler also published a curious observation about geometry in an essay titled "On the Six-Cornered Snowflake." Kepler speculated that the most efficient way to pack solid spheres of equal size in three-dimensional space was using the lattice arrangement now known as the **face-centered cubic** (Fig. 6.1). This packing results in a packing density—the fraction of the volume of the packed space occupied by the interior of the spheres—of $\frac{\pi}{3\sqrt{2}}$, and Kepler's conjecture was the statement that no other configuration of spheres can achieve (in a limiting sense when this is done over larger and larger volumes that fill up space) a higher packing density. Although intuitively plausible, even obvious-sounding to anyone who has tried to stack oranges or other spherical objects, the conjecture nonetheless proved extremely resistant to attempts by mathematicians over the ensuing centuries to prove it rigorously. In the late twentieth century, it stood as one of the most famous and longest-standing open problems in mathematics (among other markers of status, it was included as part of the 18th problem on Hilbert's famous list of 23 problems) and was finally proved by Thomas Hales [39] in 1998.

We will not discuss Hales's proof, which is very involved and does not use complex analysis; the book [38] is a good reference on this topic. However, it turns out that sphere

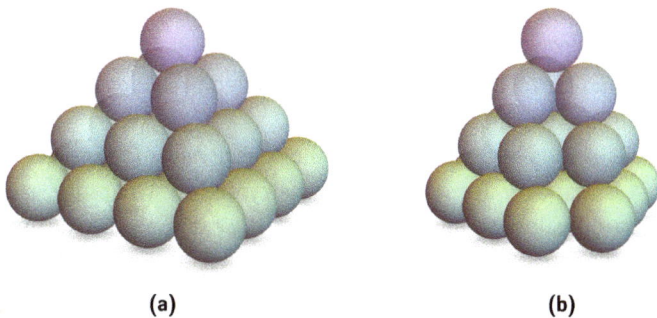

Figure 6.1: The Kepler conjecture, proved by Thomas Hales in 1998, states that the highest density for packing spheres in $\mathbb{R}^3$ is $\pi/3\sqrt{2}$. The packing density for the two lattice packings: (a) the cubic close packing (derived from the lattice known as the face-centered cubic) and (b) the hexagonal close packing.

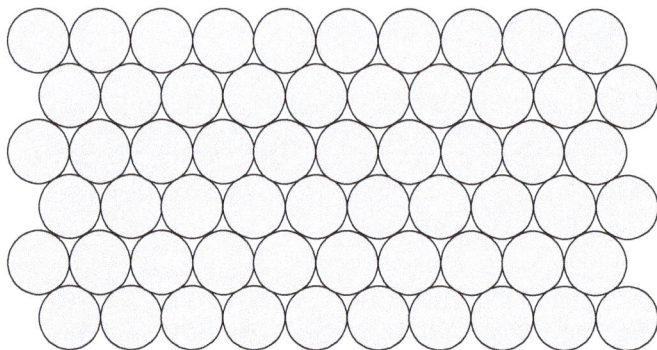

Figure 6.2: The hexagonal packing is the densest way to pack unit circles in the plane.

packing is extremely interesting to study in other dimensions as well (where "spheres" now refer to hyperspheres of appropriate dimension, and the meaning of "packing" remains the same). For example, the case of sphere packing in two dimensions (that is, circle packing) is also interesting, though it is much simpler to understand than in three dimensions and has as its solution the hexagonal lattice packing with a packing density of $\frac{\pi}{\sqrt{12}}$ (a fact that was shown, in increasing levels of generality and rigor, by Gauss in 1831, Thue in 1890, and Tóth in 1940); see Fig. 6.2. Much research in recent decades has focused on studying the question in dimensions higher than 3; see [18].

Our goal in this chapter is to explain the remarkable mathematical ideas behind the recent solution of the sphere packing problem in dimensions 8 and 24. These are currently the only dimensions apart from $d = 2, 3$ for which the problem has been solved. Specifically, we will give a detailed proof of **Viazovska's theorem**.

Theorem 6.1 (Viazovska's theorem). *The optimal sphere packing density in $\mathbb{R}^8$ is $\frac{\pi^4}{384}$.*

Theorem 6.1 was proved by Maryna Viazovska [71] in 2016.[1] Following the appearance of her groundbreaking paper, Viazovska's new insights led within days to a successful solution of the problem in dimension 24 by her and her collaborators Cohn, Kumar, Miller, and Radchenko [16]. In 2022, Viazovska was awarded the Fields Medal for these remarkable achievements and for further contributions to related problems in geometry and Fourier analysis. For more details, see [12, 13, 20, 52].

One of the remarkable aspects of the solutions to the sphere packing problem in both dimensions 8 and 24 is that they use very little geometry: in fact, what little geometrical reasoning appears only does so in connection with the explicit constructions

---

1 This statement (and our name for Theorem 6.1) are simplifications: this theorem summarizes the results and contributions of several mathematicians. However, in this writer's opinion, Viazovska's contribution being the last, as well as being inarguably most ingenious and remarkable, makes her deserving of being the eponym of the theorem.

of the optimal packings (which imply lower bounds for the packing density), whereas the proof of the matching *upper bounds* for the packing density instead draws primarily on complex analysis and the theory of modular forms, spiced up with a bit of Fourier analysis. If you have read Chapter 5, then you are well equipped to tackle this modern and quite beautiful application of complex analysis.

The E_8 lattice and sphere packing

The E_8 **sphere packing** is a packing in which each of the spheres of the packing is centered at a vertex of the so-called E_8 **lattice**, a lattice with many remarkable properties that is closely associated with (and shares a notation with) the **exceptional Lie algebra E_8**.

As the E_8 packing is an intrinsically 8-dimensional object, it is somewhat difficult to visualize what the packing "looks like." One can nonetheless gain some understanding of the qualitative behavior of the packing by considering what a single "cell" of the packing looks like — that is, a single sphere centered at the origin together with the spheres of the packing that are tangent to it. In the case of the E_8 packing, there are 240 such tangent spheres. Each of the tangent spheres is itself tangent to 56 of the other 239 spheres. This is visualized in the figure below, where the 240 spheres are represented as dots, and two dots representing spheres that are mutually tangent are connected with a line. (The positions of the dots are given by a particularly symmetric two-dimensional projection of their sphere centers in $\mathbb{R}^8$.)

A formal construction of the E_8 lattice is given in Section A.7 in the appendix.

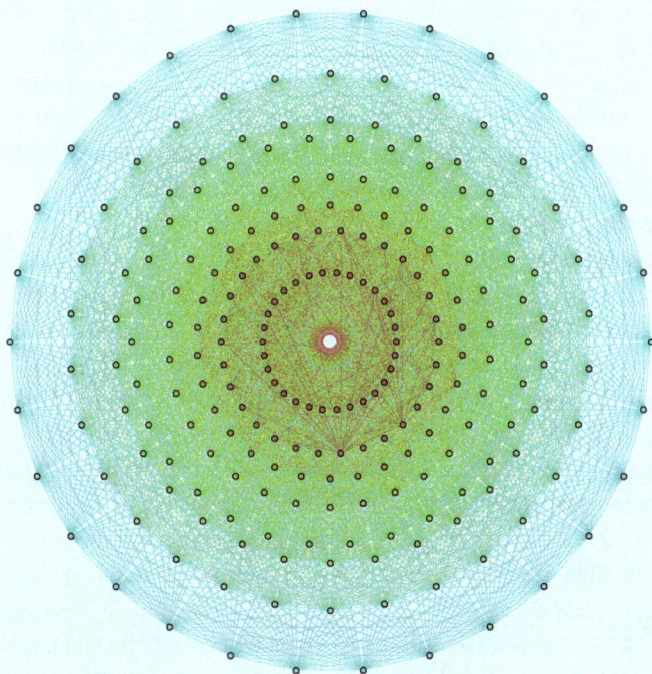

Figure 6.3: A two-dimensional projection of a packing cell in the E_8 sphere packing, which realizes the optimal sphere packing in $\mathbb{R}^8$, having a packing density of $\pi^4/384$.

6.2 A high-level overview of the proof

To understand the proof of Theorem 6.1, a bit of background is required to set up the problem for the final part of the proof, the part that involves complex analysis and is of main interest to us. Our presentation is self-contained and is split between this chapter and Appendix A. Here we give a brief overview of the full structure of the proof:

1. **Background material:** this consists of definitions and basic facts about sphere packings and lattices. This material is presented in Sections A.1–A.6 of the Appendix.
2. **Lower bound: construction of an optimal packing.** An 8-dimensional sphere packing now known to be optimal is the E_8 **sphere packing** and is based on the E_8 **lattice**; see the box on the next page. A few basic facts about this lattice will be needed, and we discuss the relevant material in Section A.7 in the Appendix. This is the "easy" part of the proof (at least in the sense that it is based on little more than elementary linear algebra), which gives a lower bound on the optimal sphere packing density.
3. **Upper bound, part I: the Cohn–Elkies bounds and magic function conjectures.** Conceptually more difficult is to prove an *upper bound* on the packing density, as that involves proving that *no* packing can have a density better than some number. Since the family of possible packings is very large (in fact, infinite-dimensional), it is not obvious how to approach this. A beautiful technique for deriving upper bounds was introduced by Cohn and Elkies [14], who discovered that the Poisson summation formula from harmonic analysis (more precisely, a multidimensional version of it for lattices) is just the right tool for the task. Their bounds, belonging to a class of bounds known as linear programming bounds, give a way of associating a numerical upper bound for the packing density with certain functions of a single (real) variable with nice properties. The problem then becomes that of optimizing the bound over the relevant family of functions in the hope of producing a sharp bound.

    Amazingly, the numerical calculations Cohn and Elkies performed for many different values of the dimension d, which gave numerical bounds that were in many cases better than those previously known, revealed that for $d = 2$, 8, and 24, their bounding technique seems to approach the value known (in the case $d = 2$) or believed at the time (in the cases $d = 8$ and 24) to equal the optimal packing density. They conjectured that in those dimensions, there exists a so-called "magic function," a function in the class of bounding functions for which the associated upper bound for the optimal sphere packing density matches the known lower bound and hence serves as a certificate that solves the sphere packing problem in that dimension. We explain the Cohn–Elkies bounding technique and their magic function conjectures in Sections A.8–A.11 of the Appendix.
4. **Upper bound, part II: Viazovska's modular form construction.** Cohn and Elkies's work reduced the sphere packing problem, at least in dimensions 8 and 24, to the problem of constructing a magic function. Viazovska discovered just the right technique for constructing the function with the desired properties in dimension 8 (and

her ideas proved also applicable to dimension 24 with minor modifications) by making an ingenious use of modular forms. Explaining the details of her construction is the main goal of this chapter.

To the reader who is completely unfamiliar with the topic of sphere packings and wishes to gain a full understanding of the proof of Theorem 6.1, a recommended path is to read Appendix A first and then proceed to reading the remainder of this chapter. Section A.7, which only deals with the explicit construction of the E_8 lattice, is not necessary to understand any other parts of the proof and may be skipped on a first reading.

6.3 Preparation: some remarks on Fourier eigenfunctions

From here on, we assume that you are familiar with the material and notation of Appendix A. The starting point for our proof is Theorem A.29, which, as explained in Section A.12, provides a kind of roadmap for constructing an E_8 magic function, based on constructing separately the Fourier-even and Fourier-odd components $\Phi_+(r)$ and $\Phi_+(r)$ associated with a hypothetical radial magic function; these functions will be constructed with the goal of manufacturing (± 1)-Fourier eigenfunctions having the prescribed set of zeros (of appropriate orders) at $\sqrt{2n}$, $n = 1, 2, \ldots$. Once these functions are constructed, they can be combined into a single radial function having the two functions as its Fourier-even and Fourier-odd components. The hope is that for the function thus constructed, the necessary conditions of Theorem A.29 will also turn out to be sufficient.

   Thus, forgetting about magic functions for the moment, our immediate goal is to construct radial Fourier eigenfunctions in 8 dimensions with the correct set of zeros. We will prove the following result.

Theorem 6.2. *There exist radial Schwartz functions $\varphi_+, \varphi_- : \mathbb{R}^8 \to \mathbb{R}$ with the following properties.*
1. *$\varphi_+(x)$ is a (+1)-Fourier eigenfunction, that is,*

$$\mathcal{F}_8[\varphi_+] = \varphi_+,$$

   where $\mathcal{F}_8$ denotes the Fourier transform in 8 dimensions (see the definition in (A.5)).
2. *$\varphi_-(x)$ is a (−1)-Fourier eigenfunction, that is,*

$$\mathcal{F}_8[\varphi_-] = -\varphi_-.$$

3. *Each of the radial profiles $\widetilde{\varphi_+}(r)$, $\widetilde{\varphi_-}(r)$ has zeros at $r = \sqrt{2n}$, $n = 1, 2, 3, \ldots$, with the zero at $\sqrt{2}$ being simple and the other zeros being of order 2.*

   Where do we begin to look for such functions? Well, probably the most famous example of such an eigenfunction is the Gaussian function

$$\gamma(x) = e^{-\pi \|x\|^2},$$

for which it follows trivially from the analogous property of the one-dimensional Gaussian,

$$\mathcal{F}_8(\gamma)(y) = \gamma(y).$$

This will be useful to us in the following way: if we let

$$\gamma_s(x) = e^{-\pi s \|x\|^2}$$

denote a *rescaled* Gaussian, then because of the scaling behavior of the Fourier transform, we have

$$\mathcal{F}_8[\gamma_s](y) = \frac{1}{s^4} \gamma_{1/s}(y). \tag{6.1}$$

This identity is valid not just for a real positive scaling parameter s, but in fact for any s in the half-plane $\mathrm{Re}(s) > 0$, since in that case, $\gamma_s(x)$ has good decay and integrability properties.

Thus we see that the rescaled Gaussian γ_s is not a Fourier eigenfunction if $s \neq 1$, but a linear combination of γ_s and $\gamma_{1/s}$ of the form $a\gamma_s + b\gamma_{1/s}$ with a, b satisfying

$$a = \pm s^4 b \tag{6.2}$$

is an eigenfunction (associated with eigenvalue ± 1 according to the choice of sign in (6.2)). More generally, we can take sums of such linear combinations involving different values of s, or even integrals with respect to s of the form

$$f(x) = \int w(s)\gamma_s(x)\, ds = \int w(s)e^{-\pi s \|x\|^2}\, ds, \tag{6.3}$$

where $w(s)$ is some weight function, and where the integration is taken over some range of values of s in the half-plane $\mathrm{Re}(s) > 0$. Under appropriate assumptions over how $w(s)$ relates to $w(1/s)$, the resulting function will be a Fourier eigenfunction. This gives a rich source of potential eigenfunctions to use for our construction.

It seems most natural to choose the interval $(0, \infty)$ as the range for the integration in (6.3); the integral (6.3) can then be thought of simply as the Laplace transform

$$\int_0^\infty w(s)e^{-\pi s z}\, ds, \tag{6.4}$$

in the variable $z = \|x\|^2$. In that case the weight function will need to satisfy $w(1/s) = \pm s^{-2} w(s)$, a condition reminiscent of one of the defining equations for a modular form.

However, this is too naive of an idea and does not work, as it does not lead to a viable path to choosing the weight function $w(s)$ in a way that causes the function $f(x)$ to have zeros at the desired radii. It turns out that a more clever choice is required that also incorporates certain *nonreal* values of the scaling parameter s (see equations (6.31) and (6.54)). Modular forms still enter the picture, but they do so in a much more subtle and surprising way. The details are given in the next two sections.

6.4 The $(+1)$-Fourier eigenfunction

In this section, we complete half of the proof of Theorem 6.2 by constructing the function $\varphi_+(x)$ and establishing its properties. The construction for $\varphi_-(x)$ is given in the next section. Both the functions $\varphi_+(x)$ and $\varphi_-(x)$ are constructed by taking the Laplace transform of two functions $U : \mathbb{H} \to \mathbb{C}$ and $V : \mathbb{H} \to \mathbb{C}$, which are given explicitly in terms of modular forms.

Let τ be a complex variable taking values in the upper half-plane, and let $q = e^{2\pi i \tau}$ as in Chapter 5. We will use the normalized versions E_4 and E_6 of the Eisenstein series G_4 and G_6 defined in (5.87)–(5.88). With these definitions, it is useful to observe that

$$E_4(\tau)^3 - E_6(\tau)^2 = \frac{1\,728}{(2\pi)^{12}} \Delta(\tau), \tag{6.5}$$

a scalar multiple of the modular discriminant (see (4.15), (4.35)).

Now define the function $U(\tau)$ by

$$U(\tau) = 108 \frac{(\tau E_4'(\tau) + 4E_4(\tau))^2}{E_4(\tau)^3 - E_6(\tau)^2}. \tag{6.6}$$

This can be expanded in the form

$$U(\tau) = 108 \left(\frac{E_4'(\tau)^2}{E_4(\tau)^3 - E_6(\tau)^2} \right) \tau^2 + 864 \left(\frac{E_4'(\tau)E_4(\tau)}{E_4(\tau)^3 - E_6(\tau)^2} \right) \tau$$
$$+ 1\,728 \left(\frac{E_4(\tau)^2}{E_4(\tau)^3 - E_6(\tau)^2} \right), \tag{6.7}$$

which will be convenient for certain calculations and highlights the structure of $U(\tau)$ as a kind of "polynomial" in τ whose "coefficients" are themselves holomorphic functions in τ that have useful modular properties and in particular are 1-periodic.

Lemma 6.3. *The function $U(\tau)$ takes real, nonnegative values on the positive imaginary axis.*

Proof. Referring to (5.87)–(5.88), it is evident that $E_4(\tau)$ and $E_6(\tau)$ take real values on the positive imaginary axis and that $E_4'(\tau)$ takes imaginary values there. Therefore (6.6) implies that $U(\tau)$ is real for $\tau = it, t > 0$. Moreover, in the fraction in (6.6), the numerator is the square of a real number (hence nonnegative) for $\tau = it$, and the denominator is a positive scalar multiple of $\Delta(it)$, which is itself a positive real number, as can be seen, e. g., from the infinite product representation (5.65). Combining these observations shows that $U(it) \geq 0$ for $t > 0$. $\square$

Lemma 6.4. $U(\tau)$ *satisfies the transformation properties*

$$U\left(-\frac{1}{\tau}\right) = \frac{1}{2\tau^2}\left(U(\tau+1) - 2U(\tau) + U(\tau-1)\right), \tag{6.8}$$

$$U\left(-\frac{1}{\tau}+1\right) = \frac{1}{\tau^2}U(\tau-1), \tag{6.9}$$

$$U\left(-\frac{1}{\tau}-1\right) = \frac{1}{\tau^2}U(\tau+1). \tag{6.10}$$

Proof. Start by noting that

$$E_4'(-1/\tau) = \tau^2\left(\frac{1}{\tau^2}E_4'(-1/\tau)\right) = \tau^2\frac{d}{d\tau}(E_4(-1/\tau)) = \tau^2\frac{d}{d\tau}(\tau^4 E_4(\tau))$$
$$= \tau^2(\tau^4 E_4'(\tau) + 4\tau^3 E_4(\tau)) = \tau^5(\tau E_4'(\tau) + 4E_4(\tau)).$$

It follows that

$$U(-1/\tau) = 108\frac{((-1/\tau)E_4'(-1/\tau) + 4E_4(-1/\tau))^2}{E_4(-1/\tau)^3 - E_6(-1/\tau)^2}$$
$$= 108\frac{((-1/\tau)\tau^5(\tau E_4'(\tau) + 4E_4(\tau)) + 4\tau^4 E_4(\tau))^2}{\tau^{12}(E_4(\tau)^3 - E_6(\tau)^2)}$$
$$= 108\frac{1}{\tau^2}\cdot\frac{E_4'(\tau)^2}{E_4(\tau)^3 - E_6(\tau)^2}. \tag{6.11}$$

On the other hand, by (6.7) and the comment above about the parenthesized expressions in that representation being 1-periodic, the discrete second difference $U(\tau+1) - 2U(\tau) + U(\tau-1)$ on the right-hand side of (6.8) is easily seen to be

$$108\left(\frac{E_4'^2}{E_4^3 - E_6^2}\right)\left((\tau+1)^2 - 2\tau^2 + (\tau-1)^2\right)$$
$$+ 864\left(\frac{E_4' E_4}{E_4^3 - E_6^2}\right)\left((\tau+1) - 2\tau + (\tau-1)\right)$$
$$+ 1728\left(\frac{E_4^2}{E_4^3 - E_6^2}\right)(1 - 2 + 1)$$
$$= 2\cdot 108\left(\frac{E_4'^2}{E_4^3 - E_6^2}\right) + 0 + 0 = 2\cdot 108\frac{E_4'^2}{E_4^3 - E_6^2}.$$

This last expression by (6.11) is equal to $2\tau^2 U(-1/\tau)$. This proves (6.8).

Next, if we denote $\tilde{U}(\tau) = \tau^2 U(-1/\tau)$, then by (6.11), $\tilde{U}(\tau)$ is also 1-periodic. Using this fact, we can write

$$U\left(-\frac{1}{\tau} + 1\right)\tau^2 = U\left(\frac{\tau - 1}{\tau}\right)\tau^2$$

$$= U\left(-1\Big/\left(-\frac{\tau}{\tau - 1}\right)\right)\left(-\frac{\tau}{\tau - 1}\right)^2 (\tau - 1)^2$$

$$= \tilde{U}\left(-\frac{\tau}{\tau - 1}\right)(\tau - 1)^2 = \tilde{U}\left(-\frac{\tau}{\tau - 1} + 1\right)(\tau - 1)^2$$

$$= \tilde{U}\left(-\frac{1}{\tau - 1}\right)(\tau - 1)^2 = U(\tau - 1),$$

which proves (6.9). Finally, (6.10) is obtained by substituting $-1/\tau$ in place of τ in (6.9). □

Lemma 6.5. *On the positive imaginary axis near $\tau = i\infty$ and $\tau = 0$, $U(\tau)$ has the asymptotic behavior*

$$U(it) = e^{2\pi t} - 240\pi t + 504 + O(t^2 e^{-2\pi t}) \quad (t \to \infty), \tag{6.12}$$

$$U(it) = O(t^2 e^{-2\pi/t}) \quad (t \to 0). \tag{6.13}$$

Proof. Using (5.87)–(5.88), the initial terms of the Taylor expansions (in powers of the variable q) of each of the parenthesized expressions in (6.7) can be readily obtained, giving the asymptotic relations, as $\tau \to i\infty$ and $q \to 0$,

$$\frac{E_4'(\tau)^2}{E_4(\tau)^3 - E_6(\tau)^2} = -\frac{400\pi^2}{3}q + O(q^2), \tag{6.14}$$

$$\frac{E_4'(\tau)E_4(\tau)}{E_4(\tau)^3 - E_6(\tau)^2} = \frac{5\pi i}{18} + O(q), \tag{6.15}$$

$$\frac{E_4(\tau)^2}{E_4(\tau)^3 - E_6(\tau)^2} = \frac{1}{1728q} + \frac{7}{24} + O(q). \tag{6.16}$$

Substituting these relations into (6.7) gives (6.12). To get (6.13), use (6.11) together with (6.14) to get that, as $t \to 0$,

$$U(it) = 108(-it)^2 \frac{E_4'(i/t)^2}{E_4(i/t)^3 - E_6(i/t)^2} = 108 \cdot \frac{400\pi^2}{3} \cdot t^2 e^{-2\pi/t} + O(t^2 e^{-4\pi/t}),$$

which proves the claim. □

Now define the holomorphic function

$$A(z) = 4i \sin^2\left(\frac{\pi z}{2}\right) \int_0^{i\infty} U(\tau)e^{\pi i \tau z} d\tau, \tag{6.17}$$

where the integral is a contour integral along the positive imaginary line. The motivation for this definition is that $\varphi_+(x)$ will later be constructed by substituting $\|x\|^2$ for z (see (6.30)). This gives a variant of the Laplace transform-based construction (6.4), but with the additional term of $\sin^2(\frac{\pi\|x\|^2}{2})$ introduced to *force* the function to have zeros at the correct points $\|x\| = \sqrt{2}, \sqrt{4}, \sqrt{6}, \ldots$. (The sine factor is squared since we want all but one of the zeros to be double zeros; recall Theorem A.29.) Some analysis is now required to verify that the idea can lead to a Fourier eigenfunction or indeed that $\varphi_+(x)$ thus defined is even a legitimate function on $\mathbb{R}^8$. We focus on the properties of $A(z)$ as a holomorphic function first before turning to a discussion of $\varphi_+(x)$ but keep the substitution $z = \|x\|^2$ in mind as you read the next few results.

Lemma 6.6. *The integral in* (6.17) *converges in the half-plane* $\mathrm{Re}(z) > 2$ *and defines a holomorphic function there.*

Proof. By Lemma 6.5 the integrand in (6.17) (with the parameterization $\tau = it$) satisfies the asymptotic bounds

$$\left|U(it)e^{-\pi tz}\right| = O(e^{-\pi(\mathrm{Re}(z)-2)t}) \quad (t \to \infty),$$
$$\left|U(it)e^{-\pi tz}\right| = O(t^2 e^{-\pi(\mathrm{Re}(z)+2)/t}) \quad (t \to 0).$$

The constant implicit in the big-O notation does not depend on z. Thus, if we write the integral in (6.17) as $I_1(z) + I_2(z)$, where $I_1(z) = \int_0^1 U(\tau)e^{\pi i\tau z}\,d\tau$ and $I_1(z) = \int_1^\infty U(\tau)e^{\pi i\tau z}\,d\tau$, then, by the standard complex analysis lemma on integrals of a family of holomorphic functions with respect to a parameter (Exercise 1.26 on p. 77), the improper integral $I_1(z)$ converges in the half-plane $\mathrm{Re}(z) > -2$ and defines a holomorphic function there. Similarly, $I_2(z)$ converges and is holomorphic in the half-plane $\mathrm{Re}(z) > 2$. $\qquad\square$

Next, we show that $A(z)$ can be continued analytically to the half-plane $\mathrm{Re}(z) > 0$, and a bit later, we will show that it can be continued analytically even beyond that half-plane. As per the usual convention in complex analysis, we continue to use the same notation $A(z)$ to denote all analytic continuations of $A(z)$.

The formula for the first analytic continuation involves integration over four paths, which we denote by Ψ_{-1}, Ψ_0, Ψ_1, and $\Psi_{i\infty}$, collectively forming the shape of an inverted pitchfork (or an inverted Greek letter Ψ), as shown in Fig. 6.4. These paths are defined as follows:

- Ψ_{-1} is the circular subarc of the unit circle leading from -1 to i;
- Ψ_1 is the circular subarc of the unit circle leading from $+1$ to i;
- Ψ_0 is the straight line segment from 0 to i;
- $\Psi_{i\infty}$ is the infinite straight line segment from i to $i\infty$.

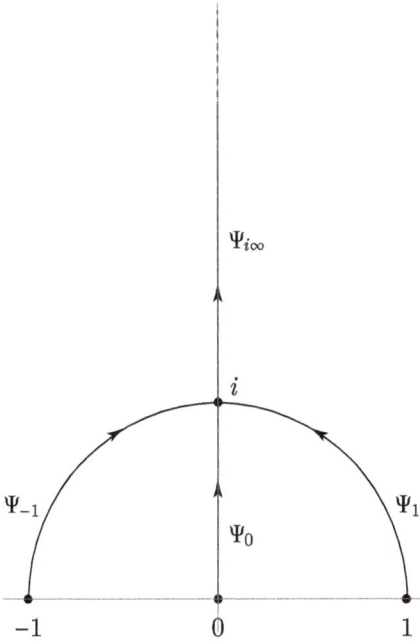

Figure 6.4: The "pitchfork paths" $\Psi_{-1}, \Psi_0, \Psi_1, \Psi_{i\infty}$.

Lemma 6.7. *The function $A(z)$ has the alternative expression*

$$A(z) = -i \int_{\Psi_{-1}} U(\tau + 1)e^{\pi i \tau z}\, d\tau - i \int_{\Psi_1} U(\tau - 1)e^{\pi i \tau z}\, d\tau$$
$$+ 2i \int_{\Psi_0} U(\tau)e^{\pi i \tau z}\, d\tau - 2i \int_{\Psi_{i\infty}} \tau^2 U(-1/\tau)e^{\pi i \tau z}\, d\tau. \tag{6.18}$$

Expression (6.18) extends the definition of $A(z)$ to a holomorphic function on the half-plane $\mathrm{Re}(z) > 0$.

Proof. Denote the right-hand side of (6.18) by $\tilde{A}(z)$, and rewrite this function as

$$\tilde{A}(z) = -i(\tilde{A}_{-1}(z) + \tilde{A}_1(z) - 2\tilde{A}_0(z) + 2\tilde{A}_{i\infty}(z)),$$

where we set

$$\tilde{A}_{-1}(z) = \int_{\Psi_{-1}} U(\tau + 1)e^{\pi i \tau z}\, d\tau, \tag{6.19}$$

$$\tilde{A}_1(z) = \int_{\Psi_1} U(\tau - 1)e^{\pi i \tau z}\, d\tau, \tag{6.20}$$

$$\tilde{A}_0(z) = \int_{\Psi_0} U(\tau) e^{\pi i \tau z} \, d\tau, \tag{6.21}$$

$$\tilde{A}_{i\infty}(z) = \int_{\Psi_{i\infty}} \tau^2 U(-1/\tau) e^{\pi i \tau z} \, d\tau. \tag{6.22}$$

Now $\tilde{A}_0(z)$ is the same as the integral $I_1(z)$ from the proof of Lemma 6.6. It was established in that proof that this integral converges to a holomorphic function in the region $\operatorname{Re}(z) > -2$. The convergence of $\tilde{A}_{i\infty}(z)$ to a holomorphic function, also in the region $\operatorname{Re}(z) > -2$, follows in a similar manner using (6.11) and (6.14).

Next, to verify the convergence of the integral $\tilde{A}_{-1}(z)$, we first rewrite it by applying a change of variables $\xi = -1/(\tau + 1)$. It is easy to check that this maps the contour Ψ_{-1} into the reverse of the straight line segment $[-\frac{1}{2} + \frac{1}{2}i, -\frac{1}{2} + i\infty)$, so we get the expression

$$\tilde{A}_{-1}(z) = - \int_{-\frac{1}{2} + \frac{1}{2}i}^{-\frac{1}{2} + i\infty} U(-1/\xi) e^{-\pi i z (\xi^{-1} + 1)} \frac{d\xi}{\xi^2}.$$

Denoting $\xi = -\frac{1}{2} + it$, where $t \geq 1/2$, we have the bounds

$$|U(-1/\xi)| = O(e^{-2\pi t})$$

(refer again to (6.11) and (6.14)) and, under the assumption that $\operatorname{Re}(z) > 0$,

$$\left| e^{-\pi i z (\xi^{-1} + 1)} \right| = \exp[\pi (\operatorname{Re}(z) \operatorname{Im}(\xi^{-1} + 1) + \operatorname{Im}(z) \operatorname{Re}(\xi^{-1} + 1))]$$

$$= \exp\left[-\pi \operatorname{Re}(z) \frac{t}{t^2 + 1/4} + \pi \operatorname{Im}(z) \frac{t^2 - 1/4}{t^2 + 1/4} \right] \leq \exp(\pi |\operatorname{Im}(z)|).$$

Therefore we conclude that given a compact set $K \subset \{\operatorname{Re}(z) > 0\}$, there is a constant $C > 0$ such that for all $z \in K$, we have

$$\int_{\Psi_{-1}} |U(\tau + 1) e^{\pi i \tau z}| \, |d\tau| = \int_{-\frac{1}{2} + \frac{1}{2}i}^{-\frac{1}{2} + i\infty} |U(-1/\xi)| \cdot \left| e^{-\pi i z (\xi^{-1} + 1)} \right| \frac{|d\xi|}{|\xi|^2}$$

$$\leq C \int_{1/2}^{\infty} e^{-2\pi t} \, dt \leq \frac{C}{2\pi}.$$

This implies the convergence of the integral to a holomorphic function in the half-plane $\operatorname{Re}(z) > 0$ by the result of Exercise 1.26.

The convergence of $\tilde{A}_1(z)$ is proved similarly to the case of $\tilde{A}_{-1}(z)$ by making the substitution $\xi = -1/(\tau - 1)$; details are left to the reader.

Having established that $\tilde{A}(z)$ is well-defined and holomorphic on $\mathrm{Re}(z) > 0$, it remains to check that it extends the definition of $A(z)$. Assume that $\mathrm{Re}(z) > 2$. First, rewrite definition (6.17) of $A(z)$ as

$$A(z) = -i\left(e^{\pi i z/2} - e^{-\pi i z/2}\right)^2 \int_0^{i\infty} U(\tau)e^{\pi i \tau z}\,d\tau$$

$$= -i\left(e^{\pi i z} - 2 + e^{-\pi i z}\right)\int_0^{i\infty} U(\tau)e^{\pi i \tau z}\,d\tau$$

$$= -i\left(\int_0^{i\infty} U(\tau)e^{\pi i(\tau+1)z}\,d\tau - 2\int_0^{i\infty} U(\tau)e^{\pi i \tau z}\,d\tau + \int_0^{i\infty} U(\tau)e^{\pi i(\tau-1)z}\,d\tau\right)$$

$$= -i\left(\int_1^{1+i\infty} U(\rho-1)e^{\pi i \rho z}\,d\rho + \int_{-1}^{-1+i\infty} U(\xi+1)e^{\pi i \xi z}\,d\xi\right.$$

$$\left. - 2\int_{\Psi_0} U(\tau)e^{\pi i \tau z}\,d\tau - 2\int_{\Psi_{i\infty}} U(\tau)e^{\pi i \tau z}\,d\tau\right), \tag{6.23}$$

where in the last step, we make the substitutions $\rho = \tau + 1$, $\xi = \tau - 1$, and for the middle integral, decompose the integral over the segment $[0, i\infty)$ into two integrals over Ψ_0 and $\Psi_{i\infty}$.

Next, observe that in (6.23), we can transform the integrals over the segments $[-1, -1 + i\infty)$ and $[1, 1 + i\infty)$ by deforming the contours: specifically, the segment $[-1, -1 + i\infty)$ can be deformed into $\Psi_{-1} + \Psi_{i\infty}$, and the segment $[1, 1 + i\infty)$ can be deformed into $\Psi_1 + \Psi_{i\infty}$. Because of the exponential decay of the integrand as $\mathrm{Im}(\tau) \to \infty$ (a fact which follows from the assumption that $\mathrm{Re}(z) > 0$, expression (6.7), and the asymptotic estimates (6.14)–(6.16)), an application of Cauchy's theorem together with an easy limiting argument shows that this deformation leaves the values of the respective integrals unchanged. The first transformed integral can therefore be rewritten as

$$\int_1^{1+i\infty} U(\rho-1)e^{\pi i \rho z}\,d\rho = \int_{\Psi_1} U(\rho-1)e^{\pi i \rho z}\,d\rho + \int_{\Psi_{i\infty}} U(\rho-1)e^{\pi i \rho z}\,d\rho,$$

and similarly the second transformed integral becomes

$$\int_{-1}^{-1+i\infty} U(\xi+1)e^{\pi i \xi z}\,d\xi = \int_{\Psi_{-1}} U(\xi+1)e^{\pi i \xi z}\,d\xi + \int_{\Psi_{i\infty}} U(\xi+1)e^{\pi i \xi z}\,d\xi.$$

Substituting these expressions into (6.23), collecting terms, and then making use of (6.8) give

$$A(z) = -i\left(\int_{\Psi_1} U(\tau - 1)e^{\pi i \tau z} \, d\tau + \int_{\Psi_{-1}} U(\tau + 1)e^{\pi i \tau z} \, d\tau \right.$$

$$\left. + \int_{\Psi_{i\infty}} (U(\tau + 1) - 2U(\tau) + U(\tau - 1))e^{\pi i \tau z} \, d\tau - 2 \int_{\Psi_0} U(\tau)e^{\pi i \tau z} \, d\tau \right)$$

$$= -i\left(\int_{\Psi_1} U(\tau - 1)e^{\pi i \tau z} \, d\tau + \int_{\Psi_{-1}} U(\tau + 1)e^{\pi i \tau z} \, d\tau \right.$$

$$\left. + 2 \int_{\Psi_{i\infty}} \tau^2 U(-1/\tau)e^{\pi i \tau z} \, d\tau - 2 \int_{\Psi_0} U(\tau)e^{\pi i \tau z} \, d\tau \right)$$

$$= -i(\tilde{A}_1(z) + \tilde{A}_{-1}(z) + 2\tilde{A}_{i\infty}(z) - 2\tilde{A}_0(z)) = \tilde{A}(z),$$

as claimed. □

Next, it is useful to derive yet another representation for $A(z)$, which continues it analytically to an even larger half-plane.

Lemma 6.8. *The function $A(z)$ is also given by the alternative expression*

$$A(z) = -4\sin^2\left(\frac{\pi z}{2}\right)\left[\frac{1}{\pi}\left(\frac{1}{z-2} - \frac{240}{z^2} + \frac{504}{z}\right)\right.$$

$$\left. + \int_0^\infty (U(it) - e^{2\pi t} + 240\pi t - 504)e^{-\pi z t} \, dt \right]. \tag{6.24}$$

The right-hand side of (6.24) defines a holomorphic function on the half-plane $\mathrm{Re}(z) > -2$ (after interpreting its values at the points $z = 0$ and $z = 2$ in a suitable limiting sense to account for removable singularities at those points) and therefore gives an analytic continuation of $A(z)$ to that half-plane.

Proof. Assume first that $\mathrm{Re}(z) > 2$. Motivated by (6.12), we write

$$A(z) = -4\sin^2\left(\frac{\pi z}{2}\right)\int_0^\infty U(it)e^{-\pi t z} \, dt$$

$$= -4\sin^2\left(\frac{\pi z}{2}\right)\int_0^\infty [(e^{2\pi t} - 240\pi t + 504)$$

$$+ (U(it) - e^{2\pi t} + 240\pi t - 504)]e^{-\pi t z} \, dt$$

$$= -4\sin^2\left(\frac{\pi z}{2}\right)\left[\int_0^\infty (e^{2\pi t} - 240\pi t + 504)e^{-\pi t z} \, dt\right.$$

$$\left. + \int_0^\infty (U(it) - e^{2\pi t} + 240\pi t - 504)e^{-\pi t z} \, dt \right].$$

Evaluating the first of the two integrals in the last expression, we obtain the representation

$$A(z) = -4\sin^2\left(\frac{\pi z}{2}\right)\left[\frac{1}{\pi(z-2)} - \frac{240}{\pi z^2} + \frac{504}{\pi z}\right.$$

$$\left. + \int_0^\infty (U(it) - e^{2\pi t} + 240\pi t - 504)e^{-\pi z t}\, dt\right].\tag{6.25}$$

Finally, observe that by (6.12) and the usual appeal to the integration lemma from Exercise 1.26, (6.25) converges to a holomorphic function in the half-plane $\operatorname{Re}(z) > -2$.   □

Lemma 6.9. *The function $A(z)$ has the special value*

$$A(0) = 240\pi.\tag{6.26}$$

Proof. $A(0)$ is the value of $A(z)$ at the removable singularity $z = 0$ of the expression in (6.24). It is easily calculated as

$$A(0) = \lim_{z\to 0}\left(-\frac{4}{\pi}\sin^2\left(\frac{\pi z}{2}\right)\left(\frac{1}{z-2} - \frac{240}{z^2} + \frac{504}{z}\right)\right)$$

$$= \lim_{z\to 0}\left(\frac{240}{z^2}\cdot\frac{4}{\pi}\sin^2\left(\frac{\pi z}{2}\right)\right) = 240\pi\lim_{z\to 0}\left(\frac{\sin(\pi z/2)}{(\pi z/2)}\right)^2 = 240\pi. \qquad\square$$

The next two lemmas establish some useful technical bounds.

Lemma 6.10. *We have the bound*

$$\int_0^\infty e^{-at-b/t}\, dt \le \frac{2}{a}e^{-\sqrt{ab}}\tag{6.27}$$

for all $a, b > 0$.

Proof. Exercise 6.1.   □

Lemma 6.11. *For any $k \ge 0$, there exist constants $C_1, C_2 > 0$ such that the kth derivative $A^{(k)}(z)$ of $A(z)$ satisfies the bound*

$$\left|A^{(k)}(z)\right| \le C_1 e^{-C_2\sqrt{\operatorname{Re}(z)}} \quad (\operatorname{Re}(z) > 3).\tag{6.28}$$

Proof. Denote $\alpha(z) = \int_0^\infty U(\tau)e^{\pi i\tau z}\, d\tau$. Then, for z with $\operatorname{Re}(z) > 3$, we have

$$\alpha^{(k)}(z) = i(-\pi)^k\int_0^\infty t^k U(it)e^{-\pi t z}\, dt = i(-\pi)^k\left(\int_0^1 + \int_1^\infty\right)t^k U(it)e^{-\pi t z}\, dt.$$

Using Lemma 6.5, we see that there exists a constant $C > 0$ such that

$$\left| a^{(k)}(z) \right| \leq C\pi^k \left(\int_0^1 t^{k+2} e^{-\pi t \, \text{Re}(z) - 2\pi/t} \, dt + \int_1^\infty t^k e^{-\pi(\text{Re}(z)-2)t} \, dt \right)$$

$$\leq C\pi^k \left(\int_0^\infty e^{-\pi t \, \text{Re}(z) - 2\pi/t} \, dt + e^{2\pi} e^{-\pi \, \text{Re}(z)} \int_1^\infty t^k e^{-\pi(\text{Re}(z)-2)(t-1)} \, dt \right).$$

In the last expression, by (6.27) the first integral is bounded from above by $c_1 e^{-c_2 \sqrt{\text{Re}(z)}}$ for some constants $c_1, c_2 > 0$. It is similarly easy to check that the second integral (including the leading multiplicative factor $e^{2\pi} e^{-\pi \, \text{Re}(z)}$) is bounded by $c_3 e^{-\pi \, \text{Re}(z)}$ for some constant $c_3 > 0$. Combining these two bounds, we get a bound of the form

$$\left| a^{(k)}(z) \right| \leq c_4 e^{-c_5 \sqrt{\text{Re}(z)}} \quad (\text{Re}(z) > 3) \tag{6.29}$$

with constants $c_4, c_5 > 0$ (possibly depending on k).

Finally, note that

$$\left| A^{(k)}(z) \right| = \left| \frac{d^k}{dz^k} \left(\sin^2\left(\frac{\pi z}{2} \right) a(z) \right) \right| = \left| \sum_{j=0}^k \binom{k}{j} a^{(j)}(z) \cdot \frac{d^{k-j}}{dz^{k-j}} \left(\sin^2\left(\frac{\pi z}{2} \right) \right) \right|$$

$$\leq \sum_{j=0}^k \binom{k}{j} \left| a^{(j)}(z) \right| \cdot \left| \frac{d^{k-j}}{dz^{k-j}} \left(\sin^2\left(\frac{\pi z}{2} \right) \right) \right|,$$

so the bound (6.29) (or more precisely, the family of bounds indexed by $k \geq 0$) also easily implies a bound of the form (6.28) for $A(z)$ for any $k \geq 0$ with constants C_1, C_2, which may depend on k. $\qquad \square$

We now use the function $A(z)$ to define a radial Fourier eigenfunction in $\mathbb{R}^8$. Define the radial function $\varphi_+ : \mathbb{R}^8 \to \mathbb{C}$ by

$$\varphi_+(x) = A(\|x\|^2). \tag{6.30}$$

By Lemma 6.7, for $x \neq 0$, this can be expressed explicitly as

$$\varphi_+(x) = -i \int_{\Psi_{-1}} U(\tau + 1) e^{\pi i \tau \|x\|^2} \, d\tau - i \int_{\Psi_1} U(\tau - 1) e^{\pi i \tau \|x\|^2} \, d\tau$$

$$+ 2i \int_{\Psi_0} U(\tau) e^{\pi i \tau \|x\|^2} \, d\tau - 2i \int_{\Psi_{i\infty}} \tau^2 U(-1/\tau) e^{\pi i \tau \|x\|^2} \, d\tau. \tag{6.31}$$

This should be thought of as the "correct" version of (6.3), in which the weight function $w(s)$ and the range for the integration are explicitly revealed.

Lemma 6.12. $\varphi_+(x)$ *is a Schwartz function.*

Proof. This follows from Lemma 6.11 and Lemma A.23. The details are left as an exercise (Exercise 6.2). □

Lemma 6.13. $\varphi_+(x)$ *is a* (+1)-*eigenfunction for the Fourier transform in* $\mathbb{R}^8$.

Proof. We evaluate the Fourier transform of φ_+ by commuting the transform operator $\mathcal{F}_8$ with the integrals in (6.31) and applying (6.1) (or rather, the generalized version of this relation that applies to complex s; see (A.12)–(A.13) in Section A.7) inside each integral. Let $y \in \mathbb{R}^d \setminus \{0\}$. Then

$$
\mathcal{F}_8[\varphi_+](y) = -i \int_{\Psi_{-1}} U(\tau+1)\mathcal{F}_8[e^{\pi i \tau \|x\|^2}](y)\, d\tau - i \int_{\Psi_1} U(\tau-1)\mathcal{F}_8[e^{\pi i \tau \|x\|^2}](y)\, d\tau
$$
$$
+ 2i \int_{\Psi_0} U(\tau)\mathcal{F}_8[e^{\pi i \tau \|x\|^2}](y)\, d\tau - 2i \int_{\Psi_{i\infty}} \tau^2 U(-1/\tau)\mathcal{F}_8[e^{\pi i \tau \|x\|^2}](y)\, d\tau
$$
$$
= -i \int_{\Psi_{-1}} U(\tau+1)\tau^{-4} e^{\pi i(-1/\tau)\|y\|^2}\, d\tau - i \int_{\Psi_1} U(\tau-1)\tau^{-4} e^{\pi i(-1/\tau)\|y\|^2}\, d\tau
$$
$$
+ 2i \int_{\Psi_0} U(\tau)\tau^{-4} e^{\pi i(-1/\tau)\|y\|^2}\, d\tau - 2i \int_{\Psi_{i\infty}} \tau^2 U(-1/\tau)\tau^{-4} e^{\pi i(-1/\tau)\|y\|^2}\, d\tau. \qquad (6.32)
$$

Now, in each of the four integrals in the last expression, make the change of variables $\rho = -1/\tau$. This change has the effect of permuting the four pitchfork paths $\Psi_{-1}, \Psi_1, \Psi_0, \Psi_{i\infty}$ according to

$$
\Psi_{-1} \longleftrightarrow \Psi_1, \quad \Psi_0 \longleftrightarrow -\Psi_{i\infty} \qquad (6.33)
$$

(where $-\Psi_{i\infty}$ refers to $\Psi_{i\infty}$ with the reverse orientation). Thus the expression in (6.32) becomes

$$
-i \int_{\Psi_1} U\left(-\frac{1}{\rho}+1\right)\rho^4 e^{\pi i \rho \|y\|^2}\frac{d\rho}{\rho^2} - i \int_{\Psi_{-1}} U\left(-\frac{1}{\rho}-1\right)\rho^4 e^{\pi i \rho \|y\|^2}\frac{d\rho}{\rho^2}
$$
$$
- 2i \int_{\Psi_{i\infty}} U\left(-\frac{1}{\rho}\right)\rho^4 e^{\pi i \rho \|y\|^2}\frac{d\rho}{\rho^2} + 2i \int_{\Psi_0} (-1/\rho)^2 U(\rho)\rho^4 e^{\pi i \rho \|y\|^2}\frac{d\rho}{\rho^2}. \qquad (6.34)
$$

By (6.9) and (6.10) this is equal to

$$
-i \int_{\Psi_1} U(\rho-1)\rho^4 e^{\pi i \rho \|y\|^2}\frac{d\rho}{\rho^2} - i \int_{\Psi_{-1}} U(\rho+1)\rho^4 e^{\pi i \rho \|y\|^2}\frac{d\rho}{\rho^2}
$$
$$
- 2i \int_{\Psi_{i\infty}} \rho^2 U\left(-\frac{1}{\rho}\right)e^{\pi i \rho \|y\|^2}\, d\rho + 2i \int_{\Psi_0} U(\rho)e^{\pi i \rho \|y\|^2}\, d\rho = \varphi_+(y).
$$

We proved the equality $\mathcal{F}_8[\varphi_+](y) = \varphi_+(y)$ for all $y \in \mathbb{R}^d \setminus \{0\}$. By continuity the claim also holds for $y = 0$. □

Lemma 6.14. *The radial profile $\widetilde{\varphi}_+(r)$ associated with $\varphi_+(x)$ has zeros at $r = \sqrt{2n}$, $n = 1, 2, 3, \ldots$. The zero at $r = \sqrt{2}$ is simple, and the zeros at $r = \sqrt{2n}$, $n \geq 2$, are of order 2.*

Proof. By (6.25), $A(z)$ has zeros at $z = 2n$, $n = 1, 2, 3, \ldots$, with the zero at $z = 2$ being simple and the zeros at $z = 2n$, $n \geq 2$ being of order 2. Since $\widetilde{\varphi}_+(r)$ is related to $A(z)$ via

$$\widetilde{\varphi}_+(r) = A(r^2),$$

the result follows. □

Suggested exercises for Section 6.4. 6.1, 6.2, 6.3, 6.4.

6.5 The (−1)-Fourier eigenfunction

Let $\theta_j(\tau), j = 2, 3, 4$, be the Jacobi thetanull functions, discussed in Subsections 5.13.1 and 5.14.3. We define

$$V(\tau) = 128 \left(\frac{\theta_3(\tau)^4 + \theta_4(\tau)^4}{\theta_2(\tau)^8} + \frac{\theta_4(\tau)^4 - \theta_2(\tau)^4}{\theta_3(\tau)^8} \right). \tag{6.35}$$

Lemma 6.15. *The function $V(\tau)$ takes real, nonnegative values on the positive imaginary axis.*

Proof. We can see from (5.54)–(5.56) and (6.35) that $V(\tau)$ is real on the positive imaginary axis. For the nonnegativity claim, it is helpful to use the connection of the theta functions $\theta_2, \theta_3, \theta_4$ to the modular lambda function $\lambda(\tau)$ (see Sections 5.13.2, 5.14.2, and 5.14.3). Using identities (5.81), we have that

$$\frac{1}{128} V(\tau) = \frac{\theta_3^4 + \theta_4^4}{\theta_2^8} + \frac{\theta_4^4 - \theta_2^4}{\theta_3^8} = \frac{1}{\theta_3^4} \cdot \frac{\theta_3^8 + \theta_3^4 \theta_4^4}{\theta_2^8} + \frac{1}{\theta_3^4} \cdot \frac{\theta_4^4 - \theta_2^4}{\theta_3^4}$$

$$= \frac{1}{\theta_3^4} \left(\frac{1}{\lambda^2} + \frac{1}{\lambda} \cdot \frac{1 - \lambda}{\lambda} + (1 - \lambda) - \lambda \right) = \frac{1}{\theta_3^4} \frac{(1 - \lambda)(2 + \lambda + 2\lambda^2)}{\lambda^2}.$$

Now note that $\lambda(it) \in (0, 1)$ for $t > 0$, as is apparent from either the second identity in (5.81) or from the infinite product representation (5.75). Since the function $x \mapsto \frac{(1-x)(2+x+2x^2)}{x^2}$ is positive for $x \in (0, 1)$, and since clearly $\theta_3(it)^4 > 0$ for $t > 0$, from the definition we get that $V(it)$ is nonnegative (in fact, positive) for $t > 0$. □

Lemma 6.16. *$V(\tau)$ satisfies the transformation properties*

$$V\left(-\frac{1}{\tau}\right) = \frac{1}{\tau^2}(V(\tau) - V(\tau + 1)) = \frac{1}{\tau^2}(V(\tau) - V(\tau - 1)), \tag{6.36}$$

$$V\left(-\frac{1}{\tau}\right) = -\frac{1}{2\tau^2}(V(\tau+1) - 2V(\tau) + V(\tau-1)), \tag{6.37}$$

$$V\left(-\frac{1}{\tau}+1\right) = -\frac{1}{\tau^2}V(\tau-1), \tag{6.38}$$

$$V\left(-\frac{1}{\tau}-1\right) = -\frac{1}{\tau^2}V(\tau+1). \tag{6.39}$$

Proof. From (6.35) and the transformation relations (5.57)–(5.59) satisfied by the functions $\theta_j(\tau)$ we get immediately that

$$V(\tau+1) = V(\tau-1) = 128\left(\frac{\theta_3(\tau)^4 + \theta_4(\tau)^4}{\theta_2(\tau)^8} + \frac{\theta_3(\tau)^4 + \theta_2(\tau)^4}{\theta_4(\tau)^8}\right), \tag{6.40}$$

$$\tau^2 V(-1/\tau) = -128\left(\frac{\theta_3(\tau)^4 + \theta_2(\tau)^4}{\theta_4(\tau)^8} + \frac{\theta_2(\tau)^4 - \theta_4(\tau)^4}{\theta_3(\tau)^8}\right), \tag{6.41}$$

which, together with (6.35), gives (6.36). Relation (6.37) then follows trivially. We also obtain from (6.41) that $\widetilde{V}(\tau) = \tau^2 V(-1/\tau)$ satisfies $\widetilde{V}(\tau+1) = -\widetilde{V}(\tau)$. This in turn implies (6.38) and (6.39) in a manner analogous to the proof of (6.9) and (6.10) from (6.8) in the previous section. □

From now on, we adopt the notation $Q = e^{\pi i \tau} = q^{1/2}$ introduced in Subsection 5.14.2. As we can see from (5.54)–(5.56) and (5.78)–(5.80), the functions θ_2^4, θ_3^4, and θ_4^4 in terms of which $V(\tau)$ is defined are all naturally expressed as power series in the variable Q, so this notation is helpful for asymptotic calculations.

Lemma 6.17. *On the positive imaginary axis near $\tau = i\infty$ and $\tau = 0$, $V(\tau)$ has the asymptotic behavior*

$$V(it) = e^{2\pi t} + 144 + O(e^{-\pi t}) \qquad (t \to \infty), \tag{6.42}$$

$$V(it) = 10\,240 t^2 e^{-\pi/t} + O(e^{-2\pi/t}) \quad (t \to 0). \tag{6.43}$$

Proof. By writing out the series expansions for $\theta_j(\tau)$ in powers of Q up to low order we find that, as $\tau \to i\infty$,

$$\theta_2(\tau)^4 = 16(Q + 4Q^3 + 6Q^5 + 8Q^7 + 13Q^9) + O(Q^{11}), \tag{6.44}$$

$$\theta_3(\tau)^4 = 1 + 8Q + 24Q^2 + 32Q^3 + 24Q^4 + 48Q^5 + 96Q^6 + O(Q^7), \tag{6.45}$$

$$\theta_4(\tau)^4 = 1 - 8Q + 24Q^2 - 32Q^3 + 24Q^4 - 48Q^5 + 96Q^6 + O(Q^7). \tag{6.46}$$

Upon substitution of these relations into (6.35), further mundane algebraic calculations give the expansion

$$V(\tau) = \frac{1}{Q^2} + 144 - 5\,120Q + 70\,524Q^2 - 626\,688Q^3 + 4\,265\,600Q^4 + O(Q^5) \tag{6.47}$$

for $V(\tau)$. This gives (6.42). A similar calculation using (6.41) gives

$$\tau^2 V(-1/\tau) = -2\,048(5Q + 612Q^3 + 23\,598Q^5) + O(Q^7), \tag{6.48}$$

easily implying (6.43) on setting $\tau = i/t$. □

Now by analogy with (6.17) define

$$B(z) = 4i \sin^2\left(\frac{\pi z}{2}\right) \int_0^{i\infty} V(\tau)e^{\pi i \tau z}\, d\tau \tag{6.49}$$

(a contour integral along the positive imaginary line).

Lemma 6.18. *The integral in (6.49) converges absolutely uniformly on compacts and defines a holomorphic function in the half-plane $\mathrm{Re}(z) > 2$.*

Proof. This follows from (6.42)–(6.43) analogously to the proof of Lemma 6.6. □

We now proceed to perform an analytic continuation of $B(z)$ to the half-plane $\mathrm{Re}(z) > -1$ in two steps that are analogous to Lemmas 6.7 and 6.8 from the previous section.

Lemma 6.19. *The function $B(z)$ has the alternative expression*

$$B(z) = -i \int_{\Psi_{-1}} V(\tau + 1)e^{\pi i \tau z}\, d\tau - i \int_{\Psi_1} V(\tau - 1)e^{\pi i \tau z}\, d\tau$$
$$+ 2i \int_{\Psi_0} V(\tau)e^{\pi i \tau z}\, d\tau + 2i \int_{\Psi_{i\infty}} \tau^2 V(-1/\tau)e^{\pi i \tau z}\, d\tau. \tag{6.50}$$

Expression (6.50) analytically continues $B(z)$ to the half-plane $\mathrm{Re}(z) > 0$.

Proof. This is similar to the proof of Lemma 6.7. As in that proof, denote the right-hand side of (6.50) by $\tilde{B}(z)$, which we represent as

$$\tilde{B}(z) = -i(\tilde{B}_{-1}(z) + \tilde{B}_1(z) - 2\tilde{B}_0(z) - 2\tilde{B}_{i\infty}(z)),$$

where

$$\tilde{B}_{-1}(z) = \int_{\Psi_{-1}} V(\tau + 1)e^{\pi i \tau z}\, d\tau,$$

$$\tilde{B}_1(z) = \int_{\Psi_1} V(\tau - 1)e^{\pi i \tau z}\, d\tau,$$

$$\tilde{B}_0(z) = \int_{\Psi_0} V(\tau)e^{\pi i \tau z}\, d\tau,$$

$$\tilde{B}_{i\infty}(z) = \int_{\Psi_{i\infty}} \tau^2 V(-1/\tau) e^{\pi i \tau z} \, d\tau.$$

The proof that the integrals converge in the half-plane $\text{Re}(z) > 0$ and are holomorphic there is similar to the analogous claim for the integrals (6.19)–(6.22) and is omitted.

We now check that $\tilde{B}(z)$ coincides with $B(z)$ where the latter is defined. Assume that $\text{Re}(z) > 2$. Rewrite definition (6.49) of $B(z)$ as

$$B(z) = -i(e^{\pi i z/2} - e^{-\pi i z/2})^2 \int_0^{i\infty} V(\tau) e^{\pi i \tau z} \, d\tau$$

$$= -i(e^{\pi i z} - 2 + e^{-\pi i z}) \int_0^{i\infty} V(\tau) e^{\pi i \tau z} \, d\tau$$

$$= -i \int_0^{i\infty} V(\tau) e^{\pi i (\tau+1)z} \, d\tau + 2i \int_0^{i\infty} V(\tau) e^{\pi i \tau z} \, d\tau - i \int_0^{i\infty} V(\tau) e^{\pi i (\tau-1)z} \, d\tau$$

$$= -i \int_1^{1+i\infty} V(\rho - 1) e^{\pi i \rho z} \, d\rho - i \int_{-1}^{-1+i\infty} V(\xi + 1) e^{\pi i \xi z} \, d\xi$$

$$+ 2i \int_{\Psi_0} V(\tau) e^{\pi i \tau z} \, d\tau + 2i \int_{\Psi_{i\infty}} V(\tau) e^{\pi i \tau z} \, d\tau. \tag{6.51}$$

Now as in the proof of Lemma 6.7, the reader can check that the straight line contours $[-1, 1, +i\infty)$ and $[-1, 1, +i\infty)$ can be deformed into the concatenated contours $\Psi_{-1} + \Psi_{i\infty}$ and $\Psi_1 + \Psi_{i\infty}$, respectively, without changing the values of the respective contour integrals. Performing this deformation transforms (6.51), after some minor rearrangement and regrouping of terms, into the relation

$$B(z) = -i \left(\int_{\Psi_1} V(\tau - 1) e^{\pi i \tau z} \, d\tau + \int_{\Psi_{-1}} V(\tau + 1) e^{\pi i \tau z} \, d\tau \right.$$

$$\left. + \int_{\Psi_{i\infty}} (V(\tau + 1) - 2V(\tau) + V(\tau - 1)) e^{\pi i \tau z} \, d\tau - 2 \int_{\Psi_0} V(\tau) e^{\pi i \tau z} \, d\tau \right),$$

whereupon, after making use of (6.37) to simplify the third of the four integrals, we finally get that

$$B(z) = -i \left(\int_{\Psi_1} V(\tau - 1) e^{\pi i \tau z} \, d\tau + \int_{\Psi_{-1}} V(\tau + 1) e^{\pi i \tau z} \, d\tau \right.$$

$$\left. - 2 \int_{\Psi_{i\infty}} \tau^2 V(-1/\tau) e^{\pi i \tau z} \, d\tau - 2 \int_{\Psi_0} V(\tau) e^{\pi i \tau z} \, d\tau \right)$$

$$= -i(\tilde{B}_1(z) + \tilde{B}_{-1}(z) - 2\tilde{B}_{i\infty}(z) - 2\tilde{B}_0(z)) = \tilde{B}(z),$$

as was to be shown. □

Lemma 6.20. *The function $B(z)$ is also given by the alternative expression*

$$B(z) = -4\sin^2\left(\frac{\pi z}{2}\right)\left[\frac{1}{\pi}\left(\frac{1}{z-2} + \frac{144}{z}\right) + \int_0^\infty (V(it) - 144 - e^{2\pi t})e^{-\pi z t}\,dt\right]. \tag{6.52}$$

Representation (6.52) analytically continues $B(z)$ to the region $\mathrm{Re}(z) > -1$ (with the obvious limiting interpretation at the points $z = 0$ and $z = 2$, which are removable singularities).

Proof. Let $\mathrm{Re}(z) > 2$. Starting from (6.49), we write

$$B(z) = -4\sin^2\left(\frac{\pi z}{2}\right)\int_0^\infty V(it)e^{-\pi z t}\,dt$$

$$= -4\sin^2\left(\frac{\pi z}{2}\right)\left[\int_0^\infty (V(it) - 144 - e^{2\pi t})e^{-\pi z t}\,dt + \int_0^\infty (144 + e^{2\pi t})e^{-\pi z t}\,dt\right]$$

$$= -4\sin^2\left(\frac{\pi z}{2}\right)\left[\frac{144}{\pi z} + \frac{1}{\pi(z-2)} + \int_0^\infty (V(it) - 144 - e^{2\pi t})e^{-\pi z t}\,dt\right].$$

Now inspect the last integral to conclude from (6.42)–(6.43) (appealing as before to the result of Exercise 1.26) that this integral converges and defines a holomorphic function on $\mathrm{Re}(z) > -1$. □

Lemma 6.21. *$B(z)$ satisfies*

$$B(0) = 0. \tag{6.53}$$

Proof. Immediate from (6.52). □

Lemma 6.22. *For any $k \geq 0$, there exist constants $C_1, C_2 > 0$ such that the kth derivative $B^{(k)}(z)$ of $A(z)$ satisfies the bound*

$$|B^{(k)}(z)| \leq C_1 e^{-C_2\sqrt{\mathrm{Re}(z)}} \quad (\mathrm{Re}(z) > 3).$$

Proof. Similar to the proof of Lemma 6.11. □

Now let $\varphi_- : \mathbb{R}^8 \to \mathbb{C}$ be the radial function defined by

$$\varphi_-(x) = B(\|x\|^2).$$

For $x \neq 0$, we can write explicitly

$$\varphi_-(x) = -i \int_{\Psi_{-1}} V(\tau+1)e^{\pi i \tau \|x\|^2}\, d\tau - i \int_{\Psi_1} V(\tau-1)e^{\pi i \tau \|x\|^2}\, d\tau$$

$$+ 2i \int_{\Psi_0} V(\tau)e^{\pi i \tau \|x\|^2}\, d\tau + 2i \int_{\Psi_{i\infty}} \tau^2 V(-1/\tau)e^{\pi i \tau \|x\|^2}\, d\tau. \tag{6.54}$$

Lemma 6.23. $\varphi_-(x)$ *is a Schwartz function.*

Proof. Analogous to the proof of Lemma 6.12. □

Lemma 6.24. $\varphi_-(x)$ *is a* (-1)-*eigenfunction for the Fourier transform in* $\mathbb{R}^8$.

Proof. This is a calculation similar to the one in the proof of Lemma 6.13. Namely, using representation (6.54) and commuting the integrals and Fourier transforms, we have for $y \in \mathbb{R}^d \setminus \{0\}$ that

$$\mathcal{F}_8[\varphi_-](y) = -i \int_{\Psi_{-1}} V(\tau+1)\mathcal{F}_8[e^{\pi i \tau \|x\|^2}](y)\, d\tau - i \int_{\Psi_1} V(\tau-1)\mathcal{F}_8[e^{\pi i \tau \|x\|^2}](y)\, d\tau$$

$$+ 2i \int_{\Psi_0} V(\tau)\mathcal{F}_8[e^{\pi i \tau \|x\|^2}](y)\, d\tau + 2i \int_{\Psi_{i\infty}} \tau^2 V(-1/\tau)\mathcal{F}_8[e^{\pi i \tau \|x\|^2}](y)\, d\tau$$

$$= -i \int_{\Psi_{-1}} V(\tau+1)\tau^{-4} e^{\pi i (-1/\tau)\|y\|^2}\, d\tau - i \int_{\Psi_1} V(\tau-1)\tau^{-4} e^{\pi i (-1/\tau)\|y\|^2}\, d\tau$$

$$+ 2i \int_{\Psi_0} V(\tau)\tau^{-4} e^{\pi i (-1/\tau)\|y\|^2}\, d\tau + 2i \int_{\Psi_{i\infty}} \tau^2 V(-1/\tau)\tau^{-4} e^{\pi i (-1/\tau)\|y\|^2}\, d\tau. \tag{6.55}$$

Now making the change of variables $\rho = -1/\tau$ as in the proof of Lemma 6.13 and recalling that the pitchfork paths get permuted as in (6.33), the expression in (6.55) becomes

$$- i \int_{\Psi_1} V\left(-\frac{1}{\rho}+1\right)\rho^4 e^{\pi i \rho \|y\|^2}\, \frac{d\rho}{\rho^2} - i \int_{\Psi_{-1}} V\left(-\frac{1}{\rho}-1\right)\rho^4 e^{\pi i \rho \|y\|^2}\, \frac{d\rho}{\rho^2}$$

$$- 2i \int_{\Psi_{i\infty}} V\left(-\frac{1}{\rho}\right)\rho^4 e^{\pi i \rho \|y\|^2}\, \frac{d\rho}{\rho^2} - 2i \int_{\Psi_0} (-1/\rho)^2 V(\rho)\rho^4 e^{\pi i \rho \|y\|^2}\, \frac{d\rho}{\rho^2}. \tag{6.56}$$

Finally, making use of (6.38)–(6.39) (the analogues of the relations (6.9)–(6.10) that were used in the proof of Lemma 6.13) gives an expression, which we easily recognize as being equal to $-\varphi_-(y)$. □

Lemma 6.25. *The radial profile* $\widetilde{\varphi}_-(r)$ *associated with* $\varphi_-(x)$ *has zeros at* $r = \sqrt{2n}$, $n = 0, 1, 2, \ldots$. *The zero at* $r = \sqrt{2}$ *is simple, and the zeros at* $r = \sqrt{2n}$, $n = 0, 2, 3, \ldots$, *are of order* 2.

Proof. It follows from (6.52) that $B(z)$ has simple zeros at $z = 0$ and $z = 2$ and double zeros at $z = 4, 6, 8, \ldots$. Since $\widetilde{\varphi}_-(r) = B(r^2)$, the result follows. □

The results of this section and the previous one, taken together, prove Theorem 6.2. This gets us most of the way toward an eventual proof of Theorem 6.1. Note that so far our analysis has treated the functions $\varphi_+(x)$ and $\varphi_-(x)$ completely separately from each other. To complete the proof of Theorem 6.1, we will need to gain some additional insight into how the two functions relate to each other or, going back to the two functions $U(\tau)$, $V(\tau)$ in terms of which $\varphi_+(x)$ and $\varphi_-(x)$ were defined, how those two functions in turn compare with each other. This is discussed in the next section.

6.6 A modular form inequality

Our goal in this section is to prove the following result.

Theorem 6.26 (Viazovska's modular form inequality). *The functions $U(\tau)$ and $V(\tau)$ satisfy the inequality*

$$U(it) < V(it) \quad (t > 0). \tag{6.57}$$

Inequality (6.57) plays a key role in the proof of Theorem 6.1; as we will see in the next section, it is needed to establish the fact that our constructed magic function candidate satisfies the nonnegativity condition in Theorem A.21.

Viazovska's original proof of Theorem 6.26 in [71] relied on computer calculations. The proof presented below, adapted from [58], offers a more direct approach.

6.6.1 Preparation

As preparation for the proof, recall the functions $\widetilde{U}(\tau)$ and $\widetilde{V}(\tau)$, which made minor appearances in the proofs of Lemmas 6.4 and 6.16. They are given by

$$\widetilde{U}(\tau) = \tau^2 U(-1/\tau) = 108\frac{(E_4')^2}{E_4^3 - E_6^2},$$

$$\widetilde{V}(\tau) = \tau^2 V(-1/\tau) = -128\left(\frac{\theta_3^4 + \theta_2^4}{\theta_4^8} + \frac{\theta_2^4 - \theta_4^4}{\theta_3^8}\right).$$

Because of the reciprocal relation between it and $i/t = -1/(it)$, inequality (6.57) is equivalent to the claim that both the inequalities $U(it) < V(it)$ and $-\widetilde{U}(it) \le -\widetilde{V}(it)$ hold for $t \ge 1$. As a further simplification, we can clear the denominators in the expressions for $U(\tau), V(\tau), \widetilde{U}(\tau), \widetilde{V}(\tau)$ by multiplying all four functions by $E_4^3 - E_6^2$ (which can also be written as $\frac{27}{4}(\theta_2\theta_3\theta_4)^8$ by (5.63) and (6.5); this function takes positive values on the positive imaginary axis). This leads us to defining the functions $F, \widetilde{F}, G, \widetilde{G}$ by

$$F(\tau) = \frac{1}{108}(E_4^3 - E_6^2)U(\tau) = (E_4')^2\tau^2 + 8E_4'E_4\tau + 16E_4^2, \tag{6.58}$$

$$\tilde{F}(\tau) = \frac{1}{108}(E_4^3 - E_6^2)\tilde{U}(\tau) = (E_4')^2, \tag{6.59}$$

$$G(\tau) = \frac{1}{108}\left(\frac{27}{4}(\theta_2\theta_3\theta_4)^8\right)V(\tau) = 8\theta_4^8(\theta_3^{12} + \theta_4^4\theta_3^8 + \theta_2^8\theta_4^4 - \theta_2^{12}), \tag{6.60}$$

$$\tilde{G}(\tau) = \frac{1}{108}\left(\frac{27}{4}(\theta_2\theta_3\theta_4)^8\right)\tilde{V}(\tau) = -8\theta_2^8(\theta_3^{12} + \theta_2^4\theta_3^8 + \theta_2^4\theta_4^8 - \theta_4^{12}). \tag{6.61}$$

The normalization by a common numerical factor of 1/108 is added to simplify some of the formulas. Our goal is now to prove the pair of inequalities

$$-\tilde{F}(it) < -\tilde{G}(it) \quad \text{if } t \geq 1, \tag{6.62}$$

$$F(it) < G(it) \quad \text{if } t \geq 1. \tag{6.63}$$

By the above remarks this will be sufficient to imply (6.57).

6.6.2 Some special values of modular forms

Our proof of inequalities (6.62)–(6.63) will rely on the numerical values of certain constants obtained from evaluating various modular forms and related functions at $\tau = i$. The relevant evaluations are given below.

Lemma 6.27 (Special values of modular forms at $\tau = i$). *We have the following identities:*

$$E_4(i) = \frac{3\Gamma(1/4)^8}{64\pi^6} \approx 1.45576, \tag{6.64}$$

$$E_4'(i) = \frac{3\Gamma(1/4)^8}{32\pi^6}i \approx 2.91152\,i, \tag{6.65}$$

$$\theta_2(i) = \frac{\Gamma(1/4)}{(2\pi)^{3/4}} \approx 0.91357, \tag{6.66}$$

$$\theta_3(i) = \frac{\Gamma(1/4)}{\sqrt{2}\,\pi^{3/4}} \approx 1.08643, \tag{6.67}$$

$$\theta_4(i) = \frac{\Gamma(1/4)}{(2\pi)^{3/4}} \approx 0.91357. \tag{6.68}$$

In these formulas, Γ is the Euler gamma function.

Sketch of proof. For the proof of (6.66)–(6.67), refer to [8, p. 325] (which appeals to results from Chapter 17 of [7]) or see alternatively [19], where these identities appear as equation (2.21) on p. 307. Evaluation (6.66) also implies (6.68) through the observation that $\theta_4(i) = \theta_2(i)$, a consequence of (5.57).

Formula (6.64) can now be shown using (6.66)–(6.68) and identity (5.60) from Chapter 5 expressing E_4 in terms of the thetanull functions.

Finally, (6.65) is obtained by combining (6.64) with the results of Exercises 5.14 and 5.18, recalling the fact (shown in Lemma 5.15) that $E_6(i) = 0$. □

The evaluations in the lemma are closely related to **Gauss's lemniscate constant**

$$\varpi = 2 \int_0^1 \frac{dx}{\sqrt{1-x^4}} = \frac{\Gamma(1/4)^2}{2\sqrt{2\pi}},$$

an important mathematical constant. See [19], [27, Sec. 6.1], [W22], [W23] for more details.

The proof of (6.62)–(6.63) given below is robust in the sense that it does not depend on the exact values given in the lemma; the inequalities we are dealing with have "slackness," so we really only need *approximate* numerical values of the five constants (6.64)–(6.68). These constants are all expressible as rapidly converging infinite series, so, as an alternative to relying on the closed-form evaluations (6.64)–(6.68), we can simply calculate the numerical values to a few digits of accuracy using a computer.

6.6.3 Proof of (6.62)

We proceed with a proof of inequality (6.62). To develop first a rough sense of why we expect such an inequality to hold, at least for large values of t, it helps to look at the expansions of the functions involved in powers of the variable Q. Those are given, as we can easily check using a computer algebra system, by

$$-\tilde{F}(\tau) = 230\,400\pi^2 Q^4 + 8\,294\,400\pi^2 Q^6 + 113\,356\,800\pi^2 Q^8 + 831\,283\,200\pi^2 Q^{10}$$
$$+ 4\,337\,971\,200\pi^2 Q^{12} + \cdots, \tag{6.69}$$
$$-\tilde{G}(\tau) = 163\,840 Q^3 + 16\,121\,856 Q^5 + 333\,250\,560 Q^7 + 3\,199\,467\,520 Q^9$$
$$+ 19\,472\,547\,840 Q^{11} + \cdots. \tag{6.70}$$

When $\tau = it$, we have $Q = e^{-\pi t}$, so a key point to note is that for large t, the dominant term in the expansion of $-\tilde{F}(it)$ decays like $e^{-4\pi t}$, whereas the dominant term in the expansion of $-\tilde{G}(it)$ decays like $e^{-3\pi t}$, so we will certainly have that $-\tilde{F}(it) < -\tilde{G}(it)$ if t is large enough.

In fact, with a bit of additional reasoning, we can show that the inequality holds for all $t \geq 1$. First, observe that the coefficients in expansion (6.69) are all nonnegative; this is immediate from (5.87) and (6.59). Second, we claim similarly that the coefficients in (6.70) are all nonnegative. To see this, note that, by the transformation properties (5.57)–(5.59) of the thetanull functions, we can represent $\tilde{G}(\tau)$ as

$$\tilde{G}(\tau) = \gamma(\tau + 1) - \gamma(\tau),$$

where $\gamma(\tau)$ is defined by

$$\gamma(\tau) = 8\theta_2^8 \theta_3^{12} + 8\theta_2^{12}\theta_3^8.$$

Now the substitution $\tau \mapsto \tau+1$ corresponds to replacing Q by $-Q$. Therefore the Q-series expansion of $-\widetilde{G}(\tau)$ has all even coefficients equal to 0 and all odd coefficients equal to twice the respective coefficients of $y(\tau)$, which are manifestly nonnegative. This proves the nonnegativity claim.

From the above remarks it now follows that the function $t \mapsto -Q^{-3}\widetilde{F}(it) = -e^{3\pi t}\widetilde{F}(it)$ is a decreasing function of t (since each term in its Q-series expansion is a nonnegative coefficient times the decreasing exponential $e^{-n\pi t}$). This implies that for $t \geq 1$, we have the bound

$$-e^{3\pi t}\widetilde{F}(it) \leq -e^{3\pi}\widetilde{F}(i) = -e^{3\pi}\left(E_4'(i)\right)^2$$

or, using (6.65),

$$-e^{3\pi t}\widetilde{F}(it) \leq e^{3\pi}\frac{9\Gamma(1/4)^{16}}{1\,024\,\pi^{12}} \approx 105\,043.78 \quad (t \geq 1). \tag{6.71}$$

On the other hand, by (6.70) and the observation about the nonnegativity of the coefficients of $-\widetilde{G}(\tau)$ we have the bound

$$-e^{3\pi t}\widetilde{G}(it) \geq 163\,840 \tag{6.72}$$

for all $t > 0$. Combining (6.71) and (6.72) gives (6.62). $\qquad\square$

6.6.4 Proof of (6.63)

As with the proof of (6.62), before tackling inequality (6.63) for the full range $t \geq 1$, it is helpful to put on our asymptotician hat and first ask the question of why we should expect the inequality to hold for large t. The answer is that the asymptotic expansions of the functions $F(it)$ and $G(it)$ are given by

$$F(it) = 16 + (-3\,840\pi t + 7\,680)Q^2 + (230\,400\pi^2 t^2 - 990\,720\pi t + 990\,720)Q^4$$
$$+ (8\,294\,400\pi^2 t^2 - 25\,205\,760\pi t + 16\,803\,840)Q^6 + \cdots, \tag{6.73}$$
$$G(it) = 16 + 1\,920Q^2 - 81\,920Q^3 + 1\,077\,120Q^4 - 8\,060\,928Q^5 + 41\,725\,440Q^6$$
$$- 166\,625\,280Q^7 + 553\,054\,080Q^8 - 1\,599\,733\,760Q^9 + \cdots, \tag{6.74}$$

where $Q = e^{-\pi t}$ as before. Here (6.74) is an ordinary Q-series expansion, whereas (6.73) is a somewhat nonstandard type of expansion that involves powers of $Q = e^{-\pi t}$, with each coefficient being itself a quadratic polynomial in t (refer to (6.58) to understand where this structure comes from).

Now the insight we get from these two expansions is that they share the same constant term 16 and that both have a next-order term proportional to Q^2 with coefficients

$7\,680 - 3\,840\pi t$ and $1\,920$, respectively. Since $7\,680 - 3\,840\pi t < 0 < 1\,920$ for $t \geq 1$, again we see that for t large, once the lower-order terms have decayed sufficiently, the relation $F(it) < G(it)$ will necessarily hold.

To turn this line of argumentation into a proof of the stronger claim that the inequality $F(it) < G(it)$ holds for all $t \geq 1$, we need to gain some measure of control over those lower-order terms, since for moderately sized t, they are not altogether negligible. This requires more subtle reasoning than that used in the proof of (6.62), since in the current case, both expansions (6.73) and+ (6.74) involve a mixture of terms with positive and negative coefficients.

Lemma 6.28. *Define*

$$W(\tau) = \theta_3^{12}\theta_2^{8} + \theta_3^{8}\theta_2^{12} + \theta_3^{12}\theta_4^{8} + \theta_3^{8}\theta_4^{12}. \tag{6.75}$$

The coefficients in the Q-series expansion of W are nonnegative.

Proof. Denote for convenience

$$Z = \theta_3^4, \quad X = \theta_2^4, \quad Y = 2Z - X.$$

Note that $\theta_4^4 = Z - X$, by (5.82). Now Z and X have Q-series expansions with nonnegative coefficients. Moreover, recalling (5.82), we see that Y can be written as $Y = Z + \theta_4^4 = \theta_3(\tau)^4 + \theta_3(\tau + 1)^4$, which implies that Y also has a Q-series expansion with nonnegative coefficients. Therefore by straightforward algebra we get that

$$W(\tau) = Z^3 X^2 + Z^2 X^3 + Z^3(Z - X)^2 + Z^2(Z - X)^3$$

$$= \left(\frac{X + Y}{2}\right)^3 X^2 + \left(\frac{X + Y}{2}\right)^2 X^3$$

$$+ \left(\frac{X + Y}{2}\right)^3 \left(\frac{-X + Y}{2}\right)^2 + \left(\frac{X + Y}{2}\right)^2 \left(\frac{-X + Y}{2}\right)^3$$

$$= \frac{1}{16}(6X^5 + 15X^4 Y + 10X^3 Y^2 + Y^5).$$

This representation clearly shows that the Q-series expansion of W also has nonnegative coefficients. □

Next, it is helpful to renormalize the functions F and G by defining the new functions

$$K(\tau) = -\frac{F(\tau) - 16}{Q^2} = -Q^{-2}(E_4')^2\tau^2 - 8Q^{-2}E_4' E_4\tau - 16Q^{-2}(E_4^2 - 1),$$

$$L(\tau) = -\frac{G(\tau) - 16}{Q^2} = -8Q^{-2}[\theta_4^8(\theta_3^{12} + \theta_4^4\theta_3^8 + \theta_2^8\theta_4^4 - \theta_2^{12}) - 2].$$

Inequality (6.63) can now be restated as the claim that $K(it) > L(it)$ for $t \geq 1$. This will follow from the combination of the following two lemmas.

Lemma 6.29. $L(it) \leq 2\,297$ for all $t \geq 1$.

Lemma 6.30. $K(it) \geq 3\,747$ for all $t \geq 1$.

Proof of Lemma 6.29. The expansion of $L(it)$ in powers of Q is easily written as

$$L(it) = -1\,920 + 81\,920Q - 1\,077\,120Q^2 + 8\,060\,928Q^3 - 41\,725\,440Q^4$$
$$+ 166\,625\,280Q^5 - 553\,054\,080Q^6 + 1\,599\,733\,760Q^7 + \cdots \tag{6.76}$$

(compare with (6.74)). Again using the substitution $\tau \mapsto \tau + 1$, we also have

$$-L(it + 1) = 1\,920 + 81\,920Q + 1\,077\,120Q^2 + 8\,060\,928Q^3 + 41\,725\,440Q^4$$
$$+ 166\,625\,280Q^5 + 553\,054\,080Q^6 + 1\,599\,733\,760Q^7 + \cdots. \tag{6.77}$$

On the other hand, using the usual properties of this substitution, we have

$$-L(\tau + 1) = Q^{-2}(G(\tau + 1) - 16) = 8Q^{-2}[\theta_3^8(\theta_4^{12} + \theta_3^4\theta_4^8 + \theta_2^8\theta_3^4 + \theta_2^{12}) - 2]$$
$$= 8Q^{-2}(W(\tau) - 2)$$

(with W defined in (6.75)). Lemma 6.28 reassures us that the coefficients in expansion (6.77) are nonnegative, and consequently the coefficients in (6.76) appear with alternating signs. Now defining

$$H(\tau) = \frac{L(\tau) - L(\tau + 1)}{2},$$

we see that $H(it)$ has the expansion

$$H(it) = 81\,920Q + 8\,060\,928Q^3 + 166\,625\,280Q^5 + 1\,599\,733\,760Q^7 + \cdots$$

with coefficients that are also nonnegative and majorize those of $L(it)$. Note moreover that the constant coefficient in $L(it)$ is $-1\,920$, whereas the constant coefficient in $H(it)$ is 0. Therefore the bound $L(it) \leq H(it) - 1\,920$ holds for all $t > 0$. In fact, since $H(it)$ is decreasing in t, we get a *constant* upper bound for $L(it)$ on the interval $[1, \infty)$, namely

$$L(it) \leq H(i) - 1\,920 \quad (t \geq 1).$$

To make this bound explicit, we express $H(\tau)$ directly in terms of thetanull functions. Appealing to (5.57)–(5.59) as before, we have

$$H(\tau) = -4Q^{-2}[\theta_4^8(\theta_3^{12} + \theta_4^4\theta_3^8 + \theta_2^8\theta_4^4 - \theta_2^{12}) - 2$$
$$- \theta_3^8(\theta_4^{12} + \theta_3^4\theta_4^8 + \theta_2^8\theta_3^4 + \theta_2^{12}) + 2]$$
$$= 4Q^{-2}(\theta_2^8\theta_3^{12} + \theta_2^{12}\theta_3^8 + \theta_2^{12}\theta_4^8 - \theta_2^8\theta_4^{12})$$
$$= 4Q^{-2}(\theta_2^8(\theta_3^{12} - \theta_4^{12}) + \theta_2^{12}(\theta_3^8 + \theta_4^8)).$$

Therefore, making use of (6.66)–(6.68), $H(i)$ can be calculated as

$$H(i) = 4e^{2\pi}\left(\frac{\Gamma(1/4)}{(2\pi)^{3/4}}\right)^{20}\left((2^{1/4})^{12} - 1 + (2^{1/4})^8 + 1\right)$$

$$= 4e^{2\pi}\frac{\Gamma(1/4)^{20}}{(2\pi)^{15}}(8+4) = 3e^{2\pi}\frac{\Gamma(1/4)^{20}}{2\,048\pi^{15}} \approx 4\,216.16.$$

We conclude that $L(it) \le 4\,217 - 1\,920 = 2\,297$ for all $t \ge 1$, as claimed. $\qquad\square$

Proof of Lemma 6.30. The asymptotic expansion for $K(it)$ is

$$K(it) = (3\,840\pi t - 7\,680) + (-230\,400\pi^2 t^2 + 990\,720\pi t - 990\,720)Q^2$$
$$+ (-8\,294\,400\pi^2 t^2 + 25\,205\,760\pi t - 16\,803\,840)Q^4 + \cdots.$$

We separate $K(it)$ into three components, defining

$$K_1(t) = 3\,840\pi t + (-230\,400\pi^2 t^2 + 990\,720\pi t - 990\,720)Q^2,$$
$$K_2(t) = Q^{-2}E_4'(it)^2 t^2 - 16Q^{-2}(E_4(it)^2 - 1) + (230\,400\pi^2 t^2 + 990\,720)Q^2,$$
$$K_3(t) = -8iQ^{-2}E_4'(it)E_4(it)t - (3\,840\pi t + 990\,720\pi t Q^2),$$

so that we have

$$K(it) = K_1(t) + K_2(t) + K_3(t).$$

The asymptotic behavior of $K_2(t)$ and $K_3(t)$ can be understood from the expansions

$$K_2(t) = -7\,680 - (8\,294\,400\pi^2 t^2 + 16\,803\,840)Q^4$$
$$- (113\,356\,800\pi^2 t^2 + 126\,819\,840)Q^6 - \cdots,$$
$$K_3(t) = 25\,205\,760\pi t Q^4 + 253\,639\,680\pi t Q^6 + 1\,500\,019\,200\pi t Q^8 + \cdots.$$

We now make the following elementary observations:
1. The function $K_1(t)$ is increasing on $[1, \infty)$.

   Proof. Assume that $t \ge 1$. Then

   $$K_1'(t) = 3\,840\pi e^{-2\pi t}(e^{2\pi t} + 120\pi^2 t^2 - 636\pi t + 774)$$
   $$\ge 3\,840\pi e^{-2\pi t}(e^{2\pi} + 120\pi^2 t^2 - 636\pi t + 774).$$

   The last expression is of the form $e^{-2\pi t}$ times a quadratic polynomial in t, which, as it is easy to check, is positive on the real line. Thus we have shown that $K_1'(t) > 0$ for $t \ge 1$, which proves the claim.

2. The function $K_2(t)$ is increasing on $[1, \infty)$.

   Proof. By inspection the expansion of $K_2(t)$ consists of the constant term $-7\,680$ plus a sum of lower-order terms, each being of the form $-(at^2 + b)e^{-n\pi t}$ for some nonnegative coefficients a, b and positive integer n. Each such term is an increasing function of t for $t \geq \frac{2}{n\pi}$, so in particular for $t \geq 1$.

3. $K_3(t) \geq 0$ for all $t > 0$.

Proof. The expansion of $K_3(t)$ has nonnegative coefficients.

Combining the above observations, we get that for $t \geq 1$,

$$K(it) \geq K_1(t) + K_2(t) \geq K_1(1) + K_2(1)$$
$$= -e^{2\pi}\left(-E_4'(i)^2 + 16E_4(i)^2 - 16\right) + 3\,840\pi + 990\,720\pi e^{-2\pi}$$
$$= -e^{2\pi}\left(\frac{9\Gamma(1/4)^{16}}{1\,024\,\pi^{12}} + 16\frac{9\Gamma(1/4)^{16}}{4\,096\,\pi^{12}} - 16\right) + 3\,840\pi + 990\,720\pi e^{-2\pi}$$
$$= -e^{2\pi}\left(\frac{45\,\Gamma(1/4)^{16}}{1\,024\,\pi^{12}} - 16\right) + 3\,840\pi + 990\,720\pi e^{-2\pi} \approx 3\,747.1,$$

as claimed. $\square$

6.7 Proof of Theorem 6.1

Define the functions

$$C(z) = A(z) + B(z),$$
$$D(z) = A(z) - B(z).$$

Lemma 6.31. *The functions $C(z)$ and $D(z)$ are holomorphic in the region $\mathrm{Re}(z) > -2$ and have the explicit representations*

$$C(z) = -4\sin^2\left(\frac{\pi z}{2}\right)\int_0^\infty (U(it) + V(it))e^{-\pi z t}\, dt \quad (\mathrm{Re}(z) > 2), \tag{6.78}$$

$$D(z) = -4\sin^2\left(\frac{\pi z}{2}\right)\int_0^\infty (U(it) - V(it))e^{-\pi z t}\, dt \quad (\mathrm{Re}(z) > 0), \tag{6.79}$$

Proof. The holomorphicity is immediate from the analytic continuation of $A(z)$ and $B(z)$ discussed in Sections 6.4–6.5. Similarly, relation (6.78) is an immediate consequence of Lemmas 6.6 and 6.18. Relation (6.79) follows as well from these lemmas *for z satisfying* $\mathrm{Re}(z) > 2$, but here we make the stronger claim that this representation remains valid in the larger half-plane $\mathrm{Re}(z) > 0$; this is related to the fact that in the analytically contin-ued representations (6.24) and (6.52), the poles $\frac{1}{z-2}$ inside the parenthesized expressions cancel each other out upon subtracting the two formulas. To make this more precise, observe that combining estimates (6.12), (6.13), (6.42), and (6.43) gives that

$$U(it) - V(it) = O(t) \quad (t \to \infty), \tag{6.80}$$

$$U(it) - V(it) = O(t^2 e^{-\pi/t}) \quad (t \to 0), \tag{6.81}$$

and this is clearly sufficient to imply the absolute convergence of the integral in (6.79), uniformly on compacts in the half-plane $\mathrm{Re}(z) > 0$. By the principle of analytic continuation, since the right-hand side of (6.79) is equal to $D(z)$ for $\mathrm{Re}(z) > 2$, it must also equal $D(z)$ on $\mathrm{Re}(z) > 0$. $\qquad\square$

Define $\varphi : \mathbb{R}^8 \to \mathbb{R}$ by

$$\varphi(x) = C(\|x\|^2) = \varphi_+(x) + \varphi_-(x).$$

By Lemmas 6.12 and 6.23, $\varphi(x)$ is a radial Schwartz function. By Lemmas 6.13 and 6.24 its Fourier transform is

$$\mathcal{F}_8[\varphi](x) = \varphi_+(x) - \varphi_-(x) = D(\|x\|^2).$$

In other words, $\varphi_+(x)$ and $\varphi_-(x)$ are the Fourier-even and Fourier-odd components in the Fourier parity decomposition of φ; see (A.20)–(A.21).

Theorem 6.32. *The function φ is a magic function for the lattice E_8. Consequently, $\Delta_{\mathrm{optimal}}(8) = \frac{\pi^4}{384}$, and the E_8 sphere packing is optimal.*

Proof. Let $\rho_0 = \sqrt{2}$. In $\mathbb{R}^8$, we have

$$\mathrm{vol}(B_{\rho_0/2}(0)) = \frac{\pi^4}{2^4 \Gamma(5)} = \frac{\pi^4}{384},$$

which is precisely the packing density of E_8 (see Theorem A.8). Therefore we need to show that φ satisfies the three conditions of Theorem A.21 with the particular value of ρ being equal to $\sqrt{2}$. Indeed, by (6.26) and (6.53) we have

$$\varphi(0) = \varphi_+(0) + \varphi_-(0) = 240\pi > 0,$$

$$\widehat{\varphi}(0) = \varphi_+(0) - \varphi_+(0) = 240\pi,$$

so the first condition is satisfied. Next, (6.78), when combined with Lemmas 6.3 and 6.15, implies that $\varphi(x) \leq 0$ for all $x \in \mathbb{R}^d$ with $\|x\| \geq \sqrt{2}$. This confirms the third condition. Finally, (6.79), together with inequality (6.57), implies that $\mathcal{F}_8[\varphi]$ is everywhere nonnegative. This is the second condition of Theorem A.21 and the final one needed to be verified. The proof that φ is a magic function for E_8 and therefore that the E_8 sphere packing is optimal for sphere packing in 8 dimensions is complete. $\qquad\square$

Suggested exercises for Section 6.7. 6.5, 6.6, 6.7.

Exercises for Chapter 6

6.1 Prove Lemma 6.10.

6.2 Prove Lemma 6.12.

6.3 Explain why the operation of commuting the Fourier transform with the integrals in (6.32) is justified.

6.4 Show that the function $A(z)$, which was analytically continued to a holomorphic function on the half-plane $\mathrm{Re}(z) > -2$ in Lemma 6.8, can in fact be continued analytically to an *entire* function.

6.5 Find the special values $\tilde{\varphi}(\sqrt{2})$, $(\tilde{\varphi})'(\sqrt{2})$, $\widehat{\tilde{\varphi}}(\sqrt{2})$, $(\widehat{\tilde{\varphi}})'(\sqrt{2})$ associated with the radial profile $\tilde{\varphi}$ of the E_8 magic function φ, the radial profile of the Fourier transform of φ, and their derivatives.

6.6 Prove that the magic function $\varphi(x)$ satisfies the following properties:

(a) $\int_{\mathbb{R}^8} \varphi(x)\, dx = 240\pi$.

(b) $\sum_{n=1}^{\infty} \sigma_3(n)\tilde{\varphi}(\sqrt{2n-1}) = 0$.

(c) $\sum_{x \in E_8} \varphi(x+y) \equiv 240\pi$ for all $y \in \mathbb{R}^8$.

6.7 **Magic function for the Leech lattice.** [16] Prove that there exists a magic function for the Leech lattice in dimension 24.

Guidance. Repeat the proof of this chapter with appropriate modifications. The function $U(\tau)$ should be replaced by

$$U_{24}(\tau) = 6\,912 \times \frac{\mu_2(\tau)\tau^2 + \mu_1(\tau)\tau + \mu_0(\tau)}{(E_4^3 - E_6^2)^2}$$

with μ_0, μ_1, μ_2 defined by

$$\mu_0(\tau) = 36(25E_6^2 - 49E_4^3),$$
$$\mu_1(\tau) = 6\pi i\big(48E_6 E_4^2 + 2(25E_6^2 - 49E_4^3)E_2\big),$$
$$\mu_2(\tau) = \pi^2\big(25E_4^4 - 49E_6^2 E_4 + 48E_6 E_4^2 E_2 + (25E_6^2 - 49E_4^3)E_2^2\big).$$

In place of $V(\tau)$, use

$$V_{24}(\tau) = 12^6 \times \frac{7\theta_4^{20}\theta_2^8 + 7\theta_4^{24}\theta_2^4 + 2\theta_4^{28}}{(E_4^3 - E_6^2)^2}.$$

See Exercise A.16 in the Appendix for the relevant properties of the Leech lattice.

A Appendix: Background on sphere packings

This appendix presents the background material on sphere packings and related notions that is necessary to understand the developments of Chapter 6. The material discussed here mostly does not involve any complex analysis (with the one notable exception being the proof of Proposition A.17 in Section A.7). Before reading this appendix, we recommend reading Sections 6.1 and 6.2 for motivation.

A.1 Sphere packings and their densities

Fix a dimension $d \geq 2$. Given $r > 0$ and $x \in \mathbb{R}^d$, denote by $B_r(x)$ the Euclidean ball of radius r centered at x. A **sphere packing** in $\mathbb{R}^d$ consists of a union of balls of equal radii with nonoverlapping interiors. We commonly denote a packing as

$$P = P(X, r) = \bigcup_{x \in X} B_r(x),$$

where $X \subset \mathbb{R}^d$ is the set of centers of the balls participating in the union, and r is their common radius. The **upper packing density** associated with a sphere packing P is

$$\Delta_P^+ = \limsup_{R \to \infty} \frac{\text{vol}(P \cap B_R(0))}{\text{vol}(B_R(0))}. \tag{A.1}$$

In the case where the limsup in (A.1) is in fact an ordinary limit, we say that P has a packing density. In that case, we denote Δ_P^+ by Δ_P and refer to this quantity simply as the packing density of P.

The **optimal packing density** of $\mathbb{R}^d$ is defined to be

$$\Delta_{\text{optimal}}(d) = \sup\{\Delta_P^+ : P \text{ is a sphere packing in } \mathbb{R}^d\}.$$

A sphere packing P in $\mathbb{R}^d$ is called **optimal** if it has a packing density and its packing density is equal to $\Delta_{\text{optimal}}(d)$.

Theorem A.1 ([35], [36, Sec. 3.viii]). *An optimal sphere packing in $\mathbb{R}^d$ exists.*

Sphere packings have a trivial scale invariance property: replacing all the balls $B_r(x)$ in a sphere packing P by their scaled copies $B_{\lambda r}(\lambda x)$ for some constant $\lambda > 0$ results in a sphere packing with the same packing density. For this reason, when proving facts about packing densities for general sphere packings, we can assume without loss of generality that a packing has some specific common sphere radius r (where r can be chosen arbitrarily for some reason of convenience).

Suggested exercises for Section A.1. A.1.

A.2 Lattices and lattice packings

A **lattice** in $\mathbb{R}^d$ is a set of points of the form

$$\Lambda = \left\{ \sum_{j=1}^{d} n_j x_j \ : \ n_1, \dots, n_d \in \mathbb{Z} \right\},$$

where $x_1, \dots, x_d$ is a linear basis for $\mathbb{R}^d$. Another notation for the same set is $\bigoplus_{j=1}^{d} \mathbb{Z} \cdot x_j$. (This may be referred to as the $\mathbb{Z}$-**span** of the vectors $x_1, \dots, x_d$. The spanning set $x_1, \dots, x_d$ is said to be a **basis** for the lattice Λ; note that it is not unique.) Given a lattice, it is easy to check that the associated union of balls $P(\Lambda, r)$ is a sphere packing if and only if $r \le r_*(\Lambda)$, where

$$r_*(\Lambda) := \frac{1}{2} \min \left\{ \left\| \sum_{j=1}^{n} n_j x_j \right\| \ : \ (n_1, \dots, n_d) \in \mathbb{Z}^d \setminus \{(0, \dots, 0)\} \right\}.$$

We refer to the sphere packing $P(\Lambda, r_*(\Lambda))$ as the **lattice sphere packing** (or lattice packing) associated with the lattice Λ and denote its packing density by δ_Λ.

It is not known whether in every dimension d there exists a lattice Λ whose associated sphere packing is optimal. This is the case in the dimensions $d = 2, 3, 8, 24$, which are the only dimensions for which the value of $\Delta_{\mathrm{optimal}}(d)$ has been established rigorously.

A.3 Periodic sphere packings

Lattice sphere packings are a particular case of a more general family of sphere packings called periodic sphere packings. These are packings that have a periodic structure associated with a lattice. More precisely, let Λ be a lattice in $\mathbb{R}^d$, let $A = \{x_1, \dots, x_m\} \subset \mathbb{R}^d$ be a finite set of points, and let $r > 0$ be a number. Assume that $\|x + x_j - x_k\| \ge 2r$ for all $1 \le j, k \le m$ and all $x \in \Lambda$, except for the case $x = 0$ and $j = k$. Then the union of balls of radius r centered around Λ-translates of the points of A is a sphere packing; that is, we define

$$P = P(X, r), \quad \text{where } X = A + \Lambda = \{x_j + y \ : \ 1 \le j \le m, \ y \in \Lambda\}. \tag{A.2}$$

A sphere packing constructed in such a way is called a **periodic sphere packing** (or periodic packing).

It is not known whether in every dimension d there exists a periodic sphere packing that is optimal. However, periodic packings are sufficiently general that they come arbitrarily close to being optimal, as the following result makes precise.

Lemma A.2 ([14, Appendix A]).

$$\Delta_{\mathrm{optimal}}(d) = \sup\{\Delta_P \ : \ P \text{ is a periodic sphere packing in } \mathbb{R}^d\}.$$

A.4 Lattice covolume

The **covolume** of a lattice $\Lambda = \bigoplus_{j=1}^{d} \mathbb{Z} \cdot x_j$, denoted $\mathrm{covol}(\Lambda)$, is defined as the absolute value of the determinant of the matrix containing the vectors $x_1, \ldots, x_d$ as its rows.

Lemma A.3. *The definition of* $\mathrm{covol}(\Lambda)$ *is independent of the choice of basis* $x_1, \ldots, x_d$ *for the lattice. Moreover, the covolume has the following geometric interpretation: it is the volume of the set*

$$\{t_1 x_1 + t_2 x_2 + \cdots + t_d x_d \ : \ t_1, \ldots, t_d \in [0,1]\}$$

*(called the **fundamental cell**, or **fundamental parallelepiped**, of the lattice associated with the basis* $x_1, \ldots, x_d$*).*

Proof. Exercise A.2. □

Lemma A.4. 1. *For a lattice* $\Lambda \subset \mathbb{R}^d$, *the packing density of the associated lattice sphere packing is given by*

$$\delta_\Lambda = \frac{\mathrm{vol}(B_{r_*(\Lambda)}(0))}{\mathrm{covol}(\Lambda)} = \frac{\pi^{d/2} r_*(\Lambda)^d}{\Gamma(\frac{d}{2} + 1) \, \mathrm{covol}(\Lambda)}. \tag{A.3}$$

2. *For a periodic sphere packing P as in* (A.2), *its packing density is*

$$\Delta_P = \frac{m \, \mathrm{vol}(B_r(0))}{\mathrm{covol}(\Lambda)} = \frac{m \pi^{d/2} r^d}{\Gamma(\frac{d}{2} + 1) \, \mathrm{covol}(\Lambda)}. \tag{A.4}$$

(In (A.3)–(A.4), Γ *denotes the Euler gamma function.)*

Proof. The second equality in each of relations (A.3) and (A.4) follows from the well-known formula for the volume of the unit ball in $\mathbb{R}^d$; see Exercise 2.3 on page 110. The proof of the additional claim relating the explicit quantities in (A.3) and (A.4) to the packing densities δ_Λ and Δ_P is left as an exercise (Exercise A.3). □

Suggested exercises for Section A.4. A.2, A.3.

A.5 Dual lattices

If Λ is a lattice in $\mathbb{R}^d$, then its **dual lattice** is the set denoted Λ^* and defined by

$$\Lambda^* = \{y \in \mathbb{R}^d \ : \ \langle x, y \rangle \in \mathbb{Z} \text{ for all } x \in \Lambda\}.$$

The fact that Λ^* is a lattice follows from Lemma A.5.

Lemma A.5. *If $B = \{x_1, \ldots, x_d\}$ is a basis for the lattice Λ, let $B^* = \{y_1, \ldots, y_d\}$ be the dual basis, when considering B as a linear basis for $\mathbb{R}^d$; that is, the vectors $y_1, \ldots, y_d$ are the unique vectors satisfying*

$$\langle x_j, y_k \rangle = \delta_{jk} \quad (1 \le j, k \le d)$$

(where δ_{jk} denotes the Kronecker delta). Then we have $\Lambda^* = \bigoplus_{j=1}^{d} \mathbb{Z} \cdot y_j$.

Proof. Exercise A.4. □

Suggested exercises for Section A.5. A.4, A.5.

A.6 The Poisson summation formula for lattices

In Chapter 2, we discussed the Poisson summation formula for functions of a single real variable (Theorem 2.6), a classical result from Fourier analysis, in the context of our proof of the functional equation of the Riemann zeta function. There is a version of the same result for functions on $\mathbb{R}^d$ involving summation over lattices. This result relates the summation of values of a nicely behaved function on $\mathbb{R}^d$ over a lattice to the summation of its Fourier transform over the dual lattice and plays an important role in the study of sphere packings.

To state the result, first recall some basic facts about Fourier transforms in d dimensions. The Fourier transform in $\mathbb{R}^d$ is the operator $\mathcal{F}_d$ taking a function $f : \mathbb{R}^d \to \mathbb{C}$ to the function $\mathcal{F}_d[f]$ given by

$$\mathcal{F}_d[f](y) = \int_{\mathbb{R}^d} f(x) \exp(-2\pi i \langle y, x \rangle) \, dx, \tag{A.5}$$

assuming appropriate integrability conditions. We also denote the Fourier transform of f by $\hat{f}$. The Fourier transform acts in a particularly nice way on **Schwartz functions**. A function $f : \mathbb{R}^d \to \mathbb{R}$ is called a Schwartz function if it satisfies

$$\sup_{x=(x_1,\ldots,x_d)\in\mathbb{R}^d} \left\| x_1^{j_1} x_2^{j_2} \cdots x_d^{j_d} \cdot \frac{\partial^{k_1} \partial^{k_2} \cdots \partial^{k_d} f}{\partial x_1^{k_1} \partial x_2^{k_2} \cdots \partial x_d^{k_d}} \right\| < \infty$$

for any integers $j_1, \ldots, j_d, k_1, \ldots, k_d \ge 0$. The following is a standard fact from analysis; see [41, p. 301] for the proof.

Proposition A.6. *The Fourier transform of a Schwartz function is also a Schwartz function.*

We can now state the Poisson summation formula for lattices.

Theorem A.7 (Poisson summation formula for lattices). *Let $\Lambda \subset \mathbb{R}^d$ be a lattice, and let $f : \mathbb{R}^d \to \mathbb{C}$ be a Schwartz function. Then*

$$\sum_{x \in \Lambda} f(x) = \frac{1}{\text{covol}(\Lambda)} \sum_{y \in \Lambda^*} \hat{f}(y). \tag{A.6}$$

Another, slightly more general, version of the Poisson summation formula for lattices is

$$\sum_{x \in \Lambda} f(x + t) = \frac{1}{\text{covol}(\Lambda)} \sum_{y \in \Lambda^*} \hat{f}(y) \exp(2\pi i \langle y, t \rangle) \quad (t \in \mathbb{R}^d). \tag{A.7}$$

In fact, equations (A.6) and (A.7) are equivalent, since (A.6) is the case $t = 0$ of (A.7), and conversely, the general case of (A.7) for arbitrary $t \in \mathbb{R}^d$ is immediately obtained from (A.6) on applying that relation to the function $g(x) = f(x + t)$.

Proof of Theorem A.7. Exercise A.6. □

Suggested exercises for Section A.6. A.6.

A.7 Construction of the lattice E_8

The goal of this section is to construct the lattice E_8, which plays a central role in the sphere packing story. We will prove the following result.

Theorem A.8. *There exists a lattice in $\mathbb{R}^8$, denoted E_8, with the following properties:*
1. *The packing density δ_{E_8} of the sphere packing associated with E_8 is $\frac{\pi^4}{384}$.*
2. *The set of Euclidean norms of points of the lattice E_8 is*

$$\{\sqrt{2n} : n = 0, 1, 2, \ldots\}.$$

An immediate corollary of the existence of E_8 is the following conceptually important result.

Corollary A.9. *The optimal sphere packing density $\Delta_{\text{optimal}}(8)$ in 8 dimensions satisfies*

$$\Delta_{\text{optimal}}(8) \geq \frac{\pi^4}{384}.$$

Several different constructions of E_8 are known; perhaps its most natural manifestation is as the lattice spanned by the **E_8 root system**, an object associated with the **E_8 Lie algebra**, one of the so-called exceptional Lie algebras that appears in a famous classification theorem. [40, p. 238] Here we give an elementary construction of E_8, which provides a straightforward path to a proof of our claims (while offering little insight into what makes E_8 so special and interesting). Define the vectors $x_1, \ldots, x_8 \in \mathbb{R}^8$ as the columns of the matrix

$$M = \begin{pmatrix} 2 & -1 & 0 & 0 & 0 & 0 & 0 & \frac{1}{2} \\ 0 & 1 & -1 & 0 & 0 & 0 & 0 & \frac{1}{2} \\ 0 & 0 & 1 & -1 & 0 & 0 & 0 & \frac{1}{2} \\ 0 & 0 & 0 & 1 & -1 & 0 & 0 & \frac{1}{2} \\ 0 & 0 & 0 & 0 & 1 & -1 & 0 & \frac{1}{2} \\ 0 & 0 & 0 & 0 & 0 & 1 & -1 & \frac{1}{2} \\ 0 & 0 & 0 & 0 & 0 & 0 & 1 & \frac{1}{2} \\ 0 & 0 & 0 & 0 & 0 & 0 & 0 & \frac{1}{2} \end{pmatrix},$$

and define

$$E_8 = \bigoplus_{j=1}^{8} \mathbb{Z} \cdot x_j.$$

Lemma A.10. E_8 is a lattice with basis $x_1, \ldots, x_8$, and its covolume is 1.

Proof. The x_j are clearly linearly independent, so E_8 is indeed a lattice in $\mathbb{R}^8$, and $\mathrm{covol}(E_8) = \det(M) = 1.$ $\qquad\square$

Lemma A.11. *The lattice E_8 has an alternative representation as*

$$E_8 = \left\{ (y_1, \ldots, y_8) \in \mathbb{Z}^8 : \sum_{j=1}^{8} y_j \equiv 0 \pmod{2} \right\}$$

$$\cup \left\{ (y_1, \ldots, y_8) \in \left(\mathbb{Z} + \frac{1}{2}\right)^8 : \sum_{j=1}^{8} y_j \equiv 0 \pmod{2} \right\}. \tag{A.8}$$

Proof. Denote the two sets participating in the union on the right-hand side of (A.8) by I_8 and J_8, respectively. By inspection, $x_1, \ldots, x_7 \in I_8$, $x_8 \in J_8$, and $I_8 \cup J_8$ is closed under the taking of linear combinations with integer coefficients. This shows that $E_8 \subset I_8 \cup J_8$. Conversely, if $y = (y_1, \ldots, y_8) \in I_8$, then we can write

$$y = \sum_{j=1}^{8} a_j x_j$$

(regarding y for convenience as a column vector), where

$$\begin{pmatrix} a_1 \\ a_2 \\ a_3 \\ a_4 \\ a_5 \\ a_6 \\ a_7 \\ a_8 \end{pmatrix} = M^{-1}y = \begin{pmatrix} \frac{1}{2} & \frac{1}{2} & \frac{1}{2} & \frac{1}{2} & \frac{1}{2} & \frac{1}{2} & \frac{1}{2} & -\frac{7}{2} \\ 0 & 1 & 1 & 1 & 1 & 1 & 1 & -6 \\ 0 & 0 & 1 & 1 & 1 & 1 & 1 & -5 \\ 0 & 0 & 0 & 1 & 1 & 1 & 1 & -4 \\ 0 & 0 & 0 & 0 & 1 & 1 & 1 & -3 \\ 0 & 0 & 0 & 0 & 0 & 1 & 1 & -2 \\ 0 & 0 & 0 & 0 & 0 & 0 & 1 & -1 \\ 0 & 0 & 0 & 0 & 0 & 0 & 0 & 2 \end{pmatrix} \begin{pmatrix} y_1 \\ y_2 \\ y_3 \\ y_4 \\ y_5 \\ y_6 \\ y_7 \\ y_8 \end{pmatrix}.$$

Again by inspection, the assumption that y_j are integers satisfying $\sum_{j=1}^{8} y_j \equiv 0 \pmod 2$ immediately implies that $a_1, \ldots, a_8$ obtained in this way are themselves integers and that therefore $y = \sum_{j=1}^{8} a_j x_j \in E_8$. This shows that $I_8 \subset E_8$. To show that also $J_8 \subset E_8$, observe that if $y \in J_8$, then $y - x_8 \in I_8$, so the previous calculation shows that $y = x_8 + \sum_{j=1}^{8} a_j x_j$, where a_j are integer coefficients, and thus once again we have that $y \in E_8$. $\square$

Lemma A.12. *For any $x, y \in E_8$, we have $\langle x, y \rangle \in \mathbb{Z}$.*

Proof. For $1 \leq j, k \leq 8$, define $t_{j,k} = \langle x_j, x_k \rangle$; explicitly, the numbers $(t_{j,k})_{j,k=1}^{8}$ are the entries of the symmetric matrix

$$M^{\top}M = \begin{pmatrix} 4 & -2 & 0 & 0 & 0 & 0 & 0 & 1 \\ -2 & 2 & -1 & 0 & 0 & 0 & 0 & 0 \\ 0 & -1 & 2 & -1 & 0 & 0 & 0 & 0 \\ 0 & 0 & -1 & 2 & 1 & 0 & 0 & 0 \\ 0 & 0 & 0 & -1 & 2 & -1 & 0 & 0 \\ 0 & 0 & 0 & 0 & -1 & 2 & -1 & 0 \\ 0 & 0 & 0 & 0 & 0 & -1 & 2 & 0 \\ 1 & 0 & 0 & 0 & 0 & 0 & 0 & 2 \end{pmatrix}.$$

Now if $x, y \in E_8$, then express x, y as $x = \sum_{j=1}^{8} a_j x_j$ and $y = \sum_{k=1}^{8} b_k x_k$ with integer coefficients a_j, b_k. Then

$$\langle x, y \rangle = \sum_{j,k=1}^{8} t_{j,k} a_j b_k, \tag{A.9}$$

which is manifestly an integer. $\square$

Lemma A.13. *For any $x \in E_8$, we have $\|x\|^2 \in 2\mathbb{Z}$.*

Proof. This is immediate from (A.9) on setting $y = x$ and noting that the double sum can be rewritten as

$$\sum_{j,k=1}^{8} t_{j,k} a_j a_k = \sum_{j=1}^{8} t_{j,j} a_j^2 + 2 \sum_{1 \leq j < k \leq 8} t_{j,k} a_j a_k,$$

which is easily recognized as an even integer. $\square$

Lemma A.14. *The packing density of the sphere packing associated with E_8 is $\frac{\pi^4}{384}$.*

Proof. Since the squared norms $\|x\|^2$ for $x \in E_8$ are nonnegative even integers, the minimal norm of a nonzero vector is at least $\sqrt{2}$. On the other hand, the vector $x = (1,1,0,0,0,0,0,0) = x_1 + x_2$ is one specific vector in E_8 with that norm, so $\sqrt{2}$ is in fact precisely the minimal nonzero norm. This establishes that

$$r_*(E_8) = \frac{\sqrt{2}}{2}.$$

Now using (A.3) together with the already established fact that $\mathrm{covol}(E_8) = 1$ gives the claim. $\qquad \square$

Lemma A.15. $E_8 = E_8^*$.

Proof. Lemma A.12 can be reformulated as the statement that $E_8 \subseteq E_8^*$. To prove the reverse inclusion, let $y_1, \ldots, y_8 \in \mathbb{R}^d$ denote the elements of the dual basis to $x_1, \ldots, x_8$. These are simply the rows of M^{-1} (or if they are thought of as column vectors, then the columns of $(M^{-1})^\top$). Now observe the somewhat trivial matrix equation

$$(M^{-1})^\top = M(M^{-1}(M^{-1})^\top)$$

$$= M \begin{pmatrix} 14 & 24 & 20 & 16 & 12 & 8 & 4 & -7 \\ 24 & 42 & 35 & 28 & 21 & 14 & 7 & -12 \\ 20 & 35 & 30 & 24 & 18 & 12 & 6 & -10 \\ 16 & 28 & 24 & 20 & 15 & 10 & 5 & -8 \\ 12 & 21 & 18 & 15 & 12 & 8 & 4 & -6 \\ 8 & 14 & 12 & 10 & 8 & 6 & 3 & -4 \\ 4 & 7 & 6 & 5 & 4 & 3 & 2 & -2 \\ -7 & -12 & -10 & -8 & -6 & -4 & -2 & 4 \end{pmatrix}.$$

For each $1 \leq j \leq 8$, the jth column y_j of the matrix $(M^{-1})^\top$ can be expressed as a linear combination of $x_1, \ldots, x_8$ with coefficients taken from the jth column of the matrix $M^{-1}(M^{-1})^\top$ (e. g., $y_1 = 14x_1 + 24x_2 + 20x_3 + 16x_4 + 12x_5 + 8x_6 + 4x_7 - 7x_8$). These coefficients are all integers, and thus $y_j \in E_8$. Since $y_1, \ldots, y_8$ are a basis for E_8^* (see Lemma A.5), we have shown that $E_8^* \subseteq E_8$. This completes the proof that $E_8 = E_8^*$ $\qquad \square$

Our last remaining task for this section is to prove the second claim in Theorem A.8. We already showed that all the squared norms of E_8 lattice vectors are even integers; it remains to show that *all* positive even integers are in fact squared norms of E_8 vectors. This will follow from a much more precise statement. Define the numbers $(a_n)_{n=0}^\infty$ by

$$a_n = \#\{x \in E_8 : \|x\|^2 = 2n\}.$$

Note that, trivially, $a_0 = 1$.

Lemma A.16. *For some constant $C > 0$, we have the bound $a_n \leq Cn^4$ for all $n \geq 1$.*

Proof. Exercise A.7. □

Proposition A.17. *We have the relation*

$$a_n = 240\sigma_3(n)$$

(where $\sigma_3(n)$ is defined in (5.2)) for all $n \geq 1$.

Remarkably, this result, which has a distinct number-theoretic flavor, can be proved using a complex-analytic argument involving modular forms. The idea is to form a kind of generating function for the squared norms of E_8 lattice vectors (known in the theory of lattices as the **theta series** of the lattice) and study its complex-analytic properties. More precisely, define a function of a complex variable τ by

$$\eta(\tau) = \sum_{x \in E_8} e^{\pi i \tau \|x\|^2} = \sum_{n=0}^{\infty} a_n e^{2\pi i n \tau}. \tag{A.10}$$

Lemma A.18. *The infinite series (A.10) converges absolutely and uniformly on compacts on the upper half-plane $\mathbb{H}$ and defines a holomorphic function there.*

Proof. By Lemma A.16,

$$\sum_{x \in E_8} \left| e^{\pi i \tau \|x\|^2} \right| \leq 1 + \sum_{n=1}^{\infty} Cn^4 e^{-2\pi n \, \mathrm{Im}(\tau)},$$

which converges uniformly in any half-plane of the form $\{\tau \ : \ \mathrm{Im}(\tau) > \kappa\}$ where $\kappa > 0$ and a fortiori on any compact subset of $\mathbb{H}$. The holomorphy follows from the standard theory (Theorem 1.39). □

Lemma A.19. *The function $\eta(\tau)$ is a modular form of weight 4.*

Proof. The equation $\eta(\tau + 1) = \eta(\tau)$ is immediate from (A.10), i. e., $\eta(\tau)$ transforms correctly under the generator T of the modular group Γ. We need to show that $\eta(\tau)$ also transforms in the correct way under the generator S, that is, that $\eta(\tau)$ satisfies the equation

$$\eta(-1/\tau) = \tau^4 \eta(\tau). \tag{A.11}$$

By Lemma 5.21 that would imply that $\eta(\tau)$ is a modular form of weight 4.

To prove (A.11), define the function $f_\tau : \mathbb{R}^8 \to \mathbb{C}$ depending on a parameter $\tau \in \mathbb{H}$ by

$$f_\tau(x) = e^{\pi i \tau \|x\|^2}. \tag{A.12}$$

In the case where τ lies on the positive imaginary axis, i.e., $\tau = it$ with $t > 0$, this is an 8-dimensional scaled Gaussian $e^{-\pi t \|x\|^2}$, which transforms under the (8-dimensional) Fourier transform as

$$\widehat{f_\tau}(y) = \tau^{-4} f_{-1/\tau}(y) = \tau^{-4} e^{\pi i(-1/\tau)\|y\|^2}. \tag{A.13}$$

(For general $\tau \in \mathbb{H}$, this equation still holds, but if you are feeling queasy about this or cannot be bothered to check it, just assume that τ is on the positive imaginary axis for now.) Applying the Poisson summation formula (A.6) and keeping in mind Lemma A.15 give

$$\sum_{x \in E_8} f_\tau(x) = \sum_{y \in E_8} \widehat{f_\tau}(y). \tag{A.14}$$

This is precisely what we need, since the left-hand side of (A.14) is equal to $\eta(\tau)$, and, by (A.13), the right-hand side is equal to $\tau^{-4} \eta(-1/\tau)$. Thus we have established (A.11). (As a final step, if you previously assumed that τ is imaginary, then now appeal to the principle of analytic continuation to argue that since the equation (A.11) holds on the positive imaginary axis, it must hold on all of $\mathbb{H}$.) □

Lemma A.20. *We have the identity*

$$\eta(\tau) = E_4(\tau) \quad (\tau \in \mathbb{H}), \tag{A.15}$$

where E_4 denotes the normalized version of the Eisenstein series G_4 defined in (5.87).

Proof. By Theorem 5.24 the vector space M_4 of modular forms of weight 4 is one-dimensional and contains $\eta(\tau)$ and $E_4(\tau)$. Thus we have

$$1 + \sum_{n=1}^\infty a_n e^{2\pi i n\tau} = K E_4(\tau) = K \cdot \left(1 + 240 \sum_{n=0}^\infty \sigma_3(n) e^{2\pi i n\tau} \right).$$

Equating the 0th Fourier coefficients on both sides gives $K = 1$, proving the claim. □

Proof of Proposition A.17. This follows immediately from (A.15), again by comparing the Fourier coefficients on both sides. □

Suggested exercises for Section A.7. A.7, A.8.

A.8 The Cohn–Elkies sphere packing bounds

Theorem A.21 (Cohn–Elkies sphere packing bounds [14]). *Let $f : \mathbb{R}^d \to \mathbb{R}$ be a Schwartz function, and let $\rho > 0$ be a number. Assume that the following conditions are satisfied:*
1. $f(0) = \widehat{f}(0) > 0$;
2. *The Fourier transform $\widehat{f}$ is real-valued, and $\widehat{f}(y) \geq 0$ for all $y \in \mathbb{R}^d$;*

3. $f(x) \leq 0$ for all $x \in \mathbb{R}^d$ such that $\|x\| \geq \rho$.

Then $\Delta_{\text{optimal}}(d)$, the optimal packing density in $\mathbb{R}^d$, satisfies

$$\Delta_{\text{optimal}}(d) \leq \text{vol}(B_{\rho/2}(0)) = \frac{\pi^{d/2} \rho^d}{2^d \Gamma(\frac{d}{2} + 1)}. \tag{A.16}$$

Proof. By Lemma A.2 it suffices to prove that $\text{vol}(B_{\rho/2}(0))$ is an upper bound for the packing density of any *periodic* sphere packing with common sphere radius $\rho/2$ (see the remark about scale invariance in Section A.1). Let P be such a packing, defined in terms of a lattice Λ and a finite set $\{x_1, \ldots, x_m\}$ as in (A.2). Recall that the fact that the common radius of the spheres in the packing is $\rho/2$ means that the Euclidean norm $\|x + x_j - x_k\|$ for any $1 \leq j, k \leq m$ and lattice point $x \in \Lambda$ is either 0 (in the case $x = 0$ and $j = k$) or is otherwise bounded from below by ρ.

Let $1 \leq j, k \leq m$. Applying the Poisson summation formula (A.7) with $t = x_j - x_k$ gives

$$\sum_{x \in \Lambda} f(x + x_j - x_k) = \frac{1}{\text{covol}(\Lambda)} \sum_{y \in \Lambda^*} \hat{f}(y) \exp(2\pi i \langle y, x_j - x_k \rangle). \tag{A.17}$$

Summing this relation over all j, k further gives that

$$\sum_{j,k=1}^{m} \sum_{x \in \Lambda} f(x + x_j - x_k)$$

$$= \frac{1}{\text{covol}(\Lambda)} \sum_{j,k=1}^{m} \sum_{y \in \Lambda^*} \hat{f}(y) \exp(2\pi i \langle y, x_j - x_k \rangle)$$

$$= \frac{1}{\text{covol}(\Lambda)} \sum_{y \in \Lambda^*} \hat{f}(y) \sum_{j,k=1}^{m} \exp(2\pi i \langle y, x_j \rangle) \overline{\exp(2\pi i \langle y, x_k \rangle)}$$

$$= \frac{1}{\text{covol}(\Lambda)} \sum_{y \in \Lambda^*} \hat{f}(y) \left(\sum_{j=1}^{m} \exp(2\pi i \langle y, x_j \rangle) \right) \overline{\left(\sum_{k=1}^{m} \exp(2\pi i \langle y, x_k \rangle) \right)}$$

$$= \frac{1}{\text{covol}(\Lambda)} \sum_{y \in \Lambda^*} \hat{f}(y) \left| \sum_{j=1}^{m} \exp(2\pi i \langle y, x_j \rangle) \right|^2. \tag{A.18}$$

The first and last expressions in this chain of relations are manifestly real numbers, and we will reach our desired conclusion by upper-bounding the former and lower-bounding the latter. Specifically, we have that

$$f(x + x_j - x_k) \quad \text{is} \quad \begin{cases} = f(0) & \text{if } x = 0 \text{ and } j = k, \\ \leq 0 & \text{otherwise,} \end{cases}$$

by our observation above about $\|x + x_j - x_k\|$ combined with the third condition in the theorem about f. Thus the leftmost expression in (A.18) is bounded from above by $mf(0)$. On the other hand, by the second condition f satisfies, the rightmost expression in (A.18) can only be made smaller by discarding all terms $y \in \Lambda \setminus \{0\}$. Thus the expression is bounded from below by $\frac{m^2}{\mathrm{covol}(\Lambda)}\widehat{f}(0) = \frac{m^2}{\mathrm{covol}(\Lambda)}f(0)$. Combining these two bounds yields the inequality

$$\mathrm{covol}(\Lambda) \geq m.$$

This is exactly what we need, since the packing density then satisfies

$$\Delta_P = \frac{m\,\mathrm{vol}(B_{\rho/2}(0))}{\mathrm{covol}\,\Lambda} \leq \mathrm{vol}(B_{\rho/2}(0)),$$

as the inequality in (A.16) claims. (The second, more explicit expression in (A.16) for the upper bound follows from the well-known formula for the volume of the unit ball in $\mathbb{R}^d$; see Exercise 2.3 on page 110.) □

A.9 Magic functions

Given a lattice $\Lambda \subset \mathbb{R}^d$ with packing density δ_Λ, a Schwartz function $f : \mathbb{R}^d \to \mathbb{R}$ is called a **magic function** for Λ if it satisfies the assumptions of Theorem A.21 with the particular value of ρ for which

$$\mathrm{vol}(B_{\rho/2}(0)) = \delta_\Lambda.$$

By Theorem A.21, if we were to prove the existence of a magic function for some specific lattice Λ, that would imply that $\Delta_{\mathrm{optimal}}(d) = \delta_\Lambda$, and that the lattice packing associated with Λ is optimal for sphere packing in $\mathbb{R}^d$, thereby resolving the sphere packing problem in dimension d.

Magic functions are a tool that seems almost too powerful (or "magic," hence the name) to exist. Indeed, heavy numerical experimentation done by Cohn and Elkies suggested that in most low dimensions they do not; but in a few special dimensions, the numerical evidence suggested that they do exist, leading to the following conjecture.

Conjecture A.22 (Cohn–Elkies [14]). *Magic functions exist for the following dimensions and lattices:*
1. *$d = 2$: the hexagonal lattice $(\mathbb{Z} \cdot (1, 0)) \oplus (\mathbb{Z} \cdot (\frac{1}{2}, \frac{\sqrt{3}}{2}))$ in $\mathbb{R}^2$;*
2. *$d = 8$: the lattice E_8;*
3. *$d = 24$: the Leech lattice (described in [18, Sec. 5.11]).*

Viazovska [71] proved the second of these conjectures by finding an explicit construction of a magic function for the lattice E_8; her proof, using complex analysis and modular forms, is the subject of Chapter 6. A subsequent construction using similar methods of a magic function for the Leech lattice in the case of 24 dimensions, proving the third conjecture, was given by Cohn et al. [16]. The first conjecture regarding the existence of a magic function for the hexagonal lattice in $\mathbb{R}^2$ remains open (as of 2023).

A.10 Radial functions and their Fourier transforms

A function $f : \mathbb{R}^d \to \mathbb{R}$ is called **radially symmetric**, or a **radial** function, if $f(x)$ depends only on the radial coordinate of x, that is, if $f(x) = f(y)$ whenever $\|x\| = \|y\|$. Clearly, $f(x)$ is radial if and only if it can be represented as

$$f(x) = \tilde{f}(\|x\|)$$

for some function $\tilde{f} : [0, \infty) \to \mathbb{R}$. The function $\tilde{f}(r)$ is determined uniquely, as $\tilde{f}(r)$ is the unique value that $f(x)$ takes on the sphere $\{x \: : \: \|x\| = r\}$. We refer to $\tilde{f}(r)$ as the **radial profile** of $f(x)$.[1]

If $f : \mathbb{R}^d \to \mathbb{R}$ is a general—not necessarily radially symmetric—function, then we can apply a standard analytic trick to $f(x)$ to obtain a radial function to perform *radial symmetrization*, that is, to average out the function over concentric spherical shells of equal radius around 0. More precisely, we define

$$f_{\mathrm{rad}}(x) = \frac{1}{s_{d-1}} \int_{S^{d-1}} f(\|x\|y)d\sigma_{d-1}(y),$$

the integral over the unit sphere $S^{d-1} = \{y \in \mathbb{R}^d \: : \: \|y\| = 1\}$ with respect to its surface area measure σ_{d-1}, normalized to be a weighted average by dividing by the total sphere surface area $s_{d-1} = \sigma_{d-1}(S^{d-1})$. We call $f_{\mathrm{rad}}(x)$ the **radially symmetrized** version of $f(x)$. Note that f is radial if and only if it coincides with its radially symmetrized version.

Lemma A.23. *Let $f : \mathbb{R}^d \to \mathbb{R}$ be a radial function. Then $f(x)$ is a Schwartz function if and only if the radial profile $\tilde{f}(r)$ satisfies the following properties:*
1. *$\tilde{f}(r)$ is the restriction to $[0, \infty)$ of an infinitely differentiable even function on $\mathbb{R}$.*
2. *$r^{-n}\tilde{f}^{(m)}(r) \xrightarrow[r \to \infty]{} 0$ for all $n, m \geq 0$.*

Proof. Exercise A.9. □

---

[1] Some authors commit the mild abuse of not making a clear distinction between a function and its radial profile, for example, by referring to them interchangeably and denoting both of them with the same symbol.

Lemma A.24. *The radially symmetrized function $f_{\mathrm{rad}}(x)$ has the following alternative expression:*

$$f_{\mathrm{rad}}(x) = \frac{1}{v_d(\mathrm{SO}(d))} \int_{\mathrm{SO}(d)} f(Ax)\, dv_d(A). \tag{A.19}$$

The meanings of the symbols in this formula are as follows: $\mathrm{SO}(d)$ *is the special orthogonal group of order d, that is, the group of $d \times d$ orthogonal matrices with determinant 1; and v_d is the Haar measure on $\mathrm{SO}(d)$, that is, the unique (up to scalar multiplication) Borel measure on $\mathrm{SO}(d)$ that is invariant under the group action, i. e., satisfies $v_d(A \cdot E) = v_d(E)$ for all $A \in \mathrm{SO}(d)$ and all Borel sets $E \subset \mathrm{SO}(d)$ (with $A \cdot E$ denoting the set of matrices $\{AB : B \in E\}$).*

Proof. Exercise A.11. □

Lemma A.25. *If $f : \mathbb{R}^d \to \mathbb{R}$ is a Schwartz function, then $\widehat{(f_{\mathrm{rad}})} = (\hat{f})_{\mathrm{rad}}$; that is, the Fourier transform of the radial symmetrization of f is equal to the radial symmetrization of the Fourier transform of f.*

Proof. If A is a $d \times d$ orthogonal matrix and $g : \mathbb{R}^d \to \mathbb{R}$, then denote by g_A the function g "rotated by the transformation A", that is,

$$g_A(x) = g(Ax) \quad (x \in \mathbb{R}^d).$$

It is trivial to check that $\widehat{(g_A)} = (\hat{g})_A$ (the Fourier transform commutes with orthogonal transformations). Now using (A.19) (applied to both f and $\hat{f}$), it follows that

$$\widehat{(f_{\mathrm{rad}})}(y) = \mathcal{F}_d\left[x \mapsto \frac{1}{v_d(\mathrm{SO}(d))} \int_{\mathrm{SO}(d)} f(Ax)\, dv_d(A)\right](y)$$

$$= \mathcal{F}_d\left[x \mapsto \frac{1}{v_d(\mathrm{SO}(d))} \int_{\mathrm{SO}(d)} f_A(x)\, dv_d(A)\right](y)$$

$$= \frac{1}{v_d(\mathrm{SO}(d))} \int_{\mathrm{SO}(d)} \widehat{(f_A)}(y)\, dv_d(A) = \frac{1}{v_d(\mathrm{SO}(d))} \int_{\mathrm{SO}(d)} (\hat{f})_A(y)\, dv_d(A)$$

$$= \frac{1}{v_d(\mathrm{SO}(d))} \int_{\mathrm{SO}(d)} \hat{f}(Ay)\, dv_d(A) = (\hat{f})_{\mathrm{rad}}(y). \qquad \square$$

From Lemma A.25 it follows in particular that the Fourier transform of a radial Schwartz function $f : \mathbb{R}^d \to \mathbb{R}$ is also a radial function. Because of this, when discussing radial functions, it is helpful to think of the Fourier transform in d dimensions as an operator acting directly on the associated radial profile. That is, if $f : \mathbb{R}^d \to \mathbb{R}$ has an associated radial profile $\tilde{f}(r)$, and $g(y) = \mathcal{F}_d[f](y)$ denotes the Fourier transform of $f(x)$

with associated radial profile $\widetilde{g}(\rho)$, then we refer to $\widetilde{g}(\rho)$ as the **radial Fourier transform** of $\widetilde{f}(r)$ and denote this as

$$\widetilde{g}(\rho) = \mathcal{F}_d^{\mathrm{rad}}[\widetilde{f}](\rho).$$

See Exercise A.13 at the end of this Appendix for more details (which are interesting but not needed for our purposes) on how this transform can be described more explicitly and some of its properties.

Radial Schwartz functions have a decomposition into "even" and "odd" parts with respect to the taking of radial Fourier transforms. This is explained in the following lemma.

Lemma A.26. *Let $f : \mathbb{R}^d \to \mathbb{R}$ be a radial Schwartz function. Then f has a unique representation of the form*

$$f = f_+ + f_-, \tag{A.20}$$

where $f_+, f_- : \mathbb{R}^d \to \mathbb{R}$ are radial Schwartz functions with

$$\mathcal{F}_d[f_+] = f_+, \quad \mathcal{F}_d[f_-] = -f_-,$$

that is, $f_\pm$ are eigenfunctions of the Fourier transform with associated eigenvalues $+1$ and -1, respectively. The Fourier transform of f is then given by

$$\mathcal{F}_d[f] = f_+ - f_-, \tag{A.21}$$

and f_+, f_- are given by

$$f_+ = \frac{f + \mathcal{F}_d[f]}{2}, \quad f_- = \frac{f - \mathcal{F}_d[f]}{2}.$$

Proof. Exercise A.12. □

We call (A.20) the **Fourier parity decomposition** for radial Schwartz functions. We call f_+ the **Fourier-even part of f** and call f_- the **Fourier-odd part of f**.

Suggested exercises for Section A.10. A.9, A.10, A.11, A.12, A.13.

A.11 Structural properties of E_8 magic functions

Theorem A.21 provides a powerful technique for proving upper bounds on the optimal packing density of $\mathbb{R}^d$. This was used in [14] to prove improved numerical upper bounds for $\Delta_{\mathrm{optimal}}(d)$. Even more intriguingly, it raises the natural question of how we can go about using the theorem to try to derive a *sharp* upper bound in any given dimension, or at least one that is best possible using the method. Needless to say, this is a highly nontrivial question. The difficulty lies in the fact that we are trying to optimize the bound (the quantity on the right-hand side of (A.16)) over a rather peculiar-looking space of functions. Without any further clues as to what sort of properties an optimizing function f might have, this is tantamount to groping in the dark.

Fortunately, in the case of 8 and 24 dimensions, Cohn and Elkies pointed out that we can infer some interesting structural properties of a hypothetical optimizer by using the additional (conjectured, at that point) knowledge that in those dimensions, the optimizers are magic functions for the E_8 and Leech lattices, respectively. Let us see what those structural properties are. We focus here on the case of 8 dimensions, where these properties turned out to be the crucial clues that ultimately led Viazovska to her construction of an E_8 magic function.

First, we can strip away one apparent layer of complexity from the optimization problem by noting that although the class of functions f we are optimizing over consists of functions on $\mathbb{R}^d$ (that is, functions of d real variables), there is no real loss of generality in assuming that the function in question is a radial function—a huge simplification, since radial functions are described in terms of their radial profile, which is a function of a *single* real variable. The idea is made precise in the following lemma.

Lemma A.27. *If $f : \mathbb{R}^d \to \mathbb{R}$ is a function satisfying the conditions of Theorem A.21 with parameter ρ, then there exists a radial Schwartz function $g : \mathbb{R}^d \to \mathbb{R}$ that satisfies the same conditions with the same value of ρ.*

Proof. Take $g = f_{\mathrm{rad}}$, the radially symmetrized version of f, which is also a Schwartz function (Exercise A.10). Using Lemma A.25, it is easy to check that g satisfies the same conditions that f satisfied, with the same value of ρ. □

A second important observation concerns a necessary condition a function must satisfy to be a magic function.

Lemma A.28. *If $f : \mathbb{R}^d \to \mathbb{R}$ is a Schwartz function that is a magic function for a lattice $\Lambda \subset \mathbb{R}^d$, then it must satisfy*

$$f(x) = 0 \quad \text{for all } x \in \Lambda \setminus \{0\} \quad \text{and}$$
$$\widehat{f}(x) = 0 \quad \text{for all } x \in \Lambda^* \setminus \{0\}.$$

Proof. First, note that we can assume without loss of generality that Λ has the property that

$$r_*(\Lambda) = \rho/2.$$

Indeed, if Λ does not satisfy this, then we can replace it by a scaled version $a\Lambda$ of itself with $a > 0$ chosen so as to cause this equation to be satisfied; the scaling does not change the value of δ_Λ, so f would still be a magic function for the rescaled lattice.

Now combining the Poisson summation formula (A.6) with the assumptions on f, we have that

$$f(0) \geq f(0) + \sum_{x \in \Lambda \setminus \{0\}} f(x) = \sum_{x \in \Lambda} f(x)$$

$$= \frac{1}{\mathrm{covol}(\Lambda)} \sum_{y \in \Lambda^*} \widehat{f}(y) = \frac{1}{\mathrm{covol}(\Lambda)} \left(\widehat{f}(0) + \sum_{y \in \Lambda^* \setminus \{0\}} \widehat{f}(y) \right)$$

$$\geq \frac{\widehat{f}(0)}{\mathrm{covol}(\Lambda)} = \frac{f(0)}{\mathrm{covol}(\Lambda)}. \tag{A.22}$$

This is equivalent to saying that $\mathrm{covol}(\Lambda) \geq 1$, which in turn is equivalent (refer to (A.3)) to the relation

$$\delta_\Lambda \leq \mathrm{vol}(B_{r_*(\Lambda)}(0)).$$

Since we assumed that f was a magic function, δ_Λ is also equal to $\mathrm{vol}(B_{\rho/2}(0))$, so a final equivalent reformulation of the inequality between the leftmost and rightmost terms in (A.22) is the statement that $\rho/2 \leq r_*(\Lambda)$. However, we started the proof by assuming that $\rho/2$ is *equal* to $r_*(\Lambda)$. This means that both (weak) inequalities in (A.22) must actually hold as *equalities*. The only way in which this can be true is if all the summation terms that were discarded to obtain those inequalities—the terms $f(x)$ for $x \in \Lambda \setminus \{0\}$ in the first inequality, which were known to be nonpositive, and the terms $\widehat{f}(y)$ for $y \in \Lambda \setminus \{0\}$ in the first inequality, which were known to be nonnegative—are necessarily 0; this was exactly the claim to prove. $\qquad\square$

Combining the above results and specializing to the case of E_8, we easily obtain the following result.

Theorem A.29 (Necessary condition for E_8 magic). *Let $f : \mathbb{R}^8 \to \mathbb{R}$ be a radial Schwartz function, and let f_+ and f_- denote the Fourier-even and Fourier-odd parts of f as in (A.20). Define the functions $\Phi, \widehat{\Phi}, \Phi_+, \Phi_- : [0, \infty) \to \mathbb{R}$ by*

$$\Phi(r) = \widetilde{f}(r) \qquad\qquad \text{(the radial profile of } f\text{),}$$

$$\widehat{\Phi}(r) = \mathcal{F}_d^{\mathrm{rad}}[\Phi](r) \qquad\qquad \text{(the radial profile of } \widehat{f}\text{),}$$

$$\Phi_+(r) = \widetilde{f}_+(r) = \frac{\Phi(r) + \widehat{\Phi}(r)}{2} \qquad \text{(the radial profile of } f_+\text{),}$$

$$\Phi_-(r) = \widetilde{f}_+(r) = \frac{\Phi(r) - \widehat{\Phi}(r)}{2} \qquad \text{(the radial profile of } f_-\text{).}$$

If f is a magic function for E_8, then the following conditions hold:
1. $\Phi(0) = \widehat{\Phi}(0) > 0$;
2. $\Phi(r), \widehat{\Phi}(r), \Phi_+(r),$ and $\Phi_-(r)$ have zeros at the points $r = \sqrt{2n}$ for $n = 1, 2, 3, \ldots$.

3. $\Phi(r)$ *does not change signs at* $r = \sqrt{2n}$ *for* $n = 2, 3, \ldots$ *(so its zeros there are of even order, assuming that it is real-analytic so that the order of the zeros is well-defined).*
4. $\widehat{\Phi}(r)$ *does not change signs at* $r = \sqrt{2n}$ *for* $n = 1, 2, 3, \ldots$ *(so its zeros there are of even order, assuming that it is real-analytic).*
5. *If* $\Phi(r)$ *does not have zeros in* $(0, \sqrt{2})$, *then it changes signs at* $r = \sqrt{2}$, *so its zero there is of odd order, assuming that it is real-analytic.*

Proof. Exercise A.14. □

Suggested exercises for Section A.11. A.14, A.15.

A.12 Summary

In this appendix, we have developed a solid framework for the study of the sphere packing problem in d dimensions, with a focus on the case of $d = 8$, from the point of view of the connections of the problem to harmonic analysis. The main tool is an analytic result, Theorem A.21, which, along with related observations such as Lemma A.27, reduces the problem to a purely analytic question: namely, can a radial function be constructed with certain special properties involving simultaneous conditions on the function and its Fourier transform?

An additional tool of importance is Theorem A.29. This result plays a motivational role in helping us think about the sphere packing problem in 8 dimensions, as it narrows down considerably the class of functions that we need to consider as hypothetical magic function candidates. Specifically, the theorem suggests that to find an E_8 magic function, we should look for a function $\Phi(r)$ of a single (radial) real variable that has the property that both $\Phi(r)$ and its radial Fourier transform $\mathcal{F}_8^{\mathrm{rad}}[\Phi]$ have zeros at the points $r = \sqrt{2}, \sqrt{4}, \sqrt{6}, \ldots$. This is a rather idiosyncratic problem quite unlike anything else mathematicians had ever seen before, and its solution eluded the researchers thinking about the problem until Viazovska came up with her breakthrough solution in 2016. Conceptually, what makes the problem hard is that it is difficult to control the zeros of a function and its Fourier transform *simultaneously*: it is straightforward to construct functions with a given set of zeros and functions whose Fourier transform has a given set of zeros, but no standard tools or ideas in (pre-2016) harmonic analysis offer much of a clue for how to do both of those things at the same time, or indeed give much insight into whether it can be done at all.

One of the conditions in Theorem A.29 offers a possible way out of this conundrum: specifically, the point of considering separately the components Φ_+ and Φ_- in the Fourier parity decomposition of Φ is that each of those components is an eigenfunction of the radial Fourier transform, and thus, if we can force it to have the required set of zeros, then its Fourier transform will *automatically* have those zeros as well. So the problem is reduced to constructing radial Fourier eigenfunctions that have zeros (with certain constraints on their orders) at $\sqrt{2n}$, $n = 1, 2, \ldots$. Of course, the condition of being a

Fourier eigenfunction is not a trivial one to satisfy either, especially when combined with the constraints on the zeros, so it is not a priori clear that this observation makes the problem anymore tractable; it seems conceivable that we have merely traded one difficult-to-satisfy condition for another.

Nonetheless, constructing Fourier eigenfunctions with the correct set of zeros turns out to be precisely the right approach. This was the path taken successfully in Viazovska's solution of the sphere packing problem in 8 dimensions; for the details, read Chapter 6, which you now have the necessary background to tackle.

Suggested exercises for Section A.12. A.16, A.17.

Exercises for Appendix A

A.1 Is an optimal sphere packing in $\mathbb{R}^d$ *unique*? Why, or why not? If not, then what can be said about the extent of the nonuniqueness?

A.2 Prove Lemma A.3.

A.3 Prove Lemma A.4.

A.4 Prove Lemma A.5.

A.5 Prove that for any lattice Λ in $\mathbb{R}^d$, $\operatorname{covol}(\Lambda^*) = \operatorname{covol}(\Lambda)^{-1}$.

A.6 Prove Theorem A.7. One possible proof proceeds in two steps: first, prove the result for the specific lattice $\Lambda = \mathbb{Z}^d$ by deducing it from the original Poisson summation formula for functions on $\mathbb{R}$. Second, derive the result in full generality by starting with the formula for $\mathbb{Z}^d$ and applying a linear coordinate change. For a more direct approach, see [15, Appendix A].

A.7 Prove Lemma A.16.

A.8 Another construction of the lattice E_8 (discussed, for example, in [12]) starts by postulating the existence of a basis $x_1, \ldots, x_8 \in \mathbb{R}^8$ whose Gram matrix (the matrix of inner products $\langle x_j, x_k \rangle$) takes the form

$$
\left(\langle x_j, x_k \rangle \right)_{j,k=1}^8 = \begin{pmatrix} 2 & -1 & 0 & 0 & 0 & 0 & 0 & 0 \\ -1 & 2 & -1 & 0 & 0 & 0 & 0 & 0 \\ 0 & -1 & 2 & -1 & -1 & 0 & 0 & 0 \\ 0 & 0 & -1 & 2 & 0 & 0 & 0 & 0 \\ 0 & 0 & -1 & 0 & 2 & -1 & 0 & 0 \\ 0 & 0 & 0 & 0 & -1 & 2 & -1 & 0 \\ 0 & 0 & 0 & 0 & 0 & -1 & 2 & -1 \\ 0 & 0 & 0 & 0 & 0 & 0 & -1 & 2 \end{pmatrix}.
$$

Prove that such a basis exists and try to redevelop the results of Section A.7 based on this construction.

A.9 Prove Lemma A.23. (See also [32, Sec. 3], [17, Subsec. 2.3].)

A.10 Prove that the radially symmetrized version of a Schwartz function is a Schwartz function.

A.11 Prove Lemma A.24.

A.12 Prove Lemma A.26.

A.13 **Radial Fourier transforms in $\mathbb{R}^d$.** [31, Sec. B.5], [45, Secs. 4.20, 4.23]

(a) Let $f : \mathbb{R}^d \to \mathbb{R}$ be a radial function with a well-defined Fourier transform. Denote $F(r) = \tilde{f}(r)$ and $G(\rho) = \tilde{\hat{f}}(\rho)$ (the radial profiles of f and $\hat{f}$, respectively). Prove that F and G are related to each other by

$$
G(\rho) = \frac{2\pi}{\rho^{d/2-1}} \int_0^\infty F(r) r^{d/2} J_{d/2-1}(2\pi\rho r) \, dr, \tag{A.23}
$$

$$F(r) = \frac{2\pi}{r^{d/2-1}} \int_0^\infty G(\rho)\rho^{d/2} J_{d/2-1}(2\pi\rho r)\, d\rho. \tag{A.24}$$

Here we use the notation $J_a(z)$ for the **Bessel function of the first kind** of index a, an entire function defined by

$$J_a(z) = \sum_{n=0}^\infty \frac{(-1)^n}{n!\,\Gamma(n+a+1)}\left(\frac{z}{2}\right)^{2n+a}$$

(see also Exercise 1.16 on p. 74). The integral transform that associates a function G on $[0,\infty)$ with another function F on $[0,\infty)$ according to (A.23) is known as the **Hankel transform**.

(b) Prove that if $f : \mathbb{R}^d \to \mathbb{R}$ is a radial square-integrable function that is an eigenfunction of the Fourier transform, that is, $\hat{F}_d(f) = \lambda f$, then $\lambda = 1$ or $\lambda = -1$.

(c) Let $a > 0$. Define the sequence of polynomials $(L_n^a(x))_{n=0}^\infty$ by the formula

$$L_n^a(x) = \sum_{k=0}^n \frac{(-1)^k}{k!}\binom{n+a}{n-k}x^k.$$

The polynomials $L_n^a(x)$ are called the **Laguerre polynomials with parameter** a. Prove that the polynomials $L_n^a(x)$ satisfy the orthogonality relation

$$\int_0^\infty L_n^a(x)L_m^a(x)e^{-x}x^a\, dx = \frac{\Gamma(n+a+1)}{n!}\delta_{mn} \qquad (m,n \geq 0).$$

Here δ_{mn} denotes the Kronecker delta.

(d) Let $d \geq 1$. Define the radial functions $\gamma_n^d(x) = G_n^d(\|x\|)$ on $\mathbb{R}^d$, $n \geq 0$, by

$$G_n^d(r) = e^{-\pi r^2} L_n^{d/2-1}(2\pi r^2)$$

$$= e^{-\pi r^2} \sum_{k=0}^n \frac{(-1)^k}{k!}\binom{n+d/2-1}{n-k}(2\pi)^k r^{2k}.$$

Prove that γ_n^d is an eigenfunction of the Fourier transform with eigenvalue $(-1)^n$.

(e) Prove that the sequence $(\gamma_n^d)_{n=0}^\infty$ forms an orthogonal basis of the subspace $L_{\mathrm{rad}}^2(\mathbb{R}^d)$ of $L^2(\mathbb{R}^d)$ consisting of radial functions. (In other words, together with the previous claim, this shows that the sequence $(\gamma_n^d)_{n=0}^\infty$ diagonalizes the restriction of the d-dimensional Fourier transform to the radial functions.)

A.14 Prove Theorem A.29.

A.15 How close are the conditions listed in Theorem A.29 to being *sufficient* for the function f to be a magic function for E_8? That is, what additional mild assumptions on $\Phi(r)$ and $\widehat{\Phi}(r)$ would guarantee that f is a magic function?

A.16 **The Leech lattice.** Prove the following analogue of Theorem A.8:

Theorem A.30 ([18, pp. 131–134]). *There exists a lattice in $\mathbb{R}^{24}$, denoted L_{24} and known as the Leech lattice, with the following properties:*

(a) *The packing density of the sphere packing associated with L_{24} is $\frac{\pi^{12}}{12!}$.*

(b) *The set of Euclidean norms of points of L_{24} is*

$$\{\sqrt{2k} \; : \; k = 0, 2, 3, 4, \ldots\}.$$

(c) *The numbers $(b_n)_{n=0}^{\infty}$ defined by*

$$b_n = \#\{x \in L_{24} \; : \; \|x\|^2 = 2n\}$$

are given explicitly by

$$b_n = \frac{65\,520}{691}(\sigma_{11}(n) - \tau(n)) \qquad (n \geq 1).$$

For the definitions of $\sigma_{11}(n)$ and $\tau(n)$, see (5.2) and (5.28).

A.17 Formulate an analogue of Theorem A.29 for the case of a magic function for the Leech lattice in 24 dimensions.

Bibliography

[1] L. V. Ahlfors. *Complex Analysis*. McGraw-Hill, New York, 3rd edition, 1979.

[2] L. V. Ahlfors. *Conformal Invariants: Topics in Geometric Function Theory*. American Mathematical Society, 2010.

[3] M. Aigner and G. Ziegler. *Proofs from THE BOOK*, 6th Ed. Springer, 2018.

[4] R. Alperin. $PSL_2(\mathbf{Z}) = \mathbf{Z}_2 * \mathbf{Z}_3$. *Am. Math. Mon.*, 100:385–386, 1993.

[5] T. M. Apostol. *Modular Functions and Dirichlet Series in Number Theory*. Springer, New York, 2nd edition, 1990.

[6] D. Beliaev. *Conformal Maps and Geometry*. World Scientific, 2018.

[7] B. C. Berndt. *Ramanujan's Notebooks, Part III*. Springer-Verlag, 1991.

[8] B. C. Berndt. *Ramanujan's Notebooks, Part V*. Springer-Verlag, 1998.

[9] S. S. Cairns. An elementary proof of the Jordan–Schoenflies theorem. *Proc. Am. Math. Soc.*, 2, 1951.

[10] J. Cardy. Critical percolation in finite geometries. *J. Phys. A*, 25:L201–L206, 1992.

[11] G. L. Cohen and G. H. Smith. A simple verification of Ilieff's conjecture for polynomials with three zeros. *Am. Math. Mon.*, 95:734–737, 1988.

[12] H. Cohn. A conceptual breakthrough in sphere packing. *Not. Am. Math. Soc.*, 64:102–115, 2017.

[13] H. Cohn. The work of Maryna Viazovska. In *Proc. Int. Cong. Math.*, volume 1, 2022.

[14] H. Cohn and N. Elkies. New upper bounds on sphere packings. I. *Ann. Math.*, 157:689–714, 2003.

[15] H. Cohn and A. Kumar. The densest lattice in twenty-four dimensions. *Electron. Res. Announc. Am. Math. Soc.*, 10:58–67, 2004.

[16] H. Cohn, A. Kumar, S. D. Miller, D. Radchenko, and M. Viazovska. The sphere packing problem in dimension 24. *Ann. Math.*, 185:1017–1033, 2017.

[17] H. Cohn, A. Kumar, S. D. Miller, D. Radchenko, and M. Viazovska. Universal optimality of the E_8 and Leech lattices and interpolation formulas. *Ann. Math.*, 196:983–1082, 2022.

[18] J. H. Conway and N. J. A. Sloane. *Sphere Packings, Lattices and Groups*. Springer, 3rd edition, 1999.

[19] D. A. Cox. The arithmetic-geometric mean of Gauss. *Enseign. Math.*, 30:275–330, 1984.

[20] D. de Laat and F. Vallentin. A breakthrough in sphere packing: the search for magic functions. *Nieuw Arch. Wiskd.*, 5/17:184–192, 2016.

[21] B. de Smit, D. Dunham, H. W. Lenstra Jr., and R. Sarhangi. Artful mathematics: the heritage of M. C. Escher. *Not. Am. Math. Soc.*, 50:446–457, 2003.

[22] M. Dewar and M. Ram Murty. An asymptotic formula for the coefficients of $j(z)$. *Int. J. Number Theory*, 9:641–652, 2013.

[23] S. Donaldson. *Riemann Surfaces*. Oxford University Press, 2011.

[24] W. Duke and O. Imamoğlu. The zeros of the Weierstrass $\wp$-function and hypergeometric series. *Math. Ann.*, 340:897–905, 2008.

[25] H. M. Edwards. *Riemann's Zeta Function*. Dover Publications, 2001.

[26] M. Eichler and D. Zagier. On the zeros of the Weierstrass $\wp$-function. *Math. Ann.*, 258:399–407, 1981.

[27] S. R. Finch. *Mathematical Constants*. Cambridge University Press, 2003.

[28] P. Flajolet and R. Sedgewick. *Analytic Combinatorics*. Cambridge University Press, 2009.

[29] T. Gannon. *Moonshine Beyond the Monster: The Bridge Connecting Algebra, Modular Forms and Physics*. Cambridge University Press, 2010.

[30] L. J. Goldstein. A history of the prime number theorem. *Am. Math. Mon.*, 80:599–615, 1973.

[31] L. Grafakos. *Classical Fourier Analysis*, 2nd Ed. Springer, 2006.

[32] L. Grafakos and G. Teschl. On Fourier transforms of radial functions and distributions. *J. Fourier Anal. Appl.*, 19:167–179, 2013.

[33] J. Gray. On the history of the Riemann mapping theorem. *Rend. Circ. Mat. Palermo*, 34:47–94, 1994.

[34] G. Grimmett. *Probability on Graphs: Random Processes on Graphs and Lattices*, 2nd Ed. Cambridge University Press, 2018.

[35] H. Groemer. Existenzsätze für Lagerungen im Euklidischen Raum. *Math. Z.*, 81:260–278, 1963.

[36] P. M. Gruber and C. G. Lekkerkerker. *Geometry of Numbers*, 2nd Ed. North Holland, 1987.

[37] X. Gu, Y. Wang, T. F. Chan, P. M. Thompson, and S.-T. Yau. Genus zero surface conformal mapping and its application to brain surface imaging. *IEEE Trans. Med. Imaging*, 23:949–958, 2004.

[38] T. Hales and S. Ferguson. *The Kepler Conjecture: The Hales–Ferguson Proof*. Springer, 2011. Ed. J. C. Lagarias.

[39] T. C. Hales. A proof of the Kepler conjecture. *Ann. Math.*, 162:1065–1185, 2005.

[40] B. C. Hall. *Lie Groups, Lie Algebras, and Representations: An Elementary Introduction*, Second Ed. Springer, 2015.

[41] J. K. Hunter and B. Nachtergaele. *Applied Analysis*. World Scientific, 2001.

[42] R. Karam. Why are complex numbers needed in quantum mechanics? Some answers for the introductory level. *Am. J. Phys.*, 88:39–45, 2020.

[43] P. Kleban and D. Zagier. Crossing probabilities and modular forms. *J. Stat. Phys.*, 113:431–454, 2003.

[44] J. Korevaar. The Wiener–Ikehara theorem by complex analysis. *Proc. Am. Math. Soc.*, 134:1107–1116, 2005.

[45] N. N. Lebedev. *Special Functions & Their Applications*. Prentice-Hall, 1965.

[46] B. Mazur and W. Stein. *Prime Numbers and the Riemann Hypothesis*. Cambridge University Press, 2016.

[47] H. L. Montgomery and R. C. Vaughan. *Multiplicative Number Theory: I. Classical Theory*. Cambridge University Press, 2006.

[48] T. Needham. *Visual Complex Analysis*. Clarendon Press, 1999.

[49] D. J. Newman. Simple analytic proof of the prime number theorem. *Am. Math. Mon.*, 87:693–696, 1980.

[50] D. Niebur. A formula for Ramanujan's τ-function. *Ill. J. Math.*, 19:448–449, 1975.

[51] B. Nienhuis. Exact critical point and critical exponents of $o(n)$ models in two dimensions. *Phys. Rev. Lett.*, 49:1062–1065, 1982.

[52] A. Okounkov. The magic of 8 and 24. In *Proc. Int. Cong. Math.*, volume 1, 2022.

[53] W. M. Oliva. *Geometric Mechanics*. Springer, 2002.

[54] I. Pak. Partition bijections, a survey. *Ramanujan J.*, 12:5–75, 2006.

[55] R. Penrose and W. Rindler. *Spinors and Space-Time, Volume 1: Two-Spinor Calculus and Relativistic Fields*. Cambridge University Press, 1987.

[56] D. Romik. Roots of the derivative of a polynomial. *Am. Math. Mon.*, 112:66–68, 2005.

[57] D. Romik. On the number of n-dimensional representations of $SU(3)$, the Bernoulli numbers, and the Witten zeta function. *Acta Arith.*, 180:111–159, 2017.

[58] D. Romik. On Viazovska's modular form inequalities. 2023. Preprint, https://arxiv.org/abs/2303.13427.

[59] D. Schattsschneider. The mathematical side of M. C. Escher. *Not. Am. Math. Soc.*, 57:706–718, 2010.

[60] W. Schlag. *A Course in Complex Analysis and Riemann Surfaces*. American Mathematical Society, 2014.

[61] J. H. Silverman. *The Arithmetic of Elliptic Curves*. Springer, 2nd edition, 2009.

[62] J. H. Silverman and J. T. Tate. *Rational Points on Elliptic Curves*. Springer, 2nd edition, 2015.

[63] N. Skoruppa. A quick combinatorial proof of Eisenstein series identities. *J. Number Theory*, 43:68–73, 1993.

[64] S. Smirnov. Critical percolation in the plane: conformal invariance, Cardy's formula, scaling limits. *C. R. Acad. Sci. Paris Sér. I Math.*, 333:239–244, 2001.

[65] S. Smirnov and H. Duminil-Copin. The connective constant of the honeycomb lattice equals $\sqrt{2 + \sqrt{2}}$. *Ann. Math.*, 175:1653–1665, 2012.

[66] E. M. Stein and R. Shakarchi. *Complex Analysis*. Princeton University Press, 2003.

[67] E. M. Stein and R. Shakarchi. *Fourier Analysis: An Introduction*. Princeton University Press, 2003.

[68] P. D. Thomas. *Conformal Projections in Geodesy and Cartography*. U.S. Government Printing Office, 1952.

[69] H. Tverberg. A proof of the Jordan curve theorem. *Bull. Lond. Math. Soc.*, 12:34–38, 1980.

[70] A. Vatwani. A simple proof of the Wiener–Ikehara Tauberian theorem. *Math. Stud.*, 84:127–134, 2015.

[71] M. Viazovska. The sphere packing problem in dimension 8. *Ann. Math.*, 185:991–1015, 2017.

[72] J. L. Walsh. History of the Riemann mapping theorem. *Am. Math. Mon.*, 80:270–276, 1973.

[73] W. Werner. Lectures on two-dimensional critical percolation. 2007. https://arxiv.org/abs/0710.0856.

[74] D. Zagier. *Values of Zeta Functions and Their Applications*, pages 497–512. Birkhäuser Basel, Basel, 1994.

[75] G. Zukav. *The Dancing Wu Li Masters: An Overview of the New Physics*. Bantam Books, 1980.

Web Bibliography

[W1]   N. J. A. Sloane. Number of self-avoiding n-step walks on honeycomb lattice. The On-Line Encyclopedia of Integer Sequences. Accessed Feb. 4, 2023. URL: https://oeis.org/A001668

[W2]   Scott Aaronson. PHYS771 Lecture 9: Quantum. Accessed Feb. 4, 2023. URL: https://www.scottaaronson.com/democritus/lec9.html

[W3]   Multiple Contributors. About the complex nature of the wave function? Physics Stack Exchange. Accessed Feb. 4, 2023. URL: https://physics.stackexchange.com/questions/8062/about-the-complex-nature-of-the-wave-function

[W4]   Hendrik Lenstra, Bart de Smit, Joost Batenburg, Hans Richter, Jacqueline Hofstra, Richard Groenewegen, Lennert Buytenhek, Martin Feleus, and Willem Jan Palenstijn. Escher and the Droste effect. Accessed Mar. 3, 2023. URL: https://pub.math.leidenuniv.nl/~smitbde/escherdroste/

[W5]   Multiple Contributors. Eschers Print Gallery and the 'Droste Effect'. Accessed Mar. 3, 2023. URL: https://cindyjs.org/gallery/main/Droste/

[W6]   Multiple Contributors. Nice proof of the Jordan curve theorem? MathOverflow. Accessed Feb. 8, 2023. URL: https://mathoverflow.net/questions/8521/nice-proof-of-the-jordan-curve-theorem

[W7]   Numberphile. ASTOUNDING: $1 + 2 + 3 + 4 + 5 + \cdots = -1/12$. Jan. 9, 2014. Accessed Feb. 4, 2023. URL: https://youtu.be/w-I6XTVZXww

[W8]   Dennis Overbye. In the End, It All Adds Up to $-1/12$. New York Times, Feb. 3, 2014. Accessed Feb. 4, 2023. URL: https://www.nytimes.com/2014/02/04/science/in-the-end-it-all-adds-up-to.html

[W9]   Terence Tao. What's new: The Euler–Maclaurin formula, Bernoulli numbers, the zeta function, and real-variable analytic continuation. Apr. 10, 2010. Accessed Feb. 4, 2023. URL: https://terrytao.wordpress.com/2010/04/10/the-euler-maclaurin-formula-bernoulli-numbers-the-zeta-function-and-real-variable-analytic-continuation/

[W10] Katie Steckles and Paul Taylor. An infinite series of blog posts which sums to $-1/12$. The Aperiodical. Accessed Feb. 4, 2023. URL: https://aperiodical.com/2014/01/an-infinite-series-of-blog-posts-which-sums-to-minus-a-twelfth/

[W11] Multiple Contributors. To sum $1 + 2 + 3 + \cdots$ to $-\frac{1}{12}$. Mathematics Stack Exchange. Accessed Feb. 4, 2023. URL: https://math.stackexchange.com/questions/39802/to-sum-123-cdots-to-frac112

[W12] Multiple Contributors. $1 + 2 + 3 + 4 + \cdots$. Wikipedia: The Free Encyclopedia. Accessed Feb. 4, 2023. URL: https://en.wikipedia.org/wiki/1_+_2_+_3_+_4_+_

[W13] Multiple Contributors. Why is the Gamma function shifted from the factorial by 1? MathOverflow. Accessed Feb. 14, 2023. URL: https://mathoverflow.net/questions/20960/why-is-the-gamma-function-shifted-from-the-factorial-by-1

[W14] Riemann Hypothesis. Clay Mathematics Institute. Accessed Mar. 15, 2023. URL: https://www.claymath.org/millennium-problems/riemann-hypothesis

[W15] Multiple Contributors. Volume of an n-ball. Wikipedia: The Free Encyclopedia. Accessed Feb. 4, 2023. URL: https://en.wikipedia.org/wiki/Volume_of_an_n-ball

[W16] N. J. A. Sloane. Least common multiple (or LCM) of $\{1, 2, \ldots, n\}$. The On-Line Encyclopedia of Integer Sequences. Accessed Mar. 13, 2023. URL: https://oeis.org/A003418

[W17] Multiple Contributors. Riemann mapping theorem for homeomorphisms. MathOverflow. Accessed Mar. 8, 2023. URL: https://mathoverflow.net/questions/66048/riemann-mapping-theorem-for-homeomorphisms

[W18] Multiple Contributors. The letter $\wp$; name and origin? MathOverflow. Accessed Mar. 17, 2023. URL: https://mathoverflow.net/questions/278130/the-letter-wp-name-origin

[W19] Multiple Contributors. Modular Forms – Eichler quote. MathOverflow. Accessed Feb. 3, 2023. URL: https://mathoverflow.net/questions/6955/modular-forms-eichler-quote

[W20] Multiple Contributors. MathOverflow. Where do the product expansions of modular forms come from? Accessed Mar. 8, 2023. URL: https://mathoverflow.net/questions/129536/where-do-the-product-expansions-of-modular-forms-come-from

[W21] The OEIS Foundation Inc. On-Line Encyclopedia of Integer Sequences. Accessed Feb. 3, 2023. URL: https://oeis.org

[W22] Multiple Contributors. Lemniscate constant. Wikipedia: The Free Encyclopedia. Accessed Feb. 22, 2023. URL: https://en.wikipedia.org/wiki/Lemniscate_constant

[W23] Eric W. Weisstein. Lemniscate Constant. From MathWorld—A Wolfram Web Resource. Accessed Feb. 22, 2023. URL: https://mathworld.wolfram.com/LemniscateConstant.html

Index

E_8
– lattice 235, 271–276, 278, 279, 282
– sphere packing 235, 271
nth root function 60

analytic continuation 43
analytic function 12
anti-derivative 26
arc length integral 24
argument principle 53
Arzelá–Ascoli theorem 133, 134
automorphism 121, 122
– group 120–122
  – of the complex plane 122
  – of the Riemann sphere 123
  – of the unit disc 126–129
  – of the upper half plane 129–131

Bernoulli numbers 74, 90, 97
Bessel functions 74, 287
beta function 110, 111
biholomorphism 118

Casorati–Weierstrass theorem 52, 122
Cauchy inequalities 40
Cauchy–Riemann equations 14, 15, 18, 19, 72, 73
Cauchy's integral formula 36, 38, 40, 41
Cauchy's theorem 28, 31–36, 44, 45, 76
Chebyshev function $\psi(x)$ 100–104, 108, 109, 117
Cohn–Elkies sphere packing bounds 276
complex torus 172, 173, 178, 190, 191, 209
– classification 173, 190, 191, 209
conformal
– automorphism 121, 122
– equivalence 118, 121
– map 7, 16–18, 118, 121, 165, 168, 208
congruence subgroup $\Gamma(2)$ 228, 229
contour integral 22–28
– fundamental theorem of calculus 26
convolution 112, 113
cotangent function
– partial fraction expansion 65, 79, 194, 223

derivative 13
– logarithmic 53
digamma function 111, 112, 114

Dirichlet eta function 113
Dirichlet series 115
discriminant
– of a cubic 168–170
– of a lattice 170

Eisenstein series 158–161, 179, 180, 182, 194, 197,
          198, 211–214, 216, 219, 221, 223, 224, 230, 231
– Fourier expansion 194
– recurrence relation 160
– weight 2 219, 221, 223, 224, 230, 231
elliptic curve 146–149, 178
elliptic invariants 159
essential singularity 52
Euler gamma function 83–89, 110, 111, 113, 257, 269
– duplication formula 110
– functional equation 85, 86, 88
– infinite product formula 84, 88
– multiplication theorem 110
– reflection formula 85
Euler pentagonal number theorem 232
Euler totient function 115
Euler–Mascheroni constant 84, 111, 113, 114

Fourier transform 93, 270, 281, 286
– radial 281, 286, 287
function
– analytic 12
– doubly periodic 149, 217
– elliptic 149
– entire 13
– harmonic 19, 38
– holomorphic 12
– inverse 61
– locally injective 61
– magic 278
– meromorphic 50
– radial 279
– Schwartz 92, 270
fundamental domain 187, 190, 193
– extended 193
fundamental theorem of algebra 9, 71

gamma density 112, 113
Gauss–Lucas theorem 73
Goursat's theorem 28, 31, 33

Hankel transform 287
holomorphic function 12
homothetic lattices 171
homotopy 32–34
Hurwitz's theorem 136, 137, 140

infinite product 63–65
– formula for the cosine function 68
– formula for the sine function 65, 78
inverse function theorem 61

Jacobi theta function 91, 94, 95, 114, 115, 215
Jacobi thetanull functions 215, 216, 226, 227, 231, 232
Jacobi triple product identity 227, 232
Jordan curve 36
Jordan curve theorem 36
Jordan–Schoenflies theorem 36

Klein's J-invariant 170, 182, 194, 197, 205–209, 214, 216
– Fourier expansion 197

Laguerre polynomials 287
Laplace transform 238, 239, 242
Laplace's equation 19
lattice 149, 268
– dual 269
Laurent series 68, 80
Leech lattice 265, 278, 279, 282, 288
line integral
– fundamental theorem of calculus 24
– of the first kind 23, 25
– of the second kind 23, 24
Liouville λ-function 115
Liouville's theorem 12, 41, 76, 77
logarithm 58, 59, 78
– principal branch 58, 78
logarithmic derivative 53, 64, 68, 223

von Mangoldt function $\Lambda(n)$ 100, 101, 115
maximum modulus principle 57, 126, 127
mean value property
– for harmonic functions 38
– for holomorphic functions 38
meromorphic function 50
Möbius μ-function 115
Möbius transformation 123–125, 127, 130, 145, 182, 185–187

modular
– discriminant 170, 182, 194, 195, 197, 198, 213, 214, 216, 219, 223, 224
  – Fourier expansion 195
– form 183, 210–227
  – entire 210, 213, 214
  – weak 210
– function 199–205
– group Γ 184, 186, 208, 214, 215, 218, 228, 230
– lambda function 216, 224, 226
– variable τ 171, 182
modular surface 190
Montel's theorem 134, 140
Morera's theorem 28, 40, 45

Newman's Tauberian theorem 104, 108
normal family 134, 140

open mapping theorem 57, 58, 62

Picard's theorem 217
Poisson summation formula 92
– for lattices 270, 276, 277
pole 52
– of a holomorphic function 47
– order 47
– simple 47
power function 60
power series 20–22, 40
pre-modular form 192, 194, 195, 197, 199
– Fourier coefficients 192
– Fourier expansion 192
– weak 192, 199
prime number theorem 5, 82, 83, 99–102, 104, 108, 109, 116
prime-counting function $\pi(x)$ 82, 100
primitive 26, 32, 36
principal part 48
principle of analytic continuation 43

radius of convergence 20, 74
Ramanujan's tau function 196, 198
region 1
– doubly connected 141, 143
– enclosed by a simple closed curve 36
– simply connected 33, 36, 125, 131, 137, 139, 140
removable singularity 44, 52
residue 48
residue theorem 48, 49, 66

Riemann hypothesis 83, 109
Riemann mapping theorem 125, 131–140
Riemann sphere 50, 51, 119, 123–125
Riemann surface 51, 172, 174, 176, 178, 190, 208
Riemann zeta function 89–99, 113, 179, 180, 215
– Euler product formula 90–92
– functional equation 90, 92, 96, 113
– trivial zeros 90, 97
Riemann's removable singularity theorem 44
Rouché's theorem 55, 56, 77

Sendov's conjecture 80
singularity 52
– essential 52
– pole 52
– removable 44, 52
sphere packing 5, 233, 235, 267
– density 267, 277
– lattice 268
– periodic 268
sum of divisors function 115, 182, 197, 198

Tauberian theorem 102
Taylor's formula 22

uniform convergence on compact subsets 45

vector field 23, 76
– conservative 76
– irrotational 76
Viazovska's modular form inequality 256
Viazovska's theorem 234

Weierstrass $\wp$-function 154–159, 161–167, 180, 217,
    225, 228
– addition theorems 180
– differential equation 159
– duplication formula 180
winding number 54, 55

zero
– of a holomorphic function 46
– order 46
– simple 46

www.ingramcontent.com/pod-product-compliance
Lightning Source LLC
Chambersburg PA
CBHW061338210326
41598CB00035B/5810